GEOGRAPHIC VARIATION IN BEHAVIOR

GEOGRAPHIC VARIATION IN BEHAVIOR

Perspectives on Evolutionary Mechanisms

Edited by

Susan A. Foster

John A. Endler

New York Oxford

Oxford University Press

1999

Oxford University Press

Oxford New York
Athens Auckland Bangkok Bogotá Buenos Aires Calcutta
Cape Town Chennai Dar es Salaam Delhi Florence Hong Kong Istanbul
Karachi Kuala Lumpur Madrid Melbourne Mexico City Mumbai
Nairobi Paris São Paulo Singapore Taipei Tokyo Toronto Warsaw

and associated companies in
Berlin Ibadan

Published by Oxford University Press, Inc.
198 Madison Avenue, New York, New York 10016

Library of Congress Cataloging-in-Publication Data
Geographic variation in behavior : perspectives on evolutionary
mechanisms / edited by Susan A. Foster and John A. Endler.
p. cm.
Includes bibliographical references and index.
ISBN 0-19-508295-8
1. Animal behavior—Evolution 2. Zoology—Variation. 3. Animal
ecology. I. Foster, Susan Adlai, 1953– . II. Endler, John A., 1947– .
QL751.G4125 1999
591.5—dc21 97-36110

9 8 7 6 5 4 3 2 1

Printed in the United States of America
on acid-free paper

Preface

In this volume we have brought together evidence that behavior varies geographically, and we explore some of the richness in phenomena, interpretations, and problems that can arise in geographical studies of behavior. The idea to publish this volume originated in a symposium at the 1991 meeting of the Animal Behavior Society at the University of North Carolina, Wilmington. We were both taken by surprise at the level of interest and even astonishment generated by the symposium. The astonishment was caused not by the excellent quality of the papers so much as by the extent of geographic variation in behavior that they documented. Apparently an assumption persists that most behavioral phenotypes are "species-typical," or invariant within species. A second surprising discovery was that few of our contributors knew of one another's work. Each had been developing research programs involving geographic variation in behavior without knowledge of other related research programs, because the taxa or the behavioral questions of interest were apparently too different. Both of these discoveries, in combination with the general level of interest on the part of participants and attendees, resulted in our decision to produce this volume.

Although the decision to publish came from responses to the symposium, this volume is not the same as the symposium. Authors have been added, and all of the chapters have been substantially revised and updated. As we go to press, we believe that these chapters reflect the state of the field.

The volume is designed mainly for two audiences. The first is the student of animal behavior (behavioral ecologist, comparative psychologist, ethologist) who wishes to explore means of studying the evolution of animal behavior. The second is the evolutionary biologist who is interested in learning about the geographic variation of behavioral traits and their value in evolutionary studies. Of course, we also hope that those who already combine an interest in both fields will find the chapters in the volume informative, interesting, and exciting. Given the different backgrounds of our anticipated readers, we have tried to make the chapters speak not only to specialists in subdisciplines of evolution or behavior, but also to readers unfamiliar with the specific area of research under discussion. The methodological discussions are adequate to provide the reader with a basic understanding of important issues, but they do not substitute for the original papers when one intends to use the methods in research.

The chapters in this volume do not attempt to explore all possible methods by which behavioral evolution can be studied, but instead, as the title indicates, focus on the insights to be gathered through the study of geographic variation in behavior. Because research on geographic variation in behavior requires that population comparisons be undertaken, the research programs described in the chapters involve elements of the comparative method. However, population phylogenies are often unavailable, and creative alternatives have to be employed to understand the impact of function and phylogeny on the distribution of phenotypes. Ultimately, we hope that improved methods will enhance our ability to incorporate phylogeny into studies of geographical variation. For those interested in the value of phylogenies for understanding the evolution of behavioral traits, we recommend *Phylogenies and the Comparative Method in Animal Behavior*, edited by Emília P. Martins (1996, Oxford University Press), which we view as a companion to this volume.

We have edited the chapters rather extensively in an effort to provide a volume with the consistency and integration more often characteristic of single-authored than multiauthored volumes. Although perhaps not achieving these goals entirely, the volume benefits from the individual expertise of the authors. In addition to our efforts and those of the authors, each chapter was reviewed by a minimum of two reviewers from appropriate disciplines. The chapters are much better for their efforts. We thank these biologists for their careful and substantive reviews: Stephen Adolph, Christine Boake, Edmund Brodie III, Douglas Fraser, Sidney Gauthreaux, Anne Houde, Charles Janson, Anne Magurran, Emília Martins, Donald Owings, Nancy Reagan, Susan Riechert, Joseph Travis, William Rowland, Daniel Rubenstein, David Wilson, Jeffrey Walker, and four anonymous reviewers.

We want especially to thank our editor at Oxford University Press, Kirk Jensen, for infinite patience, and Susan Riechert for encouraging the organization of the symposium that led ultimately to the volume.

Petersham, Massachusetts
Townsville, Queensland
August 1997

Susan A. Foster
John A. Endler

Contents

Contributors

Peter Berthold
Max-Planck-Institute for Behavioral Physiology
Radolfzell
Germany
E-mail: Peter.Berthold@uni-konstanz.de

Sue Boinski
Department of Anthropology and Division of Comparative Medicine
University of Florida
Gainesville, FL
USA
E-mail: boinski@nervm.nerdc.ufl.edu

Gordon M. Burghardt
Department of Psychology
University of Tennessee
Knoxville, TN
USA
E-mail: gburghar@utk.edu

Scott P. Carroll
Center for Population Biology
University of California, Davis
Davis, CA
USA
E-mail: scarroll@unm.edu

Patrice Showers Corneli
Department of Human Genetics
University of Utah
Salt Lake City, UT
USA
E-mail: pat@linkers.uutah.edu

Richard G. Coss
Department of Psychology
University of California, Davis
Davis, CA
USA
E-mail: rgcoss@ucdavis.edu

Timothy J. Ehlinger
Department of Biological Sciences
University of Wisconsin, Milwaukee
Milwaukee, WI
USA
E-mail: ehlinger@csd.uwm.edu

John A. Endler
Department of Biological Sciences
University of California
Santa Barbara, CA
USA
E-mail: Endler@lifesci.lscf.ucsb.edu

And
Department of Zoology and Tropical Ecology
James Cook University
Townsville, Queensland
Australia

Susan A. Foster
Biology Department
Clark University
Worcester, MA
USA
E-mail: sfoster@clarku.edu

Murray J. Littlejohn
Department of Zoology
University of Melbourne
Parkville, Victoria
Australia
E-mail: Murray_Littlejohn.Zoology@muwaye.unimelb.edu.au

Anne E. Magurran
School of Environmental and Evolutionary Biology
University of St. Andrews
St. Andrews
Scotland
E-mail: Anne.Magurran@st-andrews.ac.uk

Susan E. Riechert
Department of Ecology and Evolutionary Biology
University of Tennessee
Knoxville, TN
USA
E-mail: sriecher@utk.edu

Michael J. Ryan
Department of Zoology
University of Texas
Austin, TX
USA
E-mail: mryan@mail.utexas.edu

James M. Schwartz
Department of Zoology
University of Vermont
Burlington, VT
USA
E-mail: jschwart@gnu.uvm.edu

Daniel B. Thompson
Department of Biological Sciences
University of Nevada
Las Vegas, NV
USA
E-mail: dthompsn@ccmail.nevada.edu

Paul A. Verrell
Department of Zoology
Washington State University
Pullman, WA
USA
E-mail: verrell@wsu.edu

Walter Wilczynski
Department of Psychology
University of Texas
Austin, TX
USA
E-mail: wilczynski@psy.utexas.edu

Introduction and Aims

SUSAN A. FOSTER
JOHN A. ENDLER

Behavioral ecologists and evolutionists have come to assume that behavior evolves, is mostly adaptive, and consists of a set of traits with both genetic and environmental components. If this is generally true, evolution should result in geographical variation for behavioral traits, as it has in the more often studied morphological, physiological, and molecular traits. Such variation in conventional traits has yielded new insights in evolution and function. In spite of this, many people implicitly or explicitly assume that behavior is geographically uniform. In this volume we bring together evidence that behavior does indeed vary geographically and discuss its consequences and problems in the study of geographical variation in behavior. The collection of chapters is perhaps unusual for a volume on behavior because of the diversity of behavioral traits and issues addressed. However, all are linked to one another through the use of geographically varying behavioral traits to answer questions about the evolution of behavior. This chapter is intended to provide a historical and methodological context for the other chapters. We hope that it will be particularly useful for those who do not often think in terms of evolution or population comparison. Equally, we hope our discussion of behavioral phenotypes will allay concerns about the suitability of behavioral traits for the study of evolutionary processes.

History

The study of geographic variation in behavior is implicitly comparative in nature, in that differences in behavior across geographic regions, or across discrete populations, are examined with a particular goal in mind. There exists a rich comparative tradition in classical ethology, although comparisons were made nearly exclusively between species rather than across populations. The ethological approach to comparative studies assumed two forms (Hinde and Tinbergen 1958, Tinbergen 1964). The first was aimed at assessing current adaptive value of behavioral characters. This goal was achieved by identifying behavioral differences among closely related species and interpreting these differences in an adaptive context. This approach did not require knowledge of the evolutionary history of a character because the goal was assessment of current adaptive value. Ethologists clearly recognized the need for experimental tests of adaptive hypotheses generated through comparative studies. For example, Tinbergen et al. (1962) provided a classical direct test of an adaptive hypothesis in his experiment demonstrating reduced predation on gull nestlings when eggshells were removed from the nest by parents.

A second use of comparative data by ethologists involved interpretation of behavioral differences among related species from a phylogenetic perspective. This approach had its origins in the writings of Darwin, who clearly recognized that behavioral characters were the products not just of natural selection, but also of ancestry ("*Natura non facit saltum* applies with almost equal force to instincts as to bodily organs" [1859, p. 236]). The conceptual antecedents of the comparative approach used by classical ethologists can be traced to Darwin's writings. Whitman (1899), Heinroth (1911, 1930), and Lorenz (1935, 1941) explored the value of the comparative method for understanding the evolution of behavioral phenotypes and established the field of ethology.

The comparative method proved fruitful in that ethologists developed a number of insights into the evolution of behavior, especially display behavior (reviewed in Hinde and Tinbergen 1958, Blest 1961, Baerends 1975, Foster 1995), but the method fell into disfavor in the third quarter of this century due to uncertainty about the underlying causes of behavior (the nature–nurture controversy) and due to limitations of the methods available for phylogenetic reconstruction and for examining changes in character states on a known phylogeny (Burghardt and Gittleman 1990, Foster 1995). Advances in methods of incorporating phylogenetic information into the analysis of comparative data offer new analytical power and have revitalized the use of the comparative approach as a source of evolutionary information (see Brooks and McLennan 1991, Harvey and Pagel 1991, Miles and Dunham 1993, Maddison 1994, Martins and Hansen 1996, for recent reviews). Among the examples are a number of studies of the evolution of behavioral characters (see chapters in Martins 1996 for examples).

When ethologists applied the comparative method to interspecific data, the behavioral patterns of interest often had been studied only in a single population. They were assumed to be "species-typical," or invariant within the species. At the time there was little evidence to the contrary, and variation within species was considered to be "noise" that was of little intrinsic interest (Lorenz 1970, Mayr 1976; but see Barlow 1977, Magurran 1993, and Verrell this volume, for discussion). In contrast, population and quantitative geneticists had undertaken research that demonstrated geographic variation within species, but they rarely examined behavioral traits (Endler 1977, 1986; Zink and Remsen 1986). Koref-Santibanez (1972a,b) documented clear genetically based, geographical differences in mating behavior in *Drosophila paulistorum*, but the work was not published in a journal widely read by ethologists and was never incorporated into the ethological literature. Arnold's 1977 publication in *Science* was the first on the topic to be widely read by behavioral scientists from a variety of subdisciplines. That paper, and a pair of papers he published in 1981, stimulated research on geographic variation in behavior from an evolutionary perspective (Arnold 1981a,b).

We would argue, as have others, that there not only is special value in the study of geographic variation in behavior, but also that there are special difficulties in the collection, analysis, and interpretation of such data (e.g., Arnold 1977, 1981a,b, 1992; Barlow 1981, Endler 1983, 1986; Foster and Cameron 1996). In the next sections we provide an introduction to the use of geographic differences in behavior for understanding evolutionary processes in general and the evolution of behavioral phenotypes in particular. This is followed by a brief overview of problems often encountered in such studies that should help readers who have not considered similar methodological problems to understand and evaluate concerns expressed in individual chapters.

The Value of Geographic Comparisons

Inference of Adaptive Value

As will be evident from the chapters in this book, the most common use of population comparison has been to infer adaptive value. Although most examples involve phenotypes other than behavior (Endler 1977, 1986), chapters in this volume describe several behavioral studies of this type, and reference to others will be found within the chapters. The typical method is to examine associations between character states within populations and the environmental factors to which the populations are exposed. When particular environmental characteristics are found to be associated with particular behavioral character states across populations, and the behavioral differences are shown to reflect underlying genetic differences, adaptive differentiation can be inferred. The inference is strongest when the geographic relationship between populations expressing different phenotypes is complex because this reduces the likelihood that covarying environmental factors are responsible for the behavioral differences (see Endler 1983, 1986 for discussion).

Population comparisons sometimes can provide clearer insights into the adaptive causes of differentiation than can comparisions among species (or higher taxonomic units) because populations often have been separated for less time than have species and are more likely to still reside in the habitats in which differentiation occurred (Foster et al. 1992). Thus, the potential for correctly identifying causative selective agents is greater. Equally, populations tend to differ in fewer characteristics than do species, reducing the confounding effects of covarying traits on the analysis (Lott 1991, Arnold 1992). Therefore, population comparison provides an extremely valuable means of identifying possible causes of adaptive differentiation in wild populations.

Although population comparisons can be used to develop hypotheses as to the selective causes of population differentiation, other methods are necessary to demonstrate natural selection directly and to establish causal relationships (Endler 1983, 1986). Natural selection can be demonstrated via perturbation experiments, measurement of fitness differences among trait values within cohorts, and comparisons of trait frequency distributions among age classes or life-history stages. Once natural selection has been demonstrated, knowledge of the biology of the organism and its relationship to the environment can be used to predict changes in trait frequency distributions under nonequilibrium conditions or to predict the trait distributions under equilibrium conditions (Endler 1986). Susan Riechert's chapter in this volume (chapter 1) offers a particularly good demonstration of a complete study of natural selection on behavioral traits.

Implicit in the testing of adaptation hypotheses using these techniques is the assumption of adequate underlying genetic variation to permit, at a minimum, appropriate directional change in the trait values under nonequilibrium conditions, or directional change allowing the population to reach predicted trait values at equilibrium. Certainly, there are many behavioral traits that vary substantially within species and that have the potential to evolve rapidly in response to directional selection. This is perhaps best illustrated by the success of optimality models in predicting behavior on the basis of present-day conditions alone (e.g., Stephens and Krebs 1986, Krebs and Davies 1987, Parker and Maynard Smith 1990). Rapid adaptive responses may occur most often in quantitative traits, or in cases in which loss of a behavior pattern is favored by the current selective regime (Foster et al. 1996).

Although research on geographic variation in behavior can provide exceptional insights into the ways in which natural selection can cause adaptive differentiation of behavior, not all behavioral patterns have the potential to respond to selection rapidly and completely. To fully understand trait distributions in these cases, phylogenetic correlation, which arises through common ancestry, must be examined to detect the possibility of constraint on trait evolution. Of equal concern for studies of adaptation is the possibility that local adaptation will be precluded by high levels of gene flow between geographic regions exposed to different selective environments. Both issues are discussed below.

Phylogenetic Pattern in the Evolution of Behavior

Populations of the same species share recent common ancestors and therefore also share many behavioral patterns they have inherited. In some instances character states are shared by populations because selection has maintained the ancestral state in all the populations. Alternatively, selection may have favored divergence among the populations, but evolution of the differences may have been precluded by inadequate variation in the ancestral population. Such constraint is particularly likely when the behavior patterns are complex or when their ontogeny is determined by complex developmental rules (see Martins and Hansen 1996 for a discussion). Under these circumstances, population phenotypes may not conform to adaptive expectations, and the cause may be difficult to infer (Endler 1986).

The importance of phylogenetic relationship as an explanation for the distributions of behavioral-trait values across taxa has been examined most completely using interspecific or higher order comparisons (Martins and Hansen 1996 and chapters in Martins 1996). Although similarities between an ancestor and a daughter species are expected to decline with time, phylogenetic comparative studies among higher taxonomic units provide clear evidence that behavioral phenotypes are often inherited and that their pattern of distribution reflects historical relationships (e.g., Lorenz 1950, Lauder 1986, McLennan et al. 1988, de Queiroz and Wimberger 1993, Foster 1995, chapters in Martins 1996). Certainly then, phylogenetic relationships must also be considered in evaluating patterns of behavioral evolution through population comparison—patterns that should even more strongly reflect phylogeny. Unfortunately, the paucity of differences among populations within a species can make the generation of robust phylogenetic hypotheses difficult or impossible when standard cladistic methods are used (Crandall and Templeton 1996, Foster and Cameron 1996). Reticulate evolution caused by gene flow among populations exacerbates the problem, as does the existence of intrapopulation genetic polymorphism in rapidly evolving regions of the DNA.

Gene Flow as a Constraint on the Evolution of Population Differences

When habitat boundaries are distinct and gene flow between populations is effectively absent, populations can respond rapidly to local differences in selective regimes. Freshwater populations of fish, especially those in postglacial lakes, provide excellent examples (Bell and Foster 1994, Robinson and Wilson 1994, Skúlason and Smith 1995, Smith and Skúlason 1996, Bell and Andrews (1997), Ehlinger this volume, Magurran this volume). In species such as these, population comparisons may prove an especially effective tool for identifying adaptive differentiation and its causes because gene flow does not substantially inhibit the ability of populations to respond adaptively to local selective regimes. Equally, the effects of mutational novelties and genetic drift should be more apparent under these conditions than when gene flow blurs boundaries between populations, so sufficient population replication must be made to eliminate the effects of random divergence.

An assumption of low gene flow between populations is often unrealistic. Often a species' characteristics change gradually along environmental gradients, or dispersal capabilities permit movement across inhospitable intervening habitats (e.g., Endler 1977). The spatial characteristics of habitat patches and the dispersal capabilities of the species will determine the magnitude of gene flow (for explanation and references, see Riechert this volume, Thompson this volume). The magnitude and direction of gene flow will in part determine the extent to which local populations can diverge in response to differences in local selective regimes. When gene flow between habitats is high, little adaptive differentiation is expected (for examples, see Riechert this volume, Thompson this volume).

Although the existence of gene flow across populations will likely be perceived as a disadvantage to those interested in understanding the causes of behavioral differentiation, the dynamics of the interplay between selection, gene flow, and genetic variation harbored in interacting populations is in itself fascinating. When expected behavioral differences are not detected in population comparisons, gene flow must be explored as a possible reason, as must potential genetic constraints on the ability of local populations to respond adaptively to the local conditions.

Genes and Phenotypic Plasticity

We assume throughout this volume that the population differences under discussion have an underlying genetic basis. This is essential if Darwinian evolution is to be invoked. Although in some cases the evidence is less complete than one might wish, the example is frequently the best available to illustrate a point and has therefore provided the major substance of a chapter or an important illustrative example. In many cases, however, the evidence is compelling.

A historical criticism of efforts to study behavioral evolution has been that behavior is so labile that one cannot assume genetic differences underlie behavioral differences. The most extreme forms of this argument have held that behavior evolves in a different way from other aspects of phenotype (e.g., Hailman 1982). Although it is clearly inappropriate to assume genetic variation when one observes behavioral differences across populations, the opposite assumption of no genetic differentiation is equally flawed. There is now ample evidence of heritable variation in behavior (chapters in Boake 1994) and of genetically based population differentiation of behavior (e.g., Dingle 1994, Lynch 1994, Travis 1994; chapters in this volume). Furthermore, there is little evidence suggesting differences in levels of homoplasy among behavioral, life history, morphological, and biochemical traits (Sanderson and Donoghue 1989, de Queiroz and Wimberger 1993, Wimberger and de Queiroz 1996), although within taxonomic groups, display behavior appears to contain a stronger phylogenetic signal than does non-display behavior (Foster et al. 1996).

None of this implies that environmental influences on behavior must be small or absent. Behavioral phenotypes typically reflect the interaction between genes and environment. On the one hand, selection may act on a trait directly, influencing the magnitude or nature of its expression. On the other hand, selection may produce population-specific behavioral phenotypic plasticity. In this case, selection acts on the pattern of condition-dependent expression of behavior (norm of reaction), causing populations to respond differently to the same conditions (Carroll and Corneli this volume, Ehlinger this volume, Thompson this volume). When the reaction norms are primarily influenced by natural selection, the phenotypic plasticity is itself considered to be adaptive. Norms of reaction can thus evolve in the same way as other aspects of phenotype. Although this concept has rarely been applied to the study of behavioral evolution, the chapters in this volume by Carroll and Corneli and by Thompson provide clear demonstrations of the potential value of the approach.

Population Differentiation and Speciation

Much of the research intended to elucidate the process of speciation has involved the study of fully-formed species. This is problematic in that the process is complete and cannot be studied directly. Because speciation frequently occurs in allopatry, peripatry, or parapatry, comparisons among carefully chosen populations may enable us to understand genetic and phenotypic changes occurring during the critical, earliest stages in the process of speciation that lead to restriction in the magnitude of gene flow upon secondary contact (Lewontin 1974, Endler 1977). Indeed, such comparisons are probably essential to our understanding of mechanisms of speciation.

Although a number of authors have argued that we can understand the earliest stages of speciation through population comparison (e.g., Lewontin 1974, Endler 1989, McPhail 1994), there are few such studies. They fall into one of two categories; studies of interactions (including behavioral interactions) across hybrid zones between two relatively fully-formed species (e.g., chapters in Harrison 1993, Littlejohn this volume) and studies of variation in sexual behavior across conspecific populations (for reviews, see Wilczynski and Ryan this volume, and Verrell this volume). These studies sometimes are focused primarily on understanding the evolution of elements of mating systems that facilitate gamete union, but they may also have the expressed aim of understanding the causes of speciation. In either case, for those species of animals in which behavior plays a prominent role in mediating gamete union, studies of behavioral divergence, or of behavioral interactions between incompletely isolated species, can offer insight into the process of speciation—insights that will be difficult to glean in other ways.

The Questions

When we began editing this book, we were convinced that behavior evolves in the same way as do other characters, and we did not see the demonstration of geographic variation in behavior as very surprising. Nevertheless, the first question that may require an answer for some readers is: Does behavior, even behavior normally assumed to be species typical, exhibit substantial geographic variation? Given a positive answer to this question, the next is whether behavior evolves in the same way as do other traits. For example, do genetic differences underlie geographically varying behavioral traits, and does gene flow prevent the evolution of such variation in the way it does other traits?

The distribution of genetic variation within and across populations is also of interest. Strong local selection could remove most genetic variation within populations (Fisher 1930), or genetic variation could be retained within local populations, enhancing the population's ability to respond rapidly to selection. Because some behavioral attributes are so plastic, a particularly intriguing question is whether high levels of plasticity preclude the retention of high levels of genetic variation for plastic traits, as reflected in behavioral norms of reaction, or whether genetic variation for norms of reaction is retained even when plasticity is pronounced. If the relationships between plasticity and norms of reaction vary, do they vary in a way that is predictable in the context of spatial and temporal patterns of environmental variation? Related questions are, of course, whether learning and developmental programs generating behavior evolve over the range of a species and whether any of the differences can be interpreted in an adaptive context.

The problem of evolutionary rate, especially over relatively short time frames, can also be explored particularly effectively through population comparison. In this context, we were interested in determining whether different kinds of behavioral traits tended to evolve at different rates (for example, do maintenance behavior patterns evolve more rapidly than display behavior patterns?), whether phylogenetically ancient traits decay more slowly than traits that have evolved comparatively recently when released from selection, and whether selection on behavioral traits can drive rapid speciation. Equally, the question of the relative contributions of character state loss, expression shifts, and the evolution of novelties to populations differentiation can be explored. Are the contributions of each of these kinds of traits to evolutionary change similar to those observed in higher order comparisons?

A final set of questions is related to the generation of biodiversity. A question we would have liked to answer was whether gene flow, by increasing variation in ecotones or marginal habitats, could increase behavioral diversity, and hence, evolutionary potential. The chapters here are suggestive but do not provide a clear answer. Nonetheless, they certainly do allow us to evaluate the ways in which signals involved in mate choice evolve in response to environmental conditions, and in turn, to ask how sexual selection and natural selection influence communication systems and patterns of speciation. Equally, research described in this volume allows us to address the patterns of relationship between environmental variables and geographic and genetic differentiation as related to reproductive isolation and speciation. We can ask whether these variables covary tightly and whether behavioral phenotypes involved in mate choice covary in turn.

Ideally, we would like to have been able to provide compelling answers to all these questions. However, research on the evolution of geographic variation in behavior is just beginning, and we must be satisfied with answering some of the questions, and with demonstrating, we think convincingly, the power of this approach for answering the questions we have just posed. We will revisit these questions in the concluding chapter.

Organization of the Book

The contributors' chapters have been organized into three sections. The first section focuses on methodology and illustrates approaches to the study of adaptation, gene flow, and the evolution of phenotypic plasticity. The second section offers examples of research on population differentiation, excluding that leading to speciation. Because most chapters in the first and second sections discuss methodology, and many are linked conceptually, the choice of chapters for the two sections was difficult and somewhat arbitrary. The third section relates geographical differentation of behavior to speciation.

The closing chapter offers our perspective on the value of research on geographic variation in behavior, incorporating both our assessment of the kinds of novel insights that can be achieved and the limitations of the approach. In it we discuss themes addressed by the authors and discuss areas we believe are most in need of additional research.

References

Arnold, S. J. 1977. Polymorphism and geographic variation in the feeding behavior of the garter snake, *Thamnophis elegans*. Science 197:676–678.

Arnold, S. J. 1981a. Behavioral variation in natural populations. I. Phenotypic, genetic and environmental correlations between chemoreceptive responses to prey in the garter snake, *Thamnophis elegans*. Evolution 35:489–509.

Arnold, S. J. 1981b. Behavioral variation in natural populations. II. The inheritance of a feeding response in crosses between geographic races of the garter snake, *Thamnophis elegans*. Evolution 35:510–515.

Arnold, S. J. 1992. Behavioral variation in natural populations. VI. Prey responses by two species of garter snakes in three regions of sympatry. Animal Behaviour 44:705–719.

Baerends, G. P. 1975. An evaluation of the conflict hypothesis as an explanatory principle for the evolution of displays. *In* G. Barends, C. Beer, and A. Manning, eds. Function and evolution in behavior, pp. 187–227. Clarendon Press, Oxford.

Barlow, G. W. 1977. Modal action patterns. *In* T. A. Sebeok, ed. How animals communicate, pp. 98–134. Indiana University Press, Bloomington.

Barlow, G. W. 1981. Genetics and development of behavior, with special reference to patterned motor output. *In* K. Immelmann, G. W. Barlow, L. Petrinovich, and M. Main, eds. Behavioral development: The Bielefeld Interdisciplinary Project, pp. 191–251. Cambridge University Press, Cambridge.

Bell, M. A., and C. A. Andrews. 1997. Evolutionary consequences of postglacial colonization of fresh water by primitively anadromous fishes. *In* B. Streit, T. Städler, and C. M. Lively, eds. Evolutionary ecology of freshwater animals, pp. 324–363. Birkhäuser, Basel.

Bell, M. A., and S. A. Foster. 1994. Introduction to the evolutionary biology of the threespine stickleback. *In* M. A. Bell and S. A. Foster, eds. The evolutionary biology of the threespine stickleback, pp. 1–27. Oxford University Press, Oxford.

Blest, A. D. 1961. The concept of ritualization. *In* W. H. Thorpe and O. L. Zangwill, eds. Current problems in animal behaviour, pp. 102–124. Cambridge University Press, Cambridge.

Boake, C. R. B. (ed). 1994. Quantitative genetic studies of behavioral evolution. University of Chicago Press, Chicago.

Brooks, D. R., and D. A. McLennan. 1991. Phylogeny, ecology, and behavior: A research program in comparative biology. University of Chicago Press, Chicago.

Burghardt, G. M., and J. L. Gittleman. 1990. Comparative behavior and phylogenetic analyses: New wine, old bottles. *In* M. Bekoff and D. Jamieson, eds. Interpretation and explanation in the study of animal behavior. Volume II: Explanation, evolution, and adaptation, pp. 192–225. Westview Press, Boulder, CO.

Crandall, K. A., and A. R. Templeton. 1996. Applications of intraspecific phylogenetics. *In* P. H. Harvey, A. J. Leigh Brown, J. Maynard Smith, and S. Nee, eds. New uses for new phylogenies, pp. 81–99. Oxford University Press, Oxford.

Darwin, C. 1859. The origin of species by means of natural selection. London, Murray.

de Quieroz, A., and P. H. Wimberger. 1993. The usefulness of behavior for phylogeny estimation: Levels of homoplasy in behavioral and morphological characters. Evolution 47:46–60.

Dingle, H. 1994. Genetic analyses of animal migration. *In* C. R. B. Boake, ed. Quantitative genetic studies of behavioral evolution, pp. 145–164. University of Chicago Press, Chicago.

Endler, J. A. 1977. Geographic variation, speciation, and clines. Princeton University Press, Princeton, NJ.

Endler, J. A. 1983. Testing causal hypotheses in the study of geographic variation. *In* J. Felsenstein, ed. Numerical taxonomy, pp. 424–443. Springer-Verlag, New York.

Endler, J. A. 1986. Natural selection in the wild. Princeton University Press, Princeton, NJ.

Endler, J. A. 1989. Conceptual and other problems in speciation. *In* D. Otte and J. A. Endler, eds. Speciation and its consequences, pp. 625–648. Sinauer Associates, Sunderland, MA.

Fisher, R. A. 1930. The genetical theory of natural selection. Clarendon Press, Oxford.

Foster, S. A. 1995. Constraint, adaptation, and opportunism in the design of behavioral phenotypes. Perspectives in Ethology 11:61–81.

Foster, S. A., J. A. Baker, and M. A. Bell. 1992. Phenotypic integration of life history and morphology: An example from the three-spined stickleback, *Gasterosteus aculeatus* L. Journal of Fish Biology 41(suppl.):21–35.

Foster, S. A., and S. A. Cameron. 1996. Geographic variation in behavior: A phylogenetic framework for comparative studies. *In* E. Martins, ed. Phylogenies and the comparative method in animal behavior, pp. 138–165. Oxford University Press, Oxford.

Foster, S. A., W. A. Cresko, K. P. Johnson, M. U. Tlusty, and H. E. Willmott. 1996. Patterns of homoplasy in behavioral evolution. *In* M. J. Saunders and L. Hufford, eds. Homoplasy and the evolutionary process, pp. 245–269. Academic Press, New York.

Hailman, J. P. 1982. Evolution and behavior: An iconoclastic view. *In* H. C. Plotkin, ed. Learning, development, and culture, pp. 205–254. John Wiley and Sons, New York.

Harrison, R. G. (ed). 1993. Hybrid zones and the evolutionary process. Oxford University Press, Oxford.

Harvey, P. H., and M. D. Pagel. 1991. The comparative method in evolutionary biology. Oxford University Press, Oxford.

Heinroth, O. 1911. Beitrage zur Biologie, Insbesondere Psychologie und Ethologie der Anatiden. Verhandlungen des V. Internationalen Ornithologen-Kongresses 589–702.

Heinroth, O. 1930. Über bestimmte Bewegungsweisen der Wirbeltiere. Gesellschaft naturforschender Freunde, Berlin: 333–342.

Hinde, R. A., and N. Tinbergen. 1958. The comparative study of species-specific behavior. *In* A. Roe and G. G. Simpson, eds. Behavior and evolution, pp. 251–268. Yale University Press, New Haven, CT.

Koref-Santibanez, S. 1972a. Courtship behavior in the semispecies of the superspecies *Drosophila paulistorum*. Evolution 26:108–115.

Koref-Santibanez, S. 1972b. Courtship interaction in the semispecies of *Drosophila paulistorum*. Evolution 26:326–333.

Krebs, J. R., and N. B. Davies. 1987. An introduction to behavioral ecology. Blackwell Scientific Publications, Oxford.

Lauder, G. V. 1986. Homology, analogy, and the evolution of behavior. *In* M. Nitecki and J. Kitchell, eds. The evolution of behavior, pp. 9–40. Oxford University Press, Oxford.

Lewontin, R. C. 1974. The genetic basis of evolutionary change. Columbia University Press, New York.

Lorenz, K. 1935. Der Kumpan in der Umwelt der Vogels. Journal of Ornithology 83: 137–213.

Lorenz, K. 1941. Vergleichende Bewegungsstudien an Anatinen. Journal of Ornithology 89:19–29.

Lorenz, K. 1950. The comparative method in studying innate behavior patterns. Symposium of the Society for Experimental Biology 4:221–268.

Lorenz, K. 1970. Studies in animal and human behavior. Methuen, London.

Lott, D. F. 1991. Intraspecific variation in the social systems of wild vertebrates. Cambridge University Press, Cambridge.

Lynch, C. B. 1994. Evolutionary inferences from genetic analyses of cold adaptation in laboratory and wild populations of the house mouse. *In* C. R. B. Boake, ed. Quantitative genetic studies of behavioral evolution, pp. 278–301. University of Chicago Press, Chicago.

Maddison, D. R. 1994. Phylogenetic methods for inferring the evolutionary history and processes of change in discretely valued characters. Annual Review of Entomology 39:267–292.

Magurran, A. E. 1993. Individual differences and alternative behaviours. *In* T. J. Pitcher, ed. Behavior of teleost fishes, pp. 441–477. Chapman and Hall, London.

Martins, E. P. (ed). 1996. Phylogenies and the comparative method in animal behavior. Oxford University Press, Oxford.

Martins, E. P., and T. F. Hansen. 1996. The statistical analysis of interspecific data: A review and evaluation of phylogenetic comparative methods. *In* E. P. Martins, ed. Phylogenies and the comparative method in animal behavior, pp. 22–75. Oxford University Press, Oxford.

Mayr, E. 1976. Evolution and the diversity of life. Harvard University Press, Cambridge, MA.

McLennan, D. A., D. R. Brooks, and J. D. McPhail. 1988. The benefits of communication between comparative ethology and phylogenetic systematics: A case study using gasterosteid fishes. Canadian Journal of Zoology 66:2177–2190.

McPhail, J. D. 1994. Speciation and the evolution of reproductive isolation in the sticklebacks (*Gasterosteus*) of south-western British Columbia. *In* M. A. Bell and S. A. Foster, eds. The evolutionary biology of the threespine stickleback, pp. 399–437. Oxford University Press, Oxford.

Miles, D. B., and A. E. Dunham. 1993. Historical perspectives in ecology and evolution. Annual Review of Ecology and Systematics 24:587–619.

Parker, G. A., and J. Maynard Smith. 1990. Optimality theory in evolutionary biology. Nature 348:27–33.

Robinson, B. W., and D. S. Wilson. 1994. Character release and displacement in fishes: a neglected literature. American Naturalist 144:596–627.

Sanderson, M. J., and M. J. Donoghue. 1989. Patterns of variation in levels of homoplasy. Evolution 43:1781–1795.

Skúlason, S., and T. B. Smith. 1995. Resource polymorphisms in vertebrates. Trends in Ecology and Evolution 10:366–370.

Smith, T. B., and S. Skúlason. 1996. Evolutionary significance of resource polymorphisms in fishes, amphibians, and birds. Annual Review of Ecology and Systematics 27: 111–133.

Stephens, D. W., and J. R. Krebs. 1986. Foraging theory. Princeton University Press, Princeton, NJ.

Tinbergen, N. 1964. On aims and methods of ethology. Zeitschrift fur Tierpsychologie 20: 410–433.

Tinbergen, N., G. J. Broekhuysen, F. Feekes, J. C. W. Houghton, H. Kruuk, and E. Szulc. 1962. Egg shell removal by the black-headed gull, *Larus ribidundus* L.; a behavior component of camoflage. Behavior 19:74–118.

Travis, J. 1994. Size-dependent behavioral variation and its genetic control within and among populations. *In* C. R. B. Boake, ed. Quantitative genetic studies of behavioral evolution, pp. 165–187. University of Chicago Press, Chicago.

Whitman, C. O. 1899. Animal behavior. Woods Hole Marine Biology Lectures 6:285–338.

Wimberger, P. H., and A. de Queiroz. 1996. Comparing behavioral and morphological characters as indicators of phylogeny. *In* E. P. Martins, ed. Phylogenies and the comparative method in animal behavior, pp. 206–233. Oxford University Press, Oxford.

Zink, R. M., and J. V. Remsen, Jr. 1986. Evolutionary processes and patterns of geographic variation in birds. Current Ornithology 4:1–69.

GEOGRAPHIC VARIATION IN BEHAVIOR

1

The Use of Behavioral Ecotypes in the Study of Evolutionary Processes

SUSAN E. RIECHERT

Ecotypic variation refers to differences in traits between populations that reflect adaptation to different selection pressures. The origin of the term lies in Turesson's (1922) observation that population differences in plant growth forms often breed true in a controlled environment. From experiments, Turesson (1922) concluded that much between-population variation in phenotypic traits reflects genotypic adaptation to local conditions. He described the phenomenon as ecotypic variation and the species population exhibiting a particular variant as an ecotype.

The meaning of adaptation must be examined if the phenomenon of ecotypic variation is to be understood. Readers should refer to Reeve and Sherman (1993) for an in-depth analysis of the problems surrounding various definitions of this term. Briefly, Antonovics (1987) divided evolutionary studies into two distinct classes: those that consider the influences of past events (phylogenies) and those that consider why certain traits predominate over others in an ongoing selection process. Tinbergen (1963) proposed a similar subdivision. Most definitions of adaptation incorporate these two elements in that they require a history of selective modification of a trait. For instance, Harvey and Pagel (1991) express adaptations in terms of traits derived in their phylogenetic group that have current utilitarian function.

However, ecotypic variation refers to trait differences that reflect adaptations to local conditions at one point in time. Phylogenetic constraints need not be examined except in terms of how they might limit adaptation. Therefore, when I refer to adaptation in this chapter, I will be limiting it to "a phenotypic variant that results in the highest fitness among a specified set of variants in a given environment" (Reeve and Sherman 1993). This definition is history-free. It is based on extant competing phenotypes, and thus fitness is of significance only in reference to current alternatives.

Although the definition I borrow from Reeve and Sherman (1993) for adaptation does not specify that there be an underlying genetic mechanism, optimization models assume that there is (Charlesworth 1977). Minimally, we wish to know whether sufficient genetic complexity (variability) exists for a predicted optimal solution to be reached. Note also that traits can be adaptations even if they currently exhibit no phenotypic variability because reproductive competition among alternatives may continually eliminate inferior phenotypes.

As pointed out by Reeve and Sherman (1993), this definition of adaptation permits falsification of the hypothesis that selection accounts for the frequency of a given phenotype. It is the perceived lack of our ability to falsify the prediction of adaptation that has led to much criticism of the so-called adaptationist program (e.g., Gould and Lewontin 1979).

Different kinds of behavioral traits are likely to express ecotypic differentiation to varying degrees. Some kinds of behavioral and life-history characters clearly exhibit high levels of variation within species and, as a consequence, track environmental variation well. Others apparently display stronger phylogenetic signal and less variation that is clearly ecotypic in nature (Foster et al. 1996, Gittleman 1996b, Martins and Hansen 1996). Optimization studies and comparative studies among populations aimed at identifying patterns of adaptation and its causes will clearly be most appropriately applied to the former class of characters.

We do not yet know exactly what criteria to use in distinguishing the two classes of behavioral traits, but Gittleman et al. (1996a) note that group and home range sizes in mammals show little or no correlation with phylogeny at any interval of time. More generally, display behavior tends to exhibit lower levels of homoplasy than does non-display behavior, suggesting that display behavior patterns may generally be poor subjects of studies on ecotypic variation. When the phenotype of interest is frequency of expression of the display, or ease with which it is elicited, an ecotypic approach may be more appropriate than if the interest is in the display motor pattern (Foster et al. 1996). In general, then, non-display behavioral characters that exhibit high levels of intraspecific variation are good candidates for studies of local adaptation and its causes.

In this chapter, I review uses of population variation in traits to test for adaptation and, when the hypothesis of adaptation is falsified, to consider the evolutionary processes and factors that might limit the expression of ecotypic variation. This is a critical area of research in that most of our models of biological and ecological systems assume adaptive equilibrium with respect to the local environment. However, populations may frequently not be at selective equilibrium, and model predictions will deviate from reality as a result.

The Behavioral Ecotype

Few would question whether behavior is an important component of animal adaptation to local conditions. The relevant question is whether particular behavioral traits represent ecotypes (i.e., have strong genetic components). Many workers believe that behavior is highly plastic in that it has many more alternative states than do morphological traits (Bateson 1988, Plotkin 1988, West-Eberhard 1989). Yet, a surprising number of complex behavioral traits have been found to have underlying genetic bases. Where behavior is more flexible, it is the inherited capacity to learn that enables some populations to exploit novel environments. Genes may also influence behavioral plasticity through alteration of developmental pathways (for discussion, see Thornhill and Alcock 1983, Carroll and Corneli this volume, Thompson this volume). Descriptions of several of the better understood examples of ecotypic variation in behavior follow.

Predator–Prey Interactions in Aquatic Systems

Predator–prey interactions in aquatic systems present numerous examples of behavioral ecotypes (see also Magurran in this volume). Arnold's (1981a,b) research on ecotypic variation in garter snake diets is a classic example. Arnold (1981b) reared offspring from female *Thamnophis elegans* collected from three different California populations at one inland and two coastal sites. Slugs form a major part of the diet of coastal *T. elegans*, while individuals from the inland site, which is outside of the range of the slugs, feed primarily on frogs and fish (Arnold 1981a). Analyses of the slug-eating response of newborn *T. elegans* from pureline and population crosses indicate that slug-eating has a genetic basis. Arnold concluded that the geographic variation in this behavior is maintained by selection, though more recently (1992) he detected a phylogenetic effect in the system.

Constantino and Drummond (1990) compared population differences in prey capture techniques of another garter snake, *Thamnophis melanogaster*, which is known as a fish specialist. They tested the predatory behavior of newborn snakes from a population inhabiting a lake where fast-moving fish were abundant and from a spring-fed pond that lacked fish. In this second population, slower moving leeches and tadpoles constituted the majority of the snakes' diets. Individuals representing both populations were offered live guppies (*Poecilia reticulata*, a novel prey species) as potential prey items in test trials. The snakes from the lake population made more aerial attacks on the guppies than did pond snakes and greater success was achieved with the aerial attacks than with underwater strikes by either population. This outcome was not related to differences in motivation: the lake snakes did not have shorter latencies to attack the guppies than did the pond snakes. The method of attack in this case was found to have a significant genetic component (see also Burghardt and Schwartz this volume).

Studies of ecotypic variation have also been completed on behavioral traits associated with the avoidance of predation (see Magurran chapter 7 this volume for a complete discussion of this research). Again, aquatic systems are ideal for such studies, as one can compare water bodies with and without fish predators. Levelsley and Magurran (1988) found that field-caught minnows, *Phoxinus phoxinus*, sympatric with the pike, *Esox lucius*, responded more vigorously to alarm substance secreted by conspecifics than did those that had experienced no pike predation in the wild. The individuals from habitats exposed to pike predation were also less likely to feed after the release of an alarm substance, were more likely to seek shelter, and showed less inspection behavior toward predators. Further work with this system would be required to determine whether the observed population differences in antipredator behavior have a genetic basis and thus represent ecotypes. Thus far, Magurran (1990) has demonstrated that individuals from populations sympatric with pike show greater behavioral adjustment after early experience with predators than do individuals from pike-free habitats, and she has shown that the population differences are evident in laboratory-reared fish upon their first exposure to a predator model, indicating genetic bases for the differences.

Similarly, guppies from different populations display differences in schooling tendencies that reflect the predation regimes to which they are exposed, and the differences have again proven heritable (Seghers 1974, Breden et al. 1987, Magurran and Seghers 1990; see also Magurran this volume). Schooling is a good deterrent to chasing predators such as fish.

Several studies have shown population genetic differences in the antipredator behavior of another fish, the threespine stickleback (Giles and Huntingford 1984, Huntingford and Wright 1992; see also reviews in Huntingford et al. 1994 and Magurran this volume). Giles and Huntingford (1984) tested adult fish and fry from seven localities in Scotland for their responses to model herons and live pike. Fish from high predation-risk populations showed significantly greater predator avoidance responses than did those from low-risk sites. Huntingford and Wright (1992) subsequently exposed laboratory-reared representatives of high predation- versus low predation-risk populations to simulated attacks of a model avian predator. The attacks took place whenever individuals moved into preferred feeding areas in the test arena. Although the majority of the fish learned to avoid the high-risk patch within 15 days of the start of the experiment, the fish from the high-risk site learned significantly faster than those from the low-risk site. This work and that of Magurran (1990) demonstrate that in addition to the well-defined population differences fishes exhibit in antipredator responses, mechanisms for adjusting behavior to immediate contexts exist that can also be considered as ecotypic traits. These are examples of the learning ecotypes I mentioned earlier.

Behavioral Ecotypes in Birds

Behavioral ecotypes have also been reported for birds. For instance, Kroodsma and Canady (1985) observed differences in marsh wren (*Cistothorus palustris*) singing behavior and associated neuroanatomy that reflect adaptations to sedentary versus migratory behavior patterns. Western populations of the marsh wren are sedentary and thus have longer breeding seasons than do eastern populations that are migratory. The lack of migration in the western populations leads to high population densities and smaller territories. Individuals from the western populations also have a higher rate of encounter with neighboring conspecifics than eastern marsh wrens and a higher probability of polygynous mating (50% as compared to 5% in the eastern populations).

Laboratory-reared marsh wrens from western populations also exhibited larger song repertoires, faster rates of performing these repertoires (higher recurrence intervals), and more brain tissue devoted to controlling these song characteristics compared to the migratory eastern populations. The results of these studies are consistent with the presence of underlying genetic bases to both morphological and behavioral adaptations.

Berthold and co-workers (see Berthold this volume for a review) have documented population differences in the migration routes of the palearctic bird *Sylvia atricapilla*. These differences appear to have underlying genetic bases, as naive European blackcaps from different populations exhibited different compass orientations in field cage trials. This example of a behavioral ecotype is associated with *S. atricapilla*'s use of a novel migration route that is 1500 km northwest of its traditional Mediterranean overwintering area. Berthold et al. (1992) suggest that ameliorated winter conditions favored those individuals that were physiologically predisposed to harsh breeding-ground conditions to migrate farther than other members of their former populations. Such events could have led to the apparent population divergence in migratory patterns. The researchers presented no evidence of physiological adaptation to the colder English climate, however. The new wintering grounds in Britain afford birds improved winter food supplies, shorter migratory distances, less intraspecific competition, and earlier return to spring breeding grounds due to an advanced spring photoperiod.

These examples demonstrate that behavioral traits serve to adapt populations to local conditions. If there were more interest in investigating the variance in traits rather than in identifying modal behavior, I expect that we would find behavioral ecotypes to be common phenomena. Magurran (1993; see also Magurran this volume, Verrell this volume) suggests that behavioral biologists have heretofore regarded variation as noise and thus something to be avoided. In contrast to this earlier view, we can use variability to gain insight into evolutionary processes and the factors that may limit adaptation.

Using Ecotypic Variation in the Study of Evolutionary Processes

The study of ecotypic variation in behavior can provide insight into how evolution by natural selection works. It also can show how various factors such as genetic system constraints, recent change in local environments, phylogenetic inertia (design limitations), and gene flow (interbreeding among local populations that are under different selective regimes) can interfere with selection and, consequently, limit adaptation. Unfortunately, such studies are rare because of the difficulty in obtaining the relevant data and hesitancy to seek alternatives outside of the adaptationist program.

Gene flow is a particularly interesting factor in the evolution of ecotypic variation because it represents two problems of scale: geographic and intensity. Gene flow may limit adaptation when the geographical scale over which it occurs is greater than the geographical scale at which ecological and biological processes are operating (Endler 1977, Wilson 1980, Mishler and Donoghue 1982, Damuth 1985, Eldredge 1985, Brandon 1990; see also Thompson this volume). Bernays and Graham (1988), for instance, report that gene flow is responsible for our failure to detect feeding specializations and other adaptations in herbivorous insects that associate with specific plant hosts in an often patchy environment. Sandoval (1994) reminds us that it is not merely the distance between patches that is important here, but also patch size and shape. Smaller patches will have greater levels of gene flow, and narrow strips of homogeneous habitat will experience greater gene flow than circular patches of equivalent area.

There is also the problem of intensity of selection relative to the level of gene flow. Based on theoretical models for neutral alleles, several workers (e.g., Maruyama 1972, Slatkin 1973, 1985a) have demonstrated that even very low levels of gene flow can be strong homogenizing forces. Via and Lande (1985) obtained similar results using a model that did not assume neutrality. Brandon (1990) concludes in his review that genes selected in different environments can mix during mating and thus retard or even stop adaptation to local conditions. On the other hand, some argue that strong natural selection should overcome any homogenizing effects that gene flow might produce (e.g., Ehrlich and Raven 1969). Endler (1977) contributed an algorithm that would define those conditions under which selection might override gene flow and vice versa. His estimate of the conditions of differentiation is expressed as the ratio of of the gene flow distance to the selection geographic gradient.

Two recent studies have addressed the relationship between gene flow and selection in patchy environments (Via 1991, Sandoval 1994). Via investigated the fitness performance of clones of pea aphids (*Acrythosiphon pisium*) in patches of two alternative host plants, red clover and alfalfa. Using reciprocal transplants of aphid clones derived from one or the other host plants, she found that there was considerable genetic divergence leading to significantly reduced longevities and fecundities in clones raised on "away" (foreign) as opposed to "home" host types. Nevertheless, she did observe levels of variance in performance traits in censuses of crop patches that reflected gene flow limitation of genetic differentiation. From sticky-trap sampling, Via estimated migration (potential gene flow) to be moderate and limited to the fall migratory season involving winged males and parthenogenetic females. Sandoval (1994) also investigated potential genetic differentiation of two color morphs of the walking stick (*Timema cristina*) on different host plants (the shrubs *Ceanothus spinosus*, Rhamnaceae, and *Adenostoma fasciculatum*, Rosaceae) in a patchy environment. She removed insects from selected host plants of each type and introduced equal numbers of the two morphs, striped versus solid color. She observed selection for a shrub-specific morph in this perturbation experiment, but found that gene flow destroys the spatial organization resulting from selection, presumably because adults migrate at the start of each generation, which leads to the random dispersal of offspring with respect to color pattern.

An Example of Gene Flow Limitation of Behavioral Ecotypy

I pursued another problem involving potential gene flow limitation of adaptation. In this case, the traits were behavioral traits that linked individual fitness to local competitive environments and predation risks. The following is a review of the desert spider system in which I have explored ecotypic variation in agonistic, territorial, foraging, and antipredator behavioral traits.

The Spider

Agelenopsis aperta is a funnel-web spider (Araneae, Agelenidae) that occupies arid-lands throughout the desert southwest United States (fig. 1-1). *A. aperta* is an annual and even-aged species whose life cycle is synchronized with the pattern of rainfall in its local area (fig. 1-2, Riechert 1974). Spiders mature during the hottest and driest time of the year (April–July). The offspring emerge coincident with the summer rains and associated high levels of insect prey.

Agelenopsis builds a sheet-web with an attached funnel that extends into a crack or crevice in the ground. This funnel protects the spider from unfavorable temperature extremes (Riechert and Tracy 1975) and from predation by visual predators (Riechert and Hedrick 1990). Because the web-sheet is not sticky, *A. aperta* must be at the funnel entrance of its web-sheet to locate prey contacting the web. However, this spider spends very little time out on the web-sheet itself: *A. aperta* leaves the security of the funnel retreat only when attacking prey, adding silk to the web, or interacting with encroaching conspecifics.

Selection pressures

Agelenopsis aperta typically occupies xeric habitats like the desert grassland habitat in New Mexico (referred to here as "NM grassland," (fig. 1-3), where insect

Figure 1-1 The arid-lands spider, *Agelenopsis aperta*, sits at the mouth of its funnel, monitoring its web for prey and conspecific intruders.

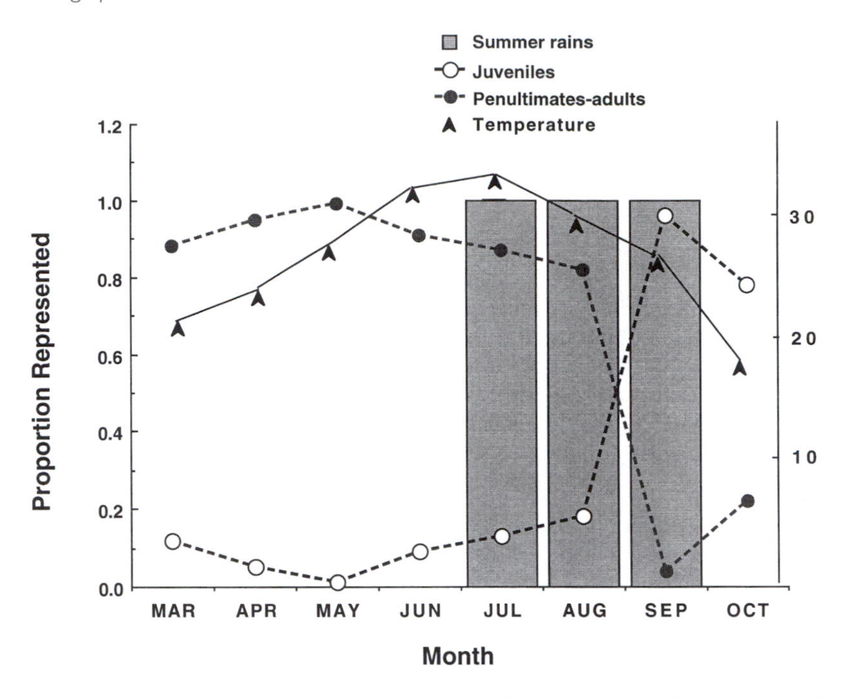

Figure 1-2 Representation of adult versus juvenile *A. aperta* in desert grassland habitat relative to summer rains and temperature over the course of the growing season of this spider (Riechert 1974). Penultimates are individuals that are one molt removed from sexual maturity. This is an annual species whose reproduction is timed to the advent of the summer rains. The spider is inactive during the cold months, November–February.

prey abundances are sufficiently low as to limit spider survival and reproductive success (table 1-1, Riechert 1981). The restrictive thermal environments afforded by these habitats further limit spider foraging activity to just a few hours a day (fig. 1-4, Riechert and Tracy 1975). Most of the time, it is either too hot (during the day) or too cold (at night) for spiders in this habitat to be active, particularly at exposed sites. Individual *A. aperta* occupying xeric habitats compete for those few web-sites that provide moderate thermal environments and higher prey availabilities (Riechert 1978a, 1979, 1981). The failure of individuals to survive to reproduction at most sites in the desert grassland habitat is evident when evaluating the contributions spiders from different-quality sites made to the gene pool in different years (fig. 1-5, Riechert 1981). While food may limit spider survival and reproductive success at the NM grassland site, predation on *A. aperta* is negligible in this habitat (Riechert and Hedrick 1990). Birds are the major predators of *A. aperta*; but their numbers are low at the desert grassland site, perhaps reflecting a shortage of perch and nesting sites as well as low levels of insect prey.

Figure 1-3 The desert grassland habitat of south-central New Mexico can support growth and reproduction among *A. aperta* on only 12% of its surface at best (Riechert 1979).

I have also investigated a population of *A. aperta* that occupies a very different habitat, a riparian woodland in the Chiricahua Mountains of southeastern Arizona (referred to here as "AZ riparian," fig. 1-6). Although the NM grassland and AZ riparian sites are at similar elevations (1400 m), a permanent spring-fed stream supports high insect numbers in the AZ riparian area (table 1-1). The stream also supports a tree canopy, which provides *A. aperta* shelter from temperature extremes. Thus, spiders in the AZ riparian area have abundant food and can forage

Table 1-1 Comparison of the availability of insect prey for *A. aperta* at three localities: New Mexico desert grassland (6 trap years), Arizona riparian (5 trap years), and Texas riparian (4 trap years).

Location	*Prey Availability (mg dry weight/trap/day)*		Proportion of Days Optimal Energy Needs Met
	Mean	SE	(≥20 mg dry weight)
New Mexico	24.9	1.1	0.35
Arizona	106.5	5.5	0.87
Texas	115.7	14.8	0.89

All items between 4 and 25 mm in body length that encountered sticky traps were included as prey. These traps were placed at web-sites after removal of the resident spiders. All comparisons showed significant mean dry weight differences, though in the cases of Arizona and Texas animals this does not reflect a significant difference in prey consumption.

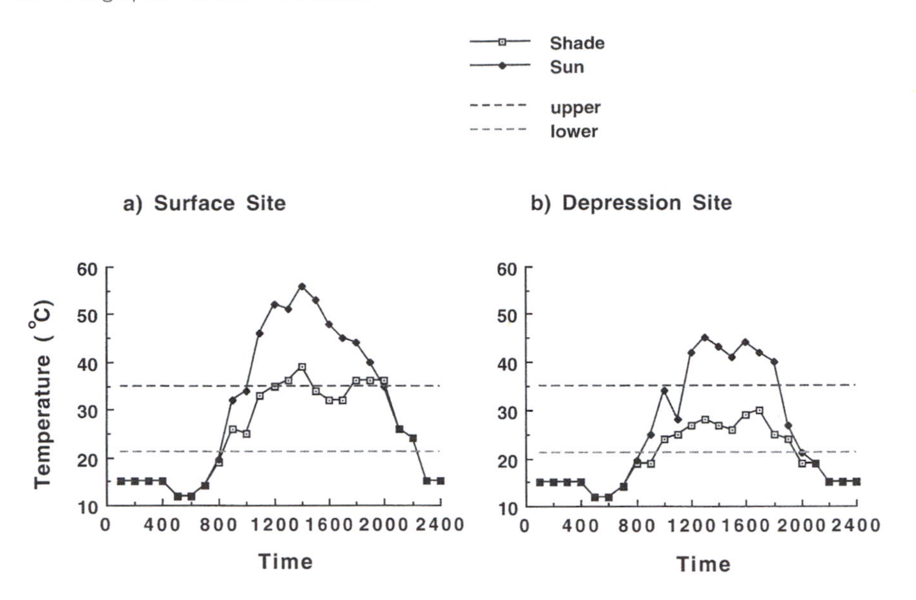

Figure 1-4 Predicted *A. aperta* body temperatures on a clear day in mid-June at (a) a surface site (low survivorship) versus (b) a depression site (highest survivorship) in the New Mexico desert grassland habitat. The two curves represent spider temperature in full sun versus full shade. Actual spider temperature when out on the web-trap will be between these two extremes, depending on the amount of sunlight to which the spider is exposed. Spider foraging is permitted only when body temperature falls between the two dashed lines, representing upper and lower limits to this spider's activity range (Riechert and Tracy 1975).

most of the day (Riechert 1979). However, in this riparian habitat where insect numbers are high and potential perch sites are abundant, bird populations are large and species representation is rich. Comparison of spider losses in quadrats exposed to bird predation versus those protected by bird netting indicate that approximately 40% of the AZ riparian local population is lost per week during the nesting season of nine species of birds in the area, although the variance is high (Riechert and Hedrick 1990). Other visual predators that commonly feed on spiders in these southwestern United States habitats, robberflies (Diptera, Asilidae) and spider wasps (Hymenoptera, Pompylidae), are not important predators on *A. aperta*. In fact, they are high profitability (low cost–benefit ratio) prey of *Agelenopsis* (Riechert 1991). In any event, the mesh size used to cover the protected quadrats did not interfere with the entrance of predaceous insects, snakes, and small rodents. These same nets are used over sticky traps to keep birds from being captured while gleaning insects from them.

Fitness-linked Behavior

The four behavioral traits that I have found to affect the individual fitness of *A. aperta* are diet choice, agonistic or contest behavior, territory size, and antipreda-

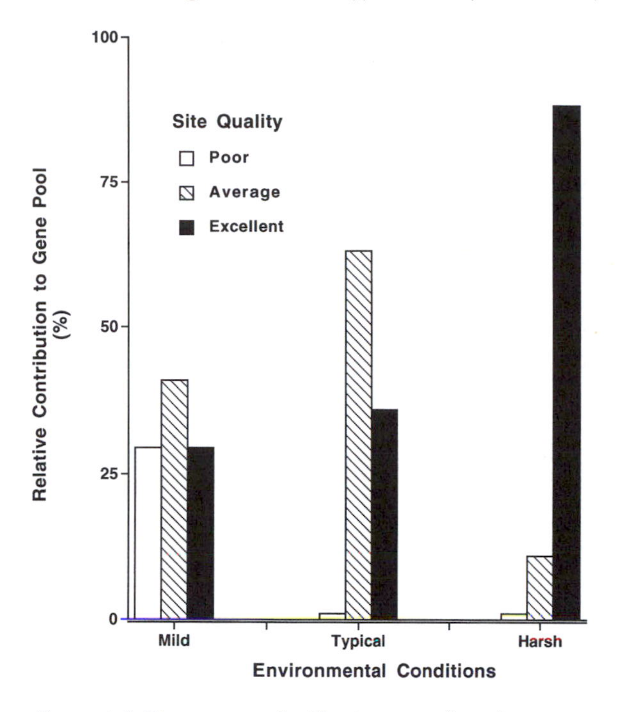

Figure 1-5 Percentage of offspring contributed to gene pool by spiders occupying different-quality sites under varying environments. Mild, typical, and harsh years refer to prey accessibility, which in turn is dependent on temperature and moisture. Typical years in this study area are characterized by winter and summer precipitation and by a hot and dry period between May and early July (see fig. 1-2). Mild years reflect greater precipitation and cloudiness, especially in late spring and early summer, while harsh years are drought years with little precipitation, no rain in the spring, and very hot and dry conditions in May, June, and July. Data represent reproductive success/individual summed for all individuals occupying sites of a given quality (Riechert 1981).

tor behavior. Diet choices, agonistic behavior, and territory sizes exhibited by individuals from the two populations reflect the differences in prey availabilities offered by arid NM grassland and mesic AZ riparian habitats: (1) NM grassland spiders attack a greater proportion of the prey encountering their webs than do AZ riparian spiders (Riechert 1991); (2) NM grassland spiders show greater levels of escalation and engage in lengthier contests over web sites than do AZ riparian spiders (Riechert 1979); and (3) NM grassland spiders maintain larger energy-based territories throughout their lives than do AZ riparian spiders (Riechert 1981).

Figure 1-6 Nearly all the riparian habitat in southeastern Arizona permits survival to reproduction for *A. aperta*. There should be low competition for available sites, but it is unexpectedly high.

Population variation in antipredator behavior reflects the higher predation pressures experienced by AZ riparian spiders: spiders from this population exhibit a greater probability of retreat into the funnel after exposure to a predatory cue and exhibit longer latencies to return to foraging once they have retreated (Riechert and Hedrick 1990).

Behavioral Ecotypes

Collaborative work with John Maynard Smith and Ann Hedrick indicates that the diet choice, agonistic or contest behavior, territory size, and antipredator behavior have strong underlying genetic bases. For instance, Hedrick and Riechert (1989) found that the differences in attack rates directed toward prey reflect longer inherent latencies to attack prey by AZ riparian spiders. Most individuals from the AZ riparian population exhibited latencies to attack prey that were two orders of magnitude greater than the latency scores of NM grassland individuals. These population differences were also observed in second-generation laboratory-reared spiders, indicating that the differences were genetic.

Comparisons of the territorial behavior of the pure population line and the crossbred AZ/NM line indicate that agonistic behavior and territory size are quantitative genetic traits that have both autosomal and sex chromosomal contributions (Riechert and Maynard Smith 1989). Finally, Riechert and Hedrick (1990) documented differences between the two populations in the time it took for spiders presented with a predatory cue to return to foraging mode. These differences were observed in both wild-caught individuals and in second-generation laboratory-reared progeny. I conclude that the two behavioral phenotypes (the more aggressive "arid-lands" versus the more fearful "riparian") of *A. aperta* are ecotypes as defined by Turesson (1922).

Deviation of the Riparian Population from Adaptive Equilibrium

Various models (evolutionarily stable strategy, ESS, and optimal foraging; see below) have been applied to the behavior of NM grassland and AZ riparian spiders to determine the extent to which these populations are at adaptive equilibrium. NM grassland spiders have been found to closely approximate ESS model predictions of contest behavior for a population that exists under high competition for web sites affording sufficient food for survival but that has negligible local rates of predation. If predation rates had been high, they would have limited spider contest and foraging activities, leading to conflicting selection pressures (Hammerstein and Riechert 1988). Desert grassland spiders also fit optimal foraging predictions of a broad diet under the pressure of severely limited prey levels (Riechert 1991). Grassland *A. aperta* show high attack rates toward all prey types encountering their webs, with the exception of small parasitic wasps, which they apparently are unable to detect (fig. 1-7). In fact, inspection of the data set indicates that most failures of grassland *A. aperta* to attempt capture of prey involved either lack of detection or an insect whose size was beyond its capture range. Because *A. aperta's* web-sheet is not sticky, these large prey readily escape. Grassland spiders do ultimately obtain many small prey to which they exhibit no immediate capture response. The spiders periodically search their webs, picking up all small prey items encountered.

Game theoretical analyses of AZ riparian *A. aperta* territorial disputes have led to the prediction that after assessing relative size, an AZ riparian spider should withdraw from a contest if it is the smaller of the two opponents in a dispute or if it is equal in weight and is the intruding individual (Hammerstein and Riechert 1988). If the spider is the larger opponent or equal in weight and the territory owner, it should display. Thus no escalation to potentially injurious behavior is predicted for this population. Yet, empirical evidence from territorial contests staged over natural web-sites indicates that AZ riparian spiders escalate to potentially injurious behavior 20% of the time when withdrawal is predicted and 49% of the time when display is predicted (Hammerstein and Riechert 1988). The foraging behavior of AZ riparian spiders also deviates from the predictions of optimal foraging theory in that individuals afforded abundant prey (AZ riparian site) should exhibit narrow diets: They should attack those prey types that provide the greatest ratio of reward/effort to a greater extent than individuals that encounter fewer prey (NM grassland spiders) (Pyke 1984). Instead, AZ riparian spiders exhibit a low attack rate toward all prey types, regardless of their profitabilities (fig. 1-7, Riechert 1991). Attacking low-profitability prey puts AZ riparian spiders at predation risk both from foraging birds and from the prey themselves (e.g., ants, predaceous bugs, and mantids).

Many years of field and laboratory measurements and several generations of models have gone into the Hammerstein and Riechert (1988) test for evolution-

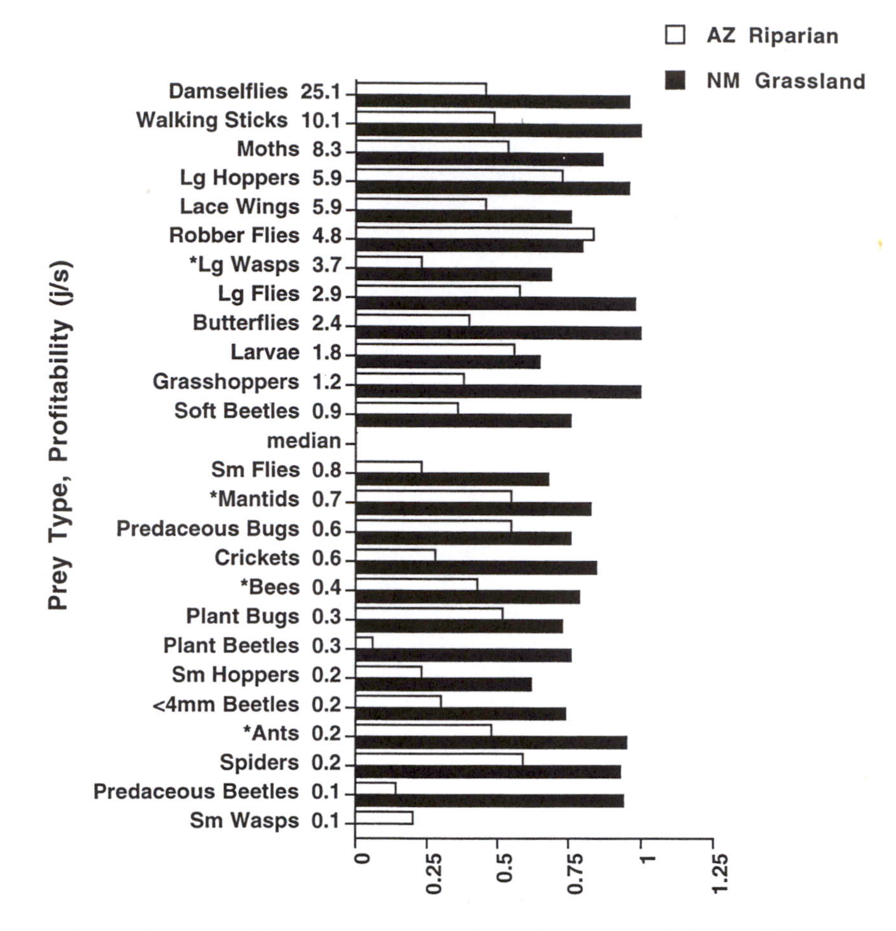

Figure 1-7 Attack rates exhibited by two populations of *A. aperta* (Arizona [AZ] riparian, New Mexico [NM] grassland) toward prey types that offer the profitabilities (capture effort relative to energetic reward) shown (Riechert 1991). Prey categories represent morphologically similar arthropods. A minimum of 25 encounters were observed for each prey category for each population.

arily stable contest strategies in the two populations. This test represents the culmination of model development by Maynard Smith, Riechert, and Hammerstein over a 10-year period. All the parameters required for the accurate estimation of strategy payoffs were measured in the field, so that we could express the value of winning or losing a single contest in terms of the future reproductive success of the particular contestants. From earlier versions of the models and empirical tests of them, we learned what information individual spiders had at different stages of these territorial disputes (Riechert 1978b, 1979, 1984; Maynard Smith 1982, Maynard Smith and Riechert 1984) as well as what each contestant communicated to its opponent (e.g., Riechert 1984). Empirical studies provided information about the relevant parameters that needed to be included in the model. For instance, for both populations, escalation led to specific probabilities of leg loss and mortality (total loss in fitness) that corresponded to size of an individual relative to that of its opponent. The loss of a leg decreased subsequent foraging rates by 10% and future success in territorial disputes by 25%. For each population, there also were parameters that had unique influences. Thus, opportunity costs (based on the probability of gaining another site in the event of losing a site) made the payoffs of winning a site high to arid grassland spiders, whereas they had little effect on the payoffs to riparian spiders. In contrast, the cost of displaying toward an opponent in a contest (largely injury and death due to predation) was high for riparian spiders, but was an order of magnitude smaller for grassland spiders (table 1-2, Hammerstein and Riechert 1988). Finally, the model predictions for AZ riparian spiders were robust in that there were marked differences in payoffs between stategies, and the empirical tests of the predictions entailed large sample sizes.

Strength is added to the conclusion that AZ riparian animals are not at adaptive equilibrium with respect to the four behavioral traits investigated by the fact that substantial individual variation (almost bimodal) is seen in all the traits. This phenomenon is exemplified in figure 1-8, which shows the latencies to return to foraging mode after a predatory cue for laboratory-reared individuals belonging to the F_2 generation. Variation of the type indicated in figure 1-8 for AZ riparian spiders is highly suggestive of population mixing.

Factors That Might Contribute to the Apparent Maladaptation of the Riparian Population

The robustness of the ESS model results, the degree of individual variability noted in AZ riparian spider behavior, and the fact that aberrant behavior is exhibited across a wide variety of traits all suggest that the failure of AZ riparian spiders to achieve an adaptive equilibrium is a real phenomenon and not a conse-

Table 1-2 Population comparisons of payoff components to spider territorial disputes: general categories and partitioning of display costs.

	New Mexico Desert Grassland	Arizona Riparian
Expected Egg-Mass Increments and Decrements (mg wet weight)		
Value of winning average site	16.7	1.6
Cost of leg loss	14.2	6.2
Cost of lethal injury	93.7	84.1
Cost of display	0.1	3.0
Decrement in Egg Production (mg wet weight)		
Energy expended	4.8×10^{-5}	2.0×10^{-6}
Loss in food intake	0.10	0.06
Loss to predation	0	2.4

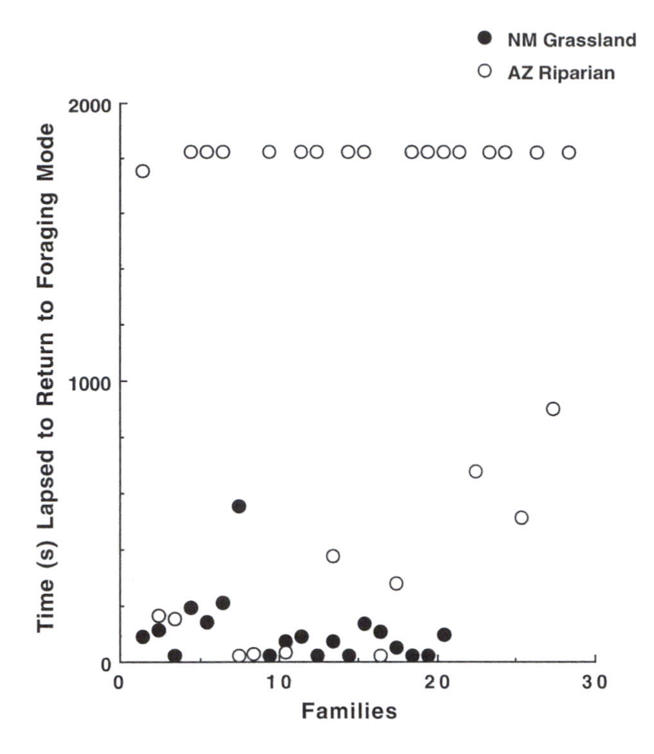

Figure 1-8 Individual scores for latency to return to the funnel after presentation of a predatory cue for F_2 generation descendents of the New Mexico (NM) desert grassland and Arizona (AZ) riparian woodland populations. Each point represents a family score obtained by sampling one individual at random from that family. The line of AZ riparian scores at the top of the graph reflects the failure of many AZ riparian spiders to return to foraging within the time alloted for the trials.

quence of inadequate model development. Repeatability estimates available for our measure of fearfulness to predatory cues (Hedrick and Riechert unpublished data) indicate that this trait, at least, is heritable and has sufficient genetic variation to permit change in response to selection.

A highly variable environment or a recent change in the local environment are often reported causes of maladaptation. Neither of these factors seem pertinent to the AZ riparian situation. Our prey availability data for the AZ riparian habitat come from 5 years of field censusing with a coefficient of variation (CV) of only 9.3% among years compared to a CV of 97.5% among years for desert grassland prey availabilities (Riechert 1993a). I attribute this lack of variability in insect levels to the close proximity of the spring-fed stream to AZ riparian spider sites. There is every indication from studies of riparian habitats that they are not of recent occurrence. Smiley et al. (1984), for instance, reported that riparian habitats in Arizona that are now wet have been accumulating sediment for the last few hundred years to few thousand years. Further, the spring that supports the riparian area at my AZ riparian study site is not downstream of an aquifer fed by a reservoir.

Predation pressure by birds in the riparian habitat selects against such behavior as maintaining excessively large territories, spending time defending the larger territories, engaging in lengthy "conventional" contests (displays without physical engagement) when withdrawal from a contest is appropriate, and attacking prey that take long to subdue relative to the consumption benefits generated. Given the marked environmental constancy and high level of predation pressure, only a few generations of selection may be required to eliminate disadvantageous behavior in *A. aperta* if sufficient genetic variability exists (Hammerstein and Riechert 1988). Repeatability estimates of *A. aperta's* antipredator behavior (Hedrick and Riechert unpublished data) indicated that high heritabilities for this trait exist in the AZ riparian population.

Based on this information, I limited further study to tests of the two alternative factors having the greatest probability of limiting adaptation given the available information: phylogenetic inertia (Riechert 1993b) and gene flow (Riechert 1993a). Under a phylogenetic inertia hypothesis, I predicted that a major change in the "wiring" of *A. aperta's* nervous system would be required to achieve a change in behavior from an arid-lands phenotype to a riparian phenotype. Because the AZ riparian population is, in fact, an island population surrounded by other *A. aperta* populations occupying more arid habitats, there is a possibility that the conditions for gene flow limitation of selection are operating as described by Sandoval (1994). Under a gene flow argument, I hypothesized that sufficient mixing and consequent interbreeding occurs between AZ riparian habitat *A. aperta* and the surrounding arid-land *A. aperta* to prevent the riparian phenotype from becoming prominent.

Phylogenetic Inertia

Testing for phylogenetic inertia requires that the traits be examined in an outgroup, a sister taxon to the taxa of interest (Wiley 1981). As an indirect test for potential phylogenetic inertia effects, I extended the study of *A. aperta* to a riparian population at the eastern extent of the species' range (along the Pedernales River in central Texas, referred to here as "TX riparian"). I consider this population to meet the outgroup requirement because it is representative of a variant of *Agelenopsis aperta* to which Wilton Ivie was preparing to assign the specific status of *Agelenopsis mexicana* when he was killed in an automobile accident. On the basis of its distinctive male genitalia, Gertsch (personal communication), has indicated that populations of *A. aperta* from central Texas and Mexico deserve at least subspecific status. He examined specimens from my Pedernales River study area and assigned them to this variant group of *A. aperta* whose specific status is in question. Gertsch described the species, *Agelenopsis aperta*, in 1934. Further, Paison and Riechert (unpublished data) have developed a phylogeny of the genus *Agelenopsis* that suggests the arid species *A. aleenae* and *A. spatula* are derived from *A. aperta* and have their origins in Mexico, presumably with *A. mexicana*. The phylogeny is based on palpal morphology and Mengel's (1964) theory of the actions of glaciation and the Great Plains as isolating barriers on speciation events and biogeography.

Because *A. aperta* in the central Texas study area is considered by systematists to deserve at least subspecific status, I used it as an outgroup by which to compare populations from New Mexico and Arizona. There are two other reasons for completing the particular study described here. This TX riparian population is not bordered by arid-lands populations, and if I falsify the hypothesis that the behavior of these riparian spiders is similar to that of the AZ riparian spiders under investigation, then I have results that support a gene flow hypothesis. Also, this population can be used to test the ESS and optimal foraging predictions developed for riparian phenotypes of *A. aperta*.

Indeed, I found that prey availabilities afforded by TX riparian and AZ riparian habitats were similar (table 1-1) and that TX riparian spiders are exposed to some predation by birds (approximately 20% of the local population is lost per week during the bird nesting season; Riechert 1993b). I also found that TX riparian spiders displayed the more fearful "riparian" phenotype predicted for the conditions just described (Riechert 1993b): they limited prey capture attempts largely to high profitability prey (table 1-3), they exhibited a low frequency of escalation in contests (none exhibited in contexts favoring withdraw and 5% in contexts favoring display), and they defended webs, but not space around them, against intrusion (no territories were maintained; table 1-4).

Because TX riparian *A. aperta* did exhibit the riparian phenotype predicted for them by ESS and optimal foraging models, I concluded that phylogenetic inertia probably did not underlie the failure of AZ riparian spiders to exhibit this pheno-

Table 1-3 Comparison of attack rates toward prey of different profitabilities for three populations of *A. aperta:* New Mexico (NM) desert grassland, Arizona (AZ) riparian, and Texas (TX) riparian.

		Capture Attempt Rates		
Prey types	Profitability Score (J/s capture effort)	NM	AZ	TX
Low Profitability				
Ants	0.2	0.96	0.59	0.31
Plant bugs	0.3	0.72	0.50	0.21
Predaceous bugs	0.6	0.53	0.67	0.00
High Profitability				
Moths	8.3	0.87	0.54	0.95
Damselflies	25.0	0.96	0.47	0.94

TX riparian spiders attacked high-profitability prey with a significantly higher frequency than they did those prey types offering lower energetic rewards, whereas NM and AZ spiders did not show significantly different attack rates between high- versus low-profitability prey types (Riechert 1993a). Note that experimental data presented in Riechert (1991) indicate that both AZ riparian and NM grassland spiders can discriminate among prey types.

Table 1-4 Results of territory-size experiments for three populations of *A. aperta*: New Mexico (NM) grassland, Arizona (AZ) riparian, and Texas (TX) riparian.

	Nearest-Neighbor Distance (cm)					
	Wild-caught Individuals			*Lab-Reared Individuals*		
	n	Mean	SE	*n*	Mean	SE
NM Grassland	13	130.5	18.5	87	33.6	6.4
AZ Riparian	24	53.6	1.1	37	51.4	6.2
TX Riparian	25	2.1	0.9	27	6.6	3.4

Protocol involved progressively moving individuals and their webs closer to neighboring conspecifics within linear enclosures until a proximity was reached that was not tolerated (see Riechert and Maynard Smith 1989 for details).

type. The study did support the gene flow hypothesis, because in the absence of potential gene flow from more arid habitats, I found this population to exhibit the behavioral traits predicted for populations under reduced competition for quality sites. The results presented for the Texas population also demonstrate the validity of our foraging and ESS model predictions.

Gene Flow

I have completed a series of studies in the AZ riparian area that conclusively demonstrate the importance of gene flow in limiting adaptation in this spider population. First, I extended the study of the AZ *A. aperta* system to three additional local populations in the Chiricahua Mountain area: desert scrub, desert riparian, and evergreen woodland (fig. 1-9). These new AZ habitats afford prey availabilities more similar to those at the NM grassland site than to those at the AZ riparian site (fig. 1-10). The *A. aperta* in these sites also show levels of escalation in contests and attack rates toward prey more similar to those predicted for NM grassland than to AZ riparian *A. aperta* (Riechert 1993a). These new habitats are, in fact, more arid than the AZ riparian habitat and the spiders inhabiting them exhibit the "arid-lands" phentoype.

I used protein electrophoresis to obtain insight into the level of gene flow that has occurred among local populations in the vicinity of the AZ riparian population. Enzymes that differ in their electrophoretic mobility as a result of allelic differences at given loci are referred to as allozymes. Allozymic variation in a population is thus indicative of genetic variation, and populations that have similar genetic structures (allele frequencies) are likely to have experienced gene flow in the past.

I applied two measures of gene flow to the allele frequency data I obtained from protein electrophoresis for local populations of *A. aperta* in the Chiricahua Mountains area: F_{st} and private allele estimates. Wright (1931) indicated that an N_m (where N = the breeding population size in a given habitat and m = the propor-

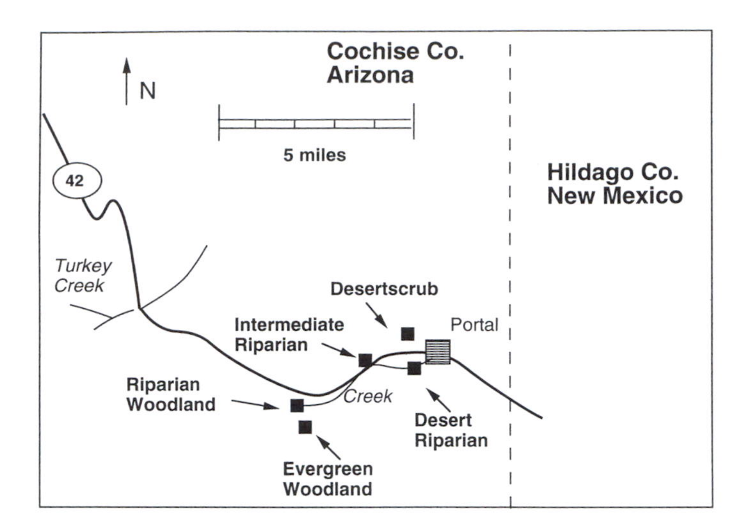

Figure 1-9 Positions of various *A. aperta* local populations in the Chiricahua Mountain area of southeastern Arizona (from Riechert 1993b).

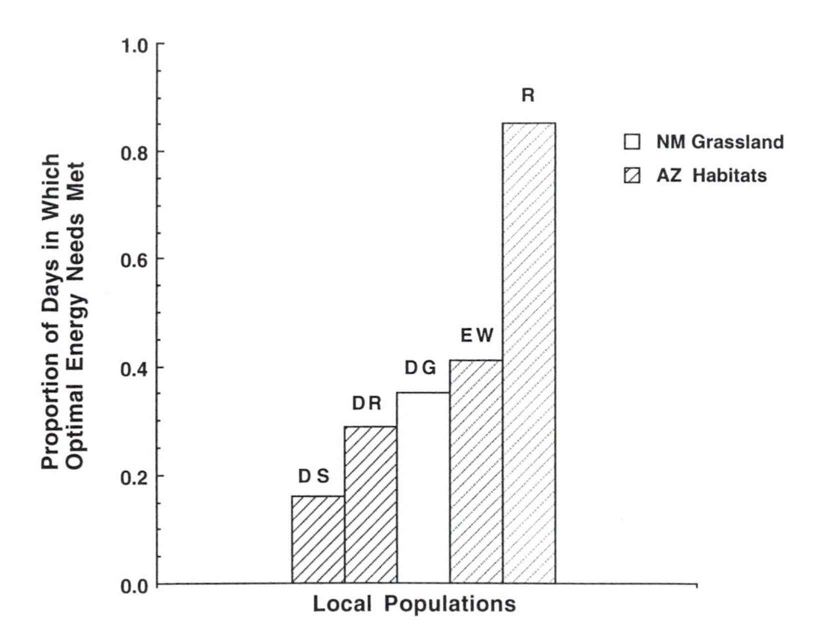

Figure 1-10 Prey available to local populations of *A. aperta* in the Chiricahua Mountain area, Cochise Co., Arizona (AZ), with New Mexico (NM) grassland included for reference. DS, desertscrub; DR, desert riparian; EW, evergreen woodland; R, AZ riparian. Optimal energy needs refers to prey levels that afford at least 20 mg dry weight of prey/day, the feeding level this spider will exhibit when given the opportunity to feed ad libitum.

tion of migrants into that habitat from other local populations) ≥ 4 indicates a panmictic population—one in which gene flow has occurred among local populations. *F* statistics are a measure of the way the genetic variance in a population is partitioned into a variance within local (sub-) populations and a variance between local populations. The fixation index, F_{st}, which is the correlation of two randomly chosen alleles in a subpopulation relative to that among alleles in the whole population, is used in the estimation of N_m (Wright 1931). Using this *F* statistic, I estimated N_m for the local AZ *A. aperta* populations to be 4.96. This value indicates that there has been general mixing of the local populations, despite the fact that *A. aperta* is not an aerial disperser. Most spiders disperse on air currents through the use of silk threads. *Agelenopsis aperta* does not exhibit this ballooning behavior; it disperses by wandering across the ground.

Most important to this particular study is not the general level of genetic mixing in the global population, but rather the level of mixing that occurs between spiders in the AZ riparian habitat and those in immediately adjoining habitats. For this purpose, Slatkin (1985b) proposes the use of an analysis based on the frequencies of alleles that are unique to particular local populations (i.e., private alleles). The so-called private allele estimates of N_m indicate that more genetic exchange occurs between local populations adjoining the AZ riparian habitat than between those more widely separated (table 1-5). This result supports the hypothesis that movement by *A. aperta* is between adjacent arid-land and riparian habitats.

I used 5 weeks of drift-fence trapping to determine the extent and direction of spider movement between the AZ riparian habitat and the immediately adjacent evergreen woodland habitat and thus to determine the extent to which gene flow might regularly occur. Ninety-seven percent of the moves of AZ riparian spiders were within the riparian habitat; 42% of the moves of evergreen woodland spiders

Table 1-5 N_m estimate of gene flow among five local populations of *A. aperta* in the Chiricahua Mountains of southeastern Arizona.

No. of Populations	$p(1)$	N_m*	Average Sample Size
5	0.0607	2.62	31.9
4	0.0540	3.17	30.8
3	0.0483	4.11	31.9
2	0.0484	4.46	34.9

Proximities of neighboring populations to the Arizona riparian population are shown in figure 1-4. N_m refers to the relationship between breeding population size and the proportion of this unit that consists of immigrants (m). At least 20 individuals from each local population were scored for each locus. Private allele estimate of Slatkin (1985b) calculated at increasing distances from Arizona riparian habitat, $p(1)$ = average frequency of alleles observed in only one population; N_m* = N_m estimate corrected for deviation of sample size from 25, and decreasing number of populations reflects the exclusion of more distant local populations from the Arizona riparian.

were into the riparian habitat. I concluded that gene flow probably regularly occurs and is predominantly from more arid habitats into the AZ riparian area. This result is predicted both from the idea that animals might preferentially move from less favorable habitats to more favorable habitats and from Sandoval's (1994) view that gene flow will be unidirectional from larger patches into smaller ones (particularly strips). The evergreen woodland habitat is extensive when compared to the narrow strip of AZ riparian habitat, which is limited to approximately 25 m on either side of a small stream.

The results of the TX riparian study lend further support to the hypothesis that gene flow limits adaptation. The riparian phenotype persists in the TX riparian population in the absence of potential gene flow from arid-lands populations (there are no surrounding arid-land populations of *A. aperta* in the area). This study also supports the validity of our optimal foraging and ESS modeling efforts. The TX riparian population closely approximates model predictions made for behavioral traits that are adapted to low competition and high levels of predation pressure.

The Interaction between Gene Flow and Selection

To test the hypothesis that gene flow restricts the adaptation of AZ riparian spiders to local conditions, I established eight 4-m^2 exclosures in the AZ riparian habitat that eliminated potential gene flow (Riechert 1993a). My aim was to determine whether a shift in behavior toward the predicted riparian phenotype would occur if selection were permitted in the absence of population mixing. After removing all resident *A. aperta*, I stocked each exclosure with recently emerged spiderlings obtained from 10 egg cases collected in the AZ riparian habitat. Approximately 80 spiderlings were released into each plot such that all exclosures received the same familial representation. I placed netting over four of the exclosures to protect the spiderlings from bird predation. Thus, all of the units excluded conspecific immigrants, hence the term *exclosure*, while half were protected from predators, and half unprotected.

After 6 months I assessed spider numbers and behavior in the exposed versus protected exclosures. I found protected exclosures contained four times as many *A. aperta* as did the exposed exclosures. Spiders remaining in the exposed exclosures tested as the more "fearful" (riparian) phenotype: on average they took longer to return to foraging after retreating to the web-funnel than did spiders from the protected exclosures (fig. 1-11, Riechert 1993a). Spiders protected from predation by birds also exhibited more variance in their "fear" scores. Finally, the spiders in the protected exclosures significantly more often won territorial contests when matched for weight against spiders from exposed exclosures (fig. 1-12). These results suggest differential mortality of the more aggressive arid-land phenotypes in the exposed exclosures. Alternatively, the differences in behavior may reflect experiential effects: spiders exposed to bird predation attempts may have developed behavior patterns appropriate to the presence of predators as a consequence of experience.

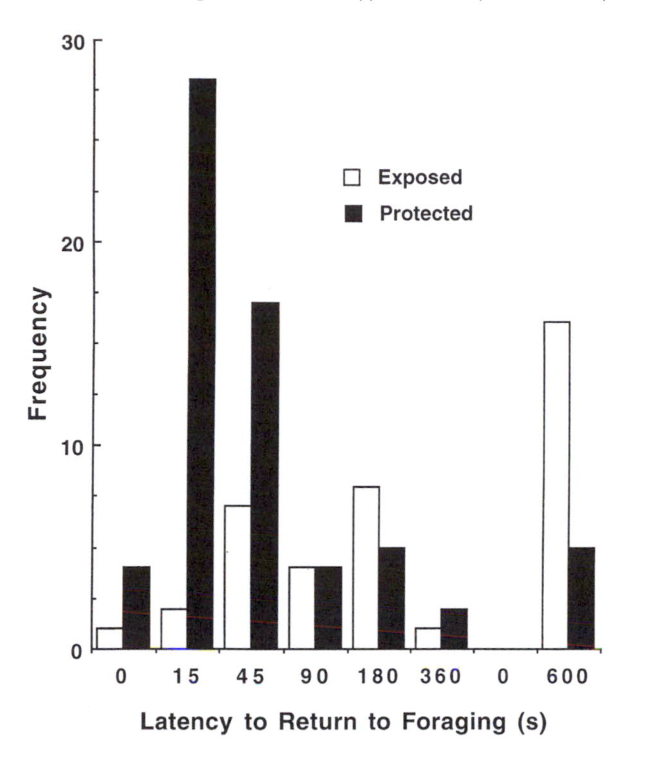

Figure 1-11 Latency to return to a foraging mode at the funnel entrance of the web after the presentation of a predatory cue for individuals collected from exclosures that were either exposed to bird predation or protected from predation for 6 months. Summaries from four replicates of each exclosure type are shown. Initially all exclosures had the same familial representation of spiders.

I tested these alternatives by collecting the offspring of the spiders subjected to the two different treatments (protected and exposed) after their emergence from egg cases the next spring. Again, I did not permit the population mixing that my movement data indicate happens each breeding season. Thus, this second test could also be used to test the hypothesis that gene flow restricts adaptation. The spiderlings from both treatments were reared in the laboratory. They were fed ad libitum and had no experience with conspecifics until they reached the penultimate stage just before maturity. At this time, individuals were randomly selected for behavioral scoring.

The results were similar to those obtained using spiders taken from the exclosures. Contests staged over webs were won by offspring of spiders from the protected exclosures (fig. 1-12). Comparing spider response in the predator cues test, the offspring from the single exposed exclosure (three of the four exposed exclosures had no second-generation *A. aperta*) exhibited greater latencies to re-

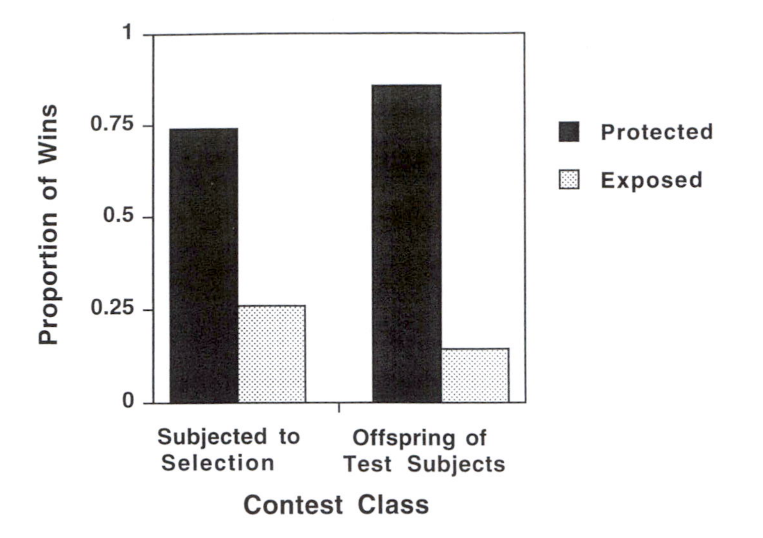

Figure 1-12 Results of pairwise contests staged between equally weighted Arizona riparian individuals, one of which represented animals exposed to predation pressure and the other animals protected from it. The two groups of animals were derived from enclosures that did not allow gene flow from more arid areas: after 6 months of selection and no potential population mixing, and offspring of matings within protected versus exposed treatments after one season of selection. The latter group was reared in the laboratory after emergence from egg cases in the enclosures.

turn to foraging after a predator cue than did any of the four test groups representing the protected treatment (table 1-6).

The results of this experiment indicate that one generation of selection is sufficient to produce a marked shift in *A. aperta* behavior in the absence of gene flow. This shift apparently is not achieved in the natural system because the levels of gene flow are sufficiently high to prevent its occurrence.

Table 1-6 Data from *t* test comparisons of predator cue test results for offspring from one exposed treatment enclosure to those from each of four protected treatment enclosures (see text for details).

	Latency to Return to Foraging (s)		
Class	Mean	SE	*p*
Exposed 1	1215.9	27.1	
Protected 1	407.2	26.0	0.00060
Protected 2	449.3	54.2	0.01500
Protected 3	161.3	21.6	0.00002
Protected 4	316.6	36.3	0.00060

Conclusions

The theme of adaptation underlies much of evolutionary thought. John Maynard Smith (1989), for instance, believes that Darwin understood that organisms were adapted before he developed his theory. Darwin then went on to conclude that adaptation is both obvious and pervasive and a concept that any theory of evolution must explain. Two recent treatments of specific systems, Trinidad guppy color patterns (Endler 1995) and Galapagos finch morphology (Grant 1986), provide ample evidence for the operation of natural selection in nature.

Nevertheless, the concept of adaptation has a long history of debate (e.g., Morgan 1903, Simpson 1953, Williams 1966, Gould and Lewontin 1979, Mayr 1983, Provine 1985, Dawkins 1986, Reeve and Sherman 1993). Critics (e.g., Gould and Lewontin 1979) maintain that adaptationists fail to consider alternative explanations: they do not apply the scientific method (attempt falsification) in studies of adaptation. There are, indeed, methodological problems associated with testing for adaptation. It is much easier to merely fault model predictions, suggest possible alternative functions for the traits, or not to publish negative data, than it is to label a trait as nonadapted or maladapted.

It is time to start pursuing this important problem with thorough empirical studies that both test the adaptiveness of traits through measurement of their fitness consequences and test among alternative explanations when a trait does not meet the criterion of Reeve and Sherman (1993) for adaptation (i.e., that it be the variant that offers the highest fitness). Traits may well be nonadaptive because (1) they are experiential responses rather than genetically programmed; (2) they are selectively neutral (e.g., the result of genetic drift); (3) they may be correlated (e.g., pleiotrophic effects of the same genes or linkage) with traits that have some adaptive function; (4) they may be anachronistic (the environment is different today from that in which the feature evolved); (5) there may be design limitations that reflect ancestral traits (phylogenetic inertia); (6) they may be limited by underlying genetic constraints (e.g., insufficient genetic complexity or lack of genetic variability); or (7) they may reflect the consequences of conflicting spatial scales (i.e., where the spatial scale at which selection is operating is smaller than that at which genetic exchange occurs).

In my long-term study of the selection pressures operating on *A. aperta*, I have used behavioral ecotypes to consider the fitness consequences of various behavioral traits as well as to consider nonadaptive alternative explanations. Through extensive ecological studies, I have identified several behavioral traits that determine individual fitness under the various selection pressures to which divergent populations are exposed.

In finding that the behavioral phenotype predicted for one population failed to obtain dominance in the population, I tested among various alternatives to natural selection. For example, one alternative might be that the traits are genetically correlated with other traits that are under different selection pressures as in pleiotropy (effected by the same set of genes). There is, in fact, evidence to suggest that the four traits I have investigated, territory size, agonistic behavior, foraging behavior and antipredator behavior, may be pleiotropic effects of the same genes or in some other way correlated (Riechert and Hedrick 1993). However, after identifying the relevant selection pressures on each, I have determined that selection on all these traits is at least in the same direction. It is not possible at this stage of my analysis of the system to determine whether selection on any one trait drives the system.

The explanation that fits the empirical data for the AZ riparian population is that gene flow from surrounding arid habitats restricts strong selection pressure for a more fearful riparian phenotype. I conclude that within a local area, gene flow can indeed underlie maladaptation. Possible responses of populations to this kind of conflict might involve restriction of the breeding range of individuals (Brandon 1990) and/or flexible behavior that monitors and responds to predation and food levels (Sultan 1987). *Agelenopsis aperta* exhibits neither of these solutions. Perhaps there are design constraints that limit the evolution of appropriate behaviors. Further population comparisons, including genetic crosses between populations, may lead to an understanding of this spider's failure to exhibit one of these adaptations.

In this chapter I have demonstrated the validity of the concept of behavioral ecotypes and offered an example of how one can conduct between-population studies of ecotypic variation to study evolutionary processes and the factors that limit adaptation. I firmly believe that one often learns more about a system from analysis of the variation inherent in it than from mean or modal behavior. Multiple populations that may or may not be under different selection pressures or levels of selection provide this desired variability. Study of them can lead to identification of significant patterns and hypotheses that might explain the underlying cause of the noted variation. At the very least, all behaviorists should use populations as replicates so that they may statistically examine the generality of their findings.

References

Antonovics, J. 1987. The evolutionary dis-synthesis: Which bottles for which wine? American Naturalist 129:321–331.

Arnold, S. J. 1981a. Behavioral variation in natural populations. I. Phenotypic, genetic and environmental correlations between chemoreceptive responses to prey in the garter snake, *Thamnophis elegans.* Evolution 35:489–509.

Arnold, S. J. 1981b. Behavioral variation in natural populations. II. The inheritance of a feeding response in crosses between geographic races of the garter snake, *Thamnophis elegans.* Evolution 35:510–515.

Arnold, S. J. 1992. Behavioral variation in natural populations. VI. Prey responses by two species of garter snakes in three regions of sympatry. Animal Behaviour 44:705–719.

Bateson, P. 1988. The active role of behaviour in evolution. *In* M. W. Ho and S. W. Fox, eds. Evolutionary processes and metaphors, pp. 191–207. Wiley, London.

Bernays, E., and M. Graham. 1988. On the evolution of host specificity in phytophagous insects. Ecology 69:886–892.

Berthhold, P., A. J. Helbig, G. Mohr, and U. Querner. 1992. Rapid microevolution of migratory behaviour in a wild bird species. Nature 360:668–670.

Brandon, R. N. 1990. Adaptation and environment. Princeton University Press, Princeton, NJ.

Breden, F., M. Scott, and E. Michel. 1987. Genetic differentiation for anti-predator behaviour in the Trinidad guppy, *Poecilia reticulata.* Animal Behaviour 35:618–620.

Charlesworth, B., 1977. Appendix. *In* D. G. Lloyd, ed. Genetic and phenotypic models of natural selection, pp. 550–560. Journal of Theoretical Biology 69:543–560.

Constantino, M. G., and H. Drummond. 1990. Population differences in fish-capturing ability of the mexican aquatic garter snake (*Thamnophis melanogaster*). Herpetologica 24:412–416.

Damuth, J. 1985. Selection among "species": A formulation in terms of natural functional units. Evolution 39:1132–1146.

Dawkins, R. 1986. The blind watchmaker. Longmans, London.

Ehrlich, P. R., and P. H. Raven. 1969. Differentiation of populations. Science 165:1228–1232.

Eldredge, N. 1985. Unfinished synthesis: Biological hierachies and modern evolutionary thought. Oxford University Press, New York.

Endler, J. A. 1977. Geographic variation, speciation, and clines. Princeton University Press, Princeton, NJ.

Endler, J. A. 1995. Multiple-trait coevolution and environmental gradients in guppies. Trends in Ecology and Evolution 10:22–29.

Foster, S. A., W. A. Cresko, P. P. Johnson, M. U. Tlusty, and H. E. Willmott. 1996 Patterns of homoplasy in behavioral evolution. *In* M. J. Sanders and L. Hufford, eds. Homoplasy and the evolutionary process, pp. 245–269. Academic Press, New York.

Giles, N., and F. A. Huntingford. 1984. Predation risk and interpopulation variation in anti-predator behaviour in the three-spine stickleback *Gasterosteus aculeatus.* Animal Behaviour 32:264–275.

Gittleman, J. L., C. G. Anderson, M. Kot, and H. Luh. 1996a. Comparative tests of evolutionary lability and rates using molecular phylogenies. *In* P. Harvey, J. Maynard Smith, and A. Leigh-Brown eds. New uses for new phylogenies, pp. 289–307. Oxford University Press, Oxford.

Gittleman, J. L., C. G. Anderson, M. Kot, and H. Luh. 1996b. Phylogenetic lability and rates of evolution: A comparison of behavioral, morphological and life history traits. *In* E. P. Martins, ed. Phylogenies and the comparative method in animal behavior, pp. 166–205. Oxford University Press, New York.

Gould, S. J., and R. C. Lewontin. 1979. The spandrels of San Marco and the Panglossian paradigm: A critique of the adaptationist programme. Proceedings of the Royal Society of London B 205:581–598.

Grant, P. R. 1986. Ecology and evolution of Darwin's finches. Princeton University Press, Princeton, NJ.

Hammerstein, P., and S. E. Riechert. 1988. Payoffs and strategies in spider territorial contests: ESS-analyses of two ecotypes. Evolutionary Ecology 2:115–138.

Harvey, P. H., and M. D. Pagel. 1991. The comparative method in evolutionary biology. Oxford University Press, Oxford.

Hedrick A. V., and S. E. Riechert. 1989. Population variation in the foraging behavior of a spider: The role of genetics. Oecologia 80:533–539.

Huntingford, F. A., and P. J. Wright. 1992. Inherited population differences in avoidance-conditioning in three-spined sticklebacks, *Gasterosteus aculeatus.* Behaviour 122: 264–273.

Huntingford, F. A., P. J. Wright, and J. F. Tierney. 1994. Adaptive variation in antipredator behaviour in threespine stickleback. *In* M. A. Bell and S. A. Foster, eds. The evolutionary biology of the threespine stickleback, pp. 277–296. Oxford University Press, Oxford.

Kroodsma, D. E., and R. A. Canady. 1985. Differences in repertoire size, singing behavior, and associated neuroanatomy among marsh wren populations have a genetic basis. Auk 102:439–446.

Levelsley, P. B., and A. E. Magurran. 1988. Population differences in the reaction of minnows to alarm substance. Journal of Fish Biology 32:699–706.

Magurran, A. E. 1990. The inheritance and development of minnow anti-predator behaviour. Animal Behaviour 39:834–842.

Magurran, A. E. 1993. Individual differences and alternative behaviours. *In* T. J. Pitcher, ed. Behaviour of teleost fishes, pp. 441–477. Chapman and Hall, London.

Magurran, A. E., and B. H. Seghers. 1990. Population differences in the schooling behaviour of newborn guppies, *Poecilia reticulata.* Ethology 84:334–342.

Maruyama, T. 1972. Distribution of gene frequencies in a geographically structured finite population. I. Distribution of neutral genes and of genes with small effect. Annals of Human Genetics 35:411–423.

Martins, E. P., and T. F. Hansen. 1996. The statistical analysis of interspecific data: A review and evaluation of phylogenetic comparative methods. *In* E. P. Martins, ed. Phylogenies and the comparative method in animal behavior, pp. 22–75. Oxford University Press, New York.

Maynard Smith, J. 1982. Evolution and the theory of games. Cambridge University Press, Cambridge.

Maynard Smith, J. 1989. Evolutionary genetics. Oxford University Press, Oxford.

Maynard Smith, J., and S. E. Riechert. 1984. A conflicting tendency model of spider agonistic behaviour: Hybrid-pure population line comparisons. Animal Behaviour 32: 564–578.

Mayr, E. 1983. How to carry out the adaptationist program? American Naturalist 121: 324–334.

Mengel, R. M. 1964. The probable history of species formation in some northern wood warblers (Parulidae). Living Bird 3:9–43.

Mishler, B. D., and M. J. Donoghue. 1982. Species concepts: A case for pluralism. Systematic Zoology 31:491–503.

Morgan, T. H. 1903. Evolution and adaptation. Macmillan, New York.

Plotkin, H. C. 1988. The role of behavior in evolution. Massachusetts Institute of Technology Press, Cambridge, MA.

Provine, W. 1985. Adaptations and the mechanisms of evolution after Darwin: a study in persistent controversies. *In* D. Kohn, ed. The Darwinian heritage, pp. 825–866. Princeton University Press, Princeton, NJ.

Pyke, G. H. 1984. Optimal foraging theory: A critical review. Annual Review of Ecology and Systematics 15:523–575.

Reeve, H. K., and P. W. Sherman. 1993. Adaptation and the goals of evolutionary research. Quarterly Review of Biology 68:1–32.

Riechert, S. E. 1974. The pattern of local web distribution in a desert spider: Mechanisms and seasonal variation. Journal of Animal Ecology 43:733–746.

Riechert, S. E. 1978a. Energy-based territoriality in populations of the desert spider, *Agelenopsis aperta* (Gertsch). Symposium of the Zoological Society of London 42:211–222.

Riechert, S. E. 1978b. Games spiders play: Behavioral variability in territorial disputes. Behavioral Ecology and Sociobiology 6:121–128.

Riechert, S. E. 1979. Games spiders play II. Resource assessment strategies. Behavioral Ecology and Sociobiology 4:1–8.

Riechert, S. E. 1981. The consequences of being territorial: spiders, a case study. American Naturalist 117:871–892.

Riechert, S. E. 1984. Games spiders play III. Cues underlying context-associated changes in agonistic behaviour. Animal Behaviour 32:1–15.

Riechert, S. E. 1991. Prey abundance versus diet breadth in a spider test system. Evolutionary Ecology 5:327–338.

Riechert, S. E. 1993a. Investigation of potential gene flow limitation of adaptation in a desert spider population. Behavioral Ecology and Sociobiology 32:355–363.

Riechert, S. E. 1993b. Test for phylogenetic inertia effects on behavioral adaptation to local environments. Behavioral Ecology and Sociobiology 32:343–348.

Riechert, S. E., and A. V. Hedrick. 1990. Levels of predation and genetically-based anti-predatory behavior in the spider, *Agelenopsis aperta*. Animal Behaviour 40:679–687.

Riechert, S. E., and A. V. Hedrick. 1993. A test for correlations among fitness-linked behavioural traits in the spider *Agelenopsis aperta* (Araneae, Agelenidae). Animal Behaviour 46:669–675.

Riechert, S. E., and J. Maynard Smith. 1989. Genetic analyses of two behavioral traits linked to individual fitness in the desert spider, *Agelenopsis aperta*. Animal Behaviour 37:624–637.

Riechert, S. E., and C. R. Tracy. 1975. Thermal balance and prey availability: bases for a model relating web-site characteristics to spider reproductive success. Ecology 56: 265–284.

Sandoval, C. P. 1994. The effects of relative geographic scales of gene flow and selection on morph frequencies in the walking-stick *Timema cristinae*. Evolution 48:1866–1879.

Seghers, B. H. 1974. Schooling behavior in the guppy (*Poecilia reticulata*): An evolutionary response to predation. Evolution 28:486–489.

Simpson, G. G. 1953. The major features of evolution. Columbia University Press, New York.

Slatkin, M. 1973. Gene flow and selection in a cline. Genetics 75:733–756.

Slatkin, M. 1985a. Gene flow in natural populations. Annual Review of Ecology and Systematics 16:393–430.

Slatkin, M. 1985b. Rare alleles as indicators of gene flow. Evolution 39:54–65.

Smiley, T. L., Nations, J. D., Pewe, T. L., and E. P. Schafer. 1984. Landscapes of Arizona: The geological story. University Press of America, Lanham, MD.

Sultan, S. E. 1987. Evolutionary implications of phenotypic plasticity in plants. Evolutionary Biology 21:127–178.

Thornhill, R., and J. Alcock. 1983. The evolution of insect mating systems. Harvard University Press, Cambridge, MA.

Tinbergen, N. 1963. On aims and methods of ethology. Zeitschrift fur Tierpsychologie 20: 410–433.

Turesson, G. 1922. The genotypic response of the plant species to the habitat. Hereditas 3:211–350.

Via, S. 1991. The genetic structure of host plant adaptation in a spatial patchwork: Demographic variability among reciprocally transplanted pea aphid clones. Evolution: 45: 827–854.

Via, S., and R. Lande. 1985. Genotype-environment interaction and the evolution of phenotypic plasticity. Evolution 39:505–523.

West-Eberhard, M. J. 1989. Phenotypic plasticity and the origins of diversity. Annual Review of Ecology and Systematics 20:249–278.

Wiley, E. O. 1981. Phylogenetics. The theory and practice of phylogenetic systematics. John Wiley and Sons, New York.

Williams, G. C. 1966. Adaptation and natural selection. Princeton University Press, Princeton, NJ.

Wilson, D. S. 1980. The natural selection of populations and communities. Benjamin/Cummings, Menlo Park, CA.

Wright, S. 1931. Evolution in mendelian populations. Genetics 16:97–159.

2

Different Spatial Scales of Natural Selection and Gene Flow

The Evolution of Behavioral Geographic Variation and Phenotypic Plasticity

DANIEL B. THOMPSON

The environmental variables hypothesized to cause behavioral adaptation are distributed across a wide array of spatial scales, from local variation in factors such as food sources and territorial encounters to regional or continental variation in factors such as seasonality and the presence of predators. Geographic variation in behavior, the topic of this book, is just one of the potential evolutionary responses to environmental variation. Because behavioral divergence among populations generated by disparate natural selection can be counterbalanced by the homogenizing influence of gene flow, adaptive geographic variation can evolve only if the spatial scale of variation in natural selection is greater than the scale of gene flow (Endler 1977, Slatkin 1978).

If geographic variation does not evolve because the spatial scale of selection is smaller than the scale of gene flow, populations may instead evolve adaptive phenotypic plasticity (Bradshaw 1965); the expression, by a single genotype, of different fitness-enhancing phenotypes in different environments. Because the same evolutionary processes operating on different spatial scales can generate behavioral geographic variation, behavioral phenotypic plasticity, or geographic variation in phenotypic plasticity, I devote this chapter to development of a hierarchical perspective for studying environmental variation and behavioral evolution. This perspective emphasizes the shared evolutionary processes and research methodologies common to different levels of spatial variation, such as the balance of gene flow, natural selection, and genetic drift, the relationship between environmental patch size and local adaptation, and the effects of historical contingencies and genetic constraints on behavioral adaptation and phenotypic plasticity.

In what follows, I review behavioral research in two unrelated taxa to illustrate the range of possible evolutionary responses to different patterns of environmental variation. First, I discuss different spatial scales of adaptation in the climbing behavior of deer mice (*Peromyscus maniculatus*) and provide a hierarchical analysis of the effects of natural selection, genetic drift, and gene flow. Second, I discuss diet-induced phenotypic plasticity in the feeding behavior of acridid grasshoppers (*Melanoplus femurrubrum* and *M. sanguinipes*) and the evolution of behavioral norms of reaction in response to local spatial and temporal variation in plant environments. In both of these examples I also discuss measurements of quantitative genetic variation and tests of alternative hypotheses that include non-adaptive processes and evolutionary constraints.

The Balance of Gene Flow and Natural Selection

Gene flow, the movement of genes and gene complexes between populations, is generally caused by individuals dispersing from one population and breeding in another. Theoretical analyses of gene flow and natural selection indicate that the interaction of these two processes sets a lower limit on the scale of differences in the environment that can generate population differentiation and geographic variation. Models of single-locus (Endler 1973, 1977; Slatkin 1973, May et al. 1975) and polygenic traits (Slatkin 1978) demonstrate that gene flow can prevent differentiation if populations are separated by distances less than a parameter termed the "characteristic length." For the case of an abrupt transition between environmental patches, the characteristic length is determined by the average gene flow distance relative to the difference in natural selection between two environments. A general prediction that holds true for most traits and types of selection is that the degree of population differentiation that evolves is proportional to the ratio of the spatial scale of differences in natural selection to the spatial scale of gene flow.

For the purposes of this chapter, I focus on populations that encounter discrete patches, or abrupt step clines, of environmental variables and experiential stimuli. For a given gene flow distance and a given difference in natural selection between two environments, the balance between selection and gene flow will be determined by the size of the environmental patches and their proximity to other patches. I use this theoretical relationship between environmental patch size and population differentiation to set up a test of the influence of natural selection and gene flow for populations sampled from a hierarchy of spatial scales of environmental variation. The basic prediction is that if natural selection of behavior varies among environments, behavioral differentiation will be detected in population samples taken from large environmental patches. In contrast, population samples from small environmental patches, adjacent to large patches of a different type, will not exhibit the same degree of behavioral differentiation due to the constraining influence of gene flow (also see Riechert this volume). This type of comparison of large and small environmental patches, and other comparisons utilizing environmental patches that vary in size, geometry, and isolation, are discussed in detail by Sandoval (1994).

Geographic Variation in Climbing Behavior of Deer Mice

In the genus *Peromyscus* (Rodentia; Cricetidae), geographic variation has been used extensively to study the evolution of behavior (e.g., Horner 1954, King 1961; Wecker 1963, King et al. 1968, King and Shea 1969, Gray 1981) and morphology (e.g., Sumner 1930; Dice 1938). The research I discuss here (Thompson 1990) builds on earlier studies that documented a widespread pattern in *Peromyscus* species of geographic variation in tree-climbing ability and external morphology (Horner 1954, Layne 1970, Dewsbury et al. 1980). The general result, documented in all studies, is that subspecies of mice sampled from forested environments have greater climbing ability and longer tails and feet than subspecies sampled from grassland and shrubland. Because all mice were reared in a common laboratory environment, the cited studies demonstrate what appears to be adaptive geographic variation in climbing behavior.

To investigate the role of natural selection and gene flow in the evolution of geographic variation in climbing behavior, I sampled replicated populations from two subspecies, *Peromyscus maniculatus rufinus* and *P. m. sonoriensis*, that occupied different-sized patches of forested and nonforested environment. The majority of mice in *P. m. rufinus* are distributed in montane conifer forest and conifer woodland in northern Arizona, USA (fig. 2-1). In contrast, most *P. m. sonoriensis* are distributed in low-elevation desert scrub devoid of trees. A small proportion of populations of *P. m. sonoriensis* can be found on isolated mountaintops in southern Nevada, in conifer woodland that is identical to the main environment of *P. m. rufinus* (fig. 2-1). In addition, there are deer mouse populations of uncertain subspecies status (Dice 1938, Hoffmeister 1986) in low-elevation desert grassland without trees that are adjacent to *P. m. rufinus* (fig. 2-1). I use the subspecific names simply to define groups of populations that are geographically clustered on a large spatial scale, with no implication that subspecies form evolutionarily distinct units (Dice 1940, Mayr 1963).

Natural Selection, Genetic Drift, and Gene Flow

I sampled 10 populations from forested and nonforested environments that varied in size. Two populations were sampled from large patches of desert scrub and two from small patches of conifer woodland within the range *of P. m. sonoriensis*. Two populations were sampled from large patches of conifer woodland and two from conifer forest within the range of *P. m. rufinus*, and two populations were sampled from small patches of desert grassland for the unknown subspecies. Replicate population samples within each environment were chosen from areas with the same density of trees. In addition, population samples were chosen in pairs with the two samples from adjacent, contrasting environments separated by similar distances (58–90 km; fig. 2-1).

From the hypothesis that natural selection of tree climbing ability occurs in forested environments, I predicted that the mean climbing ability of populations would be high in montane conifer forest, the same or slightly lower in conifer

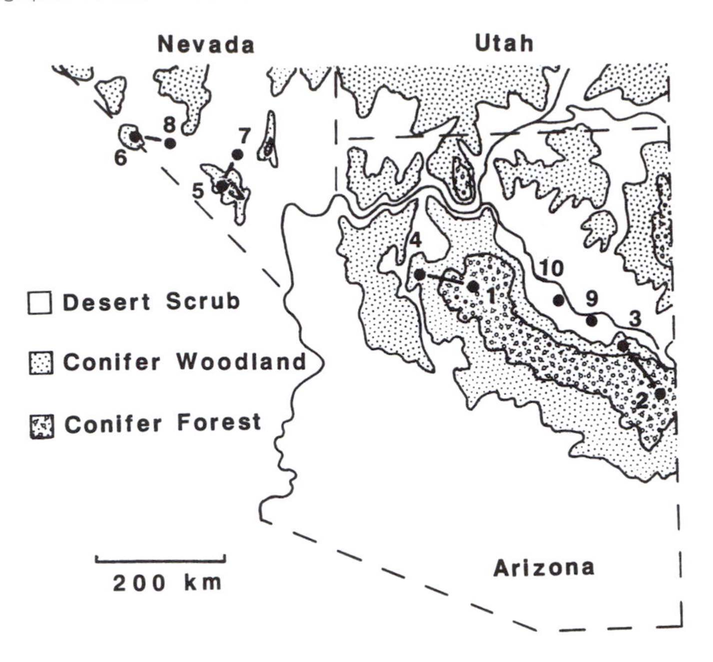

Figure 2-1 The distribution of three plant communities and the locations of 10 sites where *Peromyscus maniculatus* was sampled (Thompson 1990). Sites 1–4 are in the range of *P. m. rufinus*, and sites 5–8 are in the range of *P. m. sonorensis*. Populations in the remaining two locations are of uncertain subspecies status. Reproduced with permission from *Evolution*.

woodland, and low in desert scrub or grassland devoid of trees. This prediction is based on the observation that, for most mice, use of resources on the ground is negatively related to tree density (Horner 1954, Holbrook 1979, Layne 1970). If natural selection is the main determinant of behavior, this predicted match between climbing ability and presence of trees should be observed in samples from large and small environmental patches. If gene flow constrains local adaptation (alternative hypothesis below), I should observe a correspondence between climbing ability and environment only in samples from large environmental patches.

The null hypothesis was that evolutionary divergence among populations would be random with respect to environment due to genetic drift of alleles that affect climbing behavior. An implicit alternative that could not be distinguished from the null hypothesis was that natural selection due to an unmeasured environmental variable or natural selection of genetically correlated traits caused divergence among populations (Endler 1986).

The main alternative hypothesis for my research was that populations in small environmental patches, such as isolated mountaintops (fig. 2-1, populations 5 and 6) and small desert grassland patches (populations 9 and 10), adjacent to large patches of the contrasting environment, would be influenced by the homogenizing effects of gene flow or common history. From this hypothesis, I predicted that animals from small environmental patches would exhibit climbing abilities similar to those from the larger, adjacent patches of different habitat.

I measured the climbing ability of 266 laboratory-reared mice that were the progeny of adults captured from the 10 sampling sites (for details of methods, see Thompson 1990). The quantitative behavioral character I measured was the ability of mice to climb artificial trunks (uniformly roughened rods) of large diameter. Following Horner (1954), I scored the climbing ability of each mouse as the maximum-diameter rod it was able to climb in repeated trials. Mice were motivated to climb as a means of escape from a lighted arena. The maximum diameter is represented by the number of rods, out of nine total, that a mouse could climb. The rods of increasing diameter were arbitrarily numbered from one (6.3 mm) to nine (57.1 mm).

Mean rod-climbing scores for the two subspecies in each of the environments are diagrammed in figure 2-2. I rejected the null hypothesis of genetic drift because there was divergence in the climbing ability of populations from different environments and no significant divergence among populations within environments (table 2-1, $p = .28$). There was geographic variation in rod-climbing ability for both subspecies with the highest mean climbing scores in montane conifer forest and conifer woodland and the lowest mean climbing scores in desert grassland and desert scrub. The overall differences among the five habitat/subspecies combinations illustrated in figure 2-2 were significant (table 2-1, $p = .0002$).

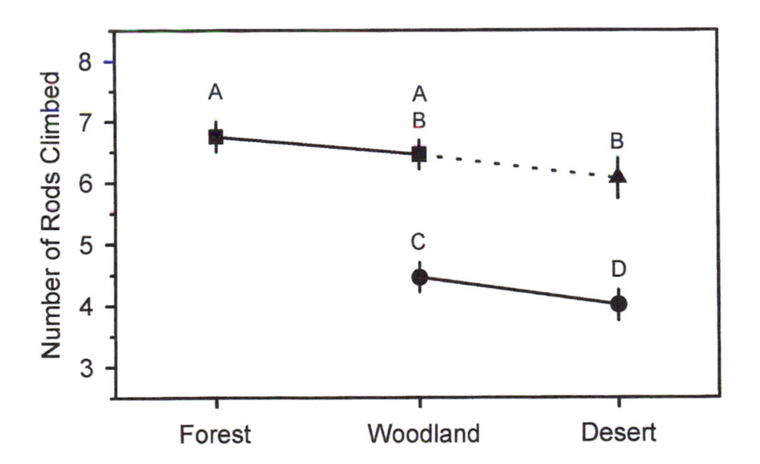

Figure 2-2 Geographic variation in the mean rod-climbing ability (see text) for populations sampled within the range of *P. m. rufinus* (square), *P. m. sonorensis* (circle), and an unknown subspecies (triangle). Means (±SE) represent two replicate populations sampled from environments in decreasing order of tree size and density: montane conifer forest, conifer woodland, and desert scrub or grassland. Means with different letters (A–D) differed significantly ($p < .05$; Student-Newman-Keuls procedure).

Table 2-1 Summary of an ANOVA of number of rods climbed by laboratory-reared progeny of *Peromyscus maniculatus*.

Source	df	F	*P*
Environment	4	66.9	.0002
Population	5	1.3	.28
Error	256		

Sampled from two replicate populations within five combinations of conifer forest, conifer woodland, and desert scrub environments that varied in size (see text). Population is nested within environment and serves as the error term in the *F* test of environment.

The general pattern of behavioral geographic variation matched the predictions of the natural selection hypothesis. However, the majority of the divergence in climbing ability between mice in forested and nonforested areas was associated with the large environmental patches (fig. 2-2; *P. m. rufinus* forest and woodland and *P. m. sonoriensis* desert). The climbing abilities of mice from desert grassland in northern Arizona were not significantly different from those of *P. m. rufinus* in adjacent woodland environments (fig. 2-2, means connected by dotted line). In addition, although the mean climbing ability in samples of *P. m. sonoriensis* from small woodland environments were significantly greater than the samples from surrounding desert scrub, mean climbing ability in the small woodland environments was significantly less than the large woodland environments of *P. m. rufinus*. These observations are consistent with the expectations of constrained local adaptation in small environmental patches.

Heritability of Climbing Ability

Evolution of geographic variation by natural selection is dependent on the precondition that there is significant heritability (proportion of phenotypic variation due to additive genetic variation) of behavior to permit population divergence (Endler 1986, Falconer 1989). Assessment of this precondition is not always meaningful because it is possible for contemporary populations to lack significant heritability for behavior that evolved in response to past natural selection (i.e., the populations had significant heritability in the past). However, the magnitude of heritability in natural populations should be determined (Arnold 1981, 1994; Mitchell-Olds and Rutledge 1986) because it will impact future behavioral evolution and because the cause of low heritability for behavior is an issue of interest in its own right (Boake 1994).

To test the precondition for a response to natural selection, I estimated the heritability of climbing behavior in a population of deer mice sampled from woodland habitat in northern Arizona (fig. 2-1, site 4). Seventy-one pairs of mice from the first laboratory-reared generation were randomly mated, and the offspring were reared in separate cages to reduce the influence of common cage environment (Thompson 1990). To calculate the heritability of climbing ability,

I used a linear regression of mean offspring climbing scores against mid-parent climbing scores (Falconer 1989). The regression coefficient provided an estimate of $h^2 = 0.352 \pm 0.077$ ($p = .005$, df = 69). Separate regressions of mean offspring climbing scores against male parent and female parent climbing scores were not significantly different from each other. Thus, there was no evidence of a maternal effect in the estimate of heritability. The presence of significant, although moderate, additive genetic variation in climbing ability in at least one population of *P. maniculatus* indicates that field populations of deer mice could evolve in response to natural selection of climbing behavior.

Insufficient Time to Respond to Natural Selection

Populations require time to respond to natural selection. Thus, associations between environmental variation and adaptive behavioral genotypes will only be observed if the populations have had sufficient time to respond to local selective regimes. There are several reasons this may not be the case for *P. maniculatus*. First, the length of time required for a response to selection is inversely related to the intensity of selection. I suspect that the differences among habitats in natural selection of climbing behavior are relatively slight. Second, the distribution of trees in the study region has changed dramatically since the onset of the Holocene. In the pluvial periods of the late Pleistocene, woodlands covered what is now desert scrub and grassland habitat in lowland areas of southern Nevada and northern Arizona (Van Devender and Spaulding 1979, Wells 1983). The high rod-climbing scores of some desert populations of *P. maniculatus* could be explained by their past exposure to natural selection in woodland environments and by a slow evolutionary response to the current lack of trees.

Third, small rodent populations are often characterized by extreme fluctuations in population size that can cause local population extinction (Terman 1968, Brown and Heski 1990). If populations in one environment have been recolonized by mice from a different environment, they may not have had sufficient time to respond to natural selection of a new level of climbing ability. Colonization can have the effect of both homogenizing gene pools and promoting differentiation through sampling drift in founder populations (Slatkin 1978, Maruyama and Kimura 1980, Wade and McCauley 1988). This type of infrequent, unidirectional genetic exchange between environments is a distinct form of gene flow that could have profound effects in reducing the occurrence of local adaptation in behavior. In summary, although contemporary gene flow between environments may constrain local adaptation (e.g., Reichert this volume), it is also possible that the patterns I have observed in *P. maniculatus* could be due to historical contingencies such as Holocene environment change or past colonization from a different environment.

Geographic Variation and Behavioral Phenotypic Plasticity

In the following sections, I consider evolution in small environmental patches that vary on a scale that is less than the distance of gene flow. In other words, the average diameter of environmental patches is less than the characteristic length set by gene flow and selection (Endler 1977, Slatkin 1978). Under these conditions, populations will remain genetically similar with respect to behavior due to the exchange of genes between environments. However, populations subject to selection in two or more environmental patches, fully connected by gene flow, may evolve adaptive phenotypic plasticity (Bradshaw 1965) or norms of reaction (Schmalhausen 1949) such that genetically similar individuals express different phenotypes in each environment. With high gene flow among environments, alleles that cause different phenotypes to develop in each environment can evolve via natural selection if the plastic phenotypic responses yield high average fitness (Via and Lande 1985).

The phenotypic plasticity that evolves in response to natural selection in one set of interconnected environments is unlikely to be adaptive in other environments that are beyond the range of gene flow. Therefore, natural selection of behavioral plasticity in geographic regions with unique combinations of environments can cause evolution of locally adapted norms of reaction and adaptive geographic variation in behavioral plasticity (e.g., Carroll and Corneli this volume). In parallel with the previous sections, I discuss experiments on grasshopper feeding behavior with reference to (1) evolution of geographic variation and phenotypic plasticity as a result of natural selection, gene flow, and genetic drift, (2) genetic variation in behavioral phenotypic plasticity as a potential constraint on evolution, and (3) the effect of temporal sequences of environments on the evolution of behavioral phenotypic plasticity.

Phenotypic Plasticity of Feeding Behavior in Grasshoppers

Grasshoppers in the genus *Melanoplus* (Orthoptera; Acrididae) encounter environments that vary on many spatial scales ranging from local differences in forage plants to regional and elevational differences in plant assemblages, temperature, and season length. On a local scale, grasshoppers encounter patches of forage plants that are relatively small, but coarse-grained in the sense that an individual usually hatches, develops, and matures in one patch of forage plants (only adults fly). Environmental patches with different plant species may differ in characteristics that affect grasshopper feeding behavior such as plant hardness (hardness is influenced by silica content and vascular bundle structure) and nutrient availability (Bernays 1986, Thompson 1992).

Both of the species I have studied, *Melanoplus femurrubrum* and *M. sanguinipes*, occur in agricultural fields and natural grass communities that vary in plant hardness. Natural selection of feeding behavior in different environments should cause the evolution of behavioral phenotypic plasticity rather than geographic variation because the spatial scale of selection is much smaller than the scale of gene flow. The first experiments I discuss indirectly test an adaptive hypothesis of behavioral phenotypic plasticity by determining if the mean plastic behavioral phenotype expressed by individuals in a given plant environment confers enhanced performance in that environment. The second set of experiments test the hypothesis that natural selection of feeding behavior reaction norms in different sets of plant environments has caused the evolution of adaptive geographic variation in behavioral plasticity.

Phenotypic Plasticity of Relative Consumption Rate

The methods I have used to induce and quantify phenotypic plasticity are described in Thompson (1992). Basically, I collected eggs from grasshoppers captured in different local plant environments. Each egg pod produced a full-sib family. After hatching, I raised one-half of each full-sib family on a soft plant diet and the other half on a hard plant diet. The soft and hard plant species differed between experiments (e.g., tables 2-2, 2-3), but the relative difference in hardness was the same. The hard plant was 2–2.5 times more difficult to cut than the soft plant. Feeding behavior experiments were conducted on grasshoppers deprived of food for 12–16 h. To quantify consumption rate (milligrams plant ingested per second), food-deprived grasshoppers were weighed, observed for 5 min to determine cumulative feeding time, and then weighed again. To calculate relative consumption rate, I divided the milligrams of plant ingested by the total feeding time and adjusted for body size. In analyzing differences in feeding behavior, I refer to "soft" or "hard" test plants (randomly assigned in a feeding trial) and "soft" or "hard" diet plants (rearing diet). Phenotypic plasticity of feeding behavior is present if there are significant differences between rearing diets in the consumption of a given type of test plant.

The prediction from an adaptive hypothesis was that relative consumption rate for grasshoppers tested on hard plants should be higher for individuals raised on hard-plant diets relative to full-sibs raised on soft-plant diets. The opposite should hold for grasshoppers tested on soft plants. The null hypothesis was that there was no behavioral plasticity, and an alternative hypothesis was that genetic drift of alleles with differential sensitivity to the environment has caused evolution of behavioral norms of reaction that do not enhance feeding performance.

For *M. femurrubrum* (from Johnson County, Iowa, USA) feeding on soft test plants, there was no significant difference between grasshoppers raised on hard-plant versus soft-plant diets (table 2-2). Thus, there was no phenotypic plasticity of relative consumption rate for grasshoppers feeding on soft plants. In contrast, for grasshoppers feeding on hard test plants, those raised on a hard-plant diet had significantly greater relative consumption rate than those raised on a soft-plant diet (table 2-2). This phenotypic plasticity in relative consumption rate led to a 37% increase in feeding rates for individuals raised on a hard-plant diet and tested on hard plants.

For *M. sanguinipes* (from Clark County, Nevada, USA) feeding on soft test plants, there was no significant difference in relative consumption rate between hard and soft rearing diets. For grasshoppers feeding on hard test plants, those raised on a hard-plant diet had significantly greater relative consumption rate than those raised on a soft-plant diet (table 2-2, fig. 2-3). The phenotypic plasticity in *M. sanguinipes* relative consumption rate led to a 43% increase in feeding performance for individuals raised on a hard-plant diet and tested on a hard plant.

In neither species was there a significant effect of the source environment

Table 2-2 Mean relative consumption rate (mg/s adjusted for body size) and standard error (in parentheses) of adult *Melanoplus femurrubrum* ($n = 90$) and *M. sanguinipes* ($n = 89$) when feeding on hard or soft test plants.

	Hard plant diet	Soft plant diet
M. femurrubrum[a]		
Hard test plant	0.049 (0.001)	0.036 (0.001)*
Soft test plant	0.071 (0.001)	0.077 (0.002)
M. sanguinipes[b]		
Hard test plant	0.056 (0.001)	0.028 (0.001)*
Soft test plant	0.088 (0.001)	0.074 (0.001)

Differences in mean consumption rate between grasshoppers raised on hard- and soft-plant diets indicate phenotypic plasticity in their feeding performance on a given test plant.

[a]In this experiment, the hard plant was rye grass (*Lolium perenne*) and the soft plant was red clover (*Trifolium repens*).

[b]In this experiment, the hard plant was buffalo grass (*Buchloe dactyloides*) and the soft plant was rye grass (*Lolium perenne*).

*$p < .05$: means significantly different between plant diets.

(plant type in the field where eggs were collected) on laboratory-reared grasshopper feeding behavior ($p > .26$). As predicted, there was no small-scale geographic differentiation in mean feeding behavior. Instead, grasshoppers of both species exhibited diet-induced behavioral plasticity that enhanced feeding performance on hard-plant diets, an observation consistent with the predictions of an adaptive hypothesis.

Geographic Variation in Relative Consumption Rate

The lesser migratory grasshopper, *M. sanguinipes*, exhibits a high degree of geographic variation in morphological traits (e.g., Gurney and Brooks 1959, Chapco 1980) and life-history traits (e.g., Dingle et al. 1990), particularly along elevational gradients in the Sierra Nevada and Rocky Mountains. The high degree of environmental heterogeneity encountered by *M. sanguinipes* across this region provides an opportunity to test for adaptive geographic variation in behavioral plasticity.

In table 2-3, I present a preliminary analysis of geographic variation in diet-induced plasticity of feeding behavior for third nymphal instar *M. sanguinipes* (L. Taylor and D. Thompson, unpublished data). Eggs were collected from grasshoppers sampled from two short-grass prairie populations in Larimer County, Colorado, USA, and two montane meadow populations in Placer and Amador Counties, California, USA. Grasshoppers hatched from the eggs were raised on hard- or soft-plant diets until the third instar and then videotaped in feeding trials with hard or soft test plants. In this experiment, the area of plant leaf consumed

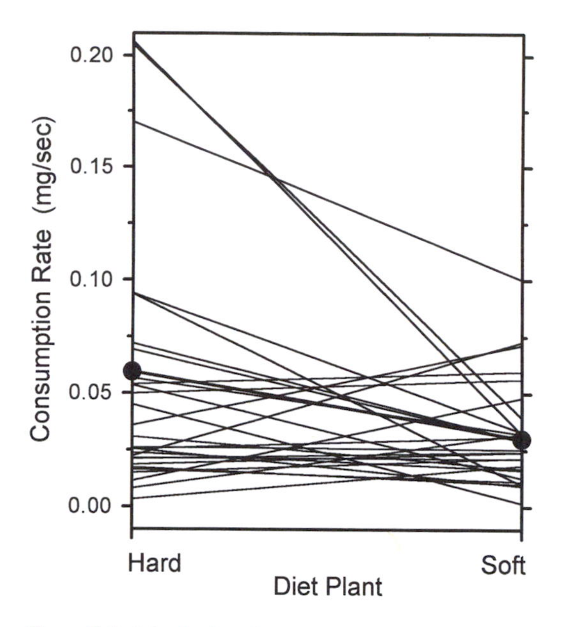

Figure 2-3 Diet-induced phenotypic plasticity in relative consumption rate (mg/s adjusted for body size) of hard test plants (buffalo grass) for *Melanoplus sanguinipes*. Each line connects the mean relative consumption rate of hard test plants for full-sibs raised on hard (buffalo grass) or soft (rye grass) plant diets. The mean phenotypic plasticity for 30 full-sib families, represented by the filled circles and dark line, is significant (ANOVA, $p = .01$).

per incision and the incision rate were measured directly to estimate relative consumption rate (square millimeter leaf per second).

Geographic variation in phenotypic plasticity can be quantified in an ANOVA with plant diet, geographic region, and geographic region × diet interaction as the sources of variation. Plant diet provides a test of diet-induced plasticity in relative consumption rate averaged across geographic regions, and geographic region provides a test of regional genetic divergence in mean relative consumption rate averaged across plant diet. The geographic region × diet interaction provides a test of geographic variation in diet-induced phenotypic plasticity. If this interaction is significant, populations sampled from different geographic regions have different plastic behavioral responses to the environments and, therefore, genetically divergent norms of reaction. If two or more populations are sampled within each region, the ANOVA is nested and the F test of geographic region is constructed with the mean square of population nested within geographic region as the error term. In addition, the F test of geographic region × diet is constructed with the mean square of population within geographic region × diet as the error term.

Table 2-3. Mean relative consumption rate (mm^2 leaf/s adjusted for body size) and standard error (in parentheses) of third nymphal instar *Melanoplus sanguinipes* ($n = 264$) sampled from Colorado and California, USA (see text) when feeding on hard or soft test plants.

	Hard-plant diet	Soft-plant diet
Hard test plant[a]		
Colorado	0.038 (0.003)	0.036 (0.005)
California	0.014 (0.002)	0.025 (0.002)*
Soft test plant		
Colorado	0.096 (0.009)	0.064 (0.004)*
California	0.078 (0.014)	0.086 (0.005)

Differences in mean consumption rate between grasshoppers raised on hard and soft plant diets indicate phenotypic plasticity in their feeding performance on a given test plant.

[a]In this experiment, the hard plant was western wheatgrass (*Agropyron smithii*) and the soft plant was hairy vetch (*Vicia villosa*).

*$p < .05$; means significantly different between plant diets.

A nested ANOVA of third nymphal instar grasshoppers feeding on hard test plants revealed no significant effect of plant diet ($p = .22$) or geographic region × diet ($p = .37$), but a significant effect of geographic region ($p = .001$). The divergence between geographic regions was due to the significantly higher mean relative consumption rate, averaged across plant diet, of grasshoppers from Colorado (mean = 0.037) relative to California (mean = 0.070; table 2-3). Grasshoppers from Colorado had slightly lower mean relative consumption rate of hard plants when they were raised on hard-plant diets (table 2-3), and grasshoppers from California exhibited no behavioral phenotypic plasticity. These feeding performance results are not consistent with the predictions of an adaptive hypothesis.

For grasshoppers feeding on soft test plants, there was a significant effect of plant diet ($p = .001$), geographic region ($p = .001$), and geographic region × diet ($p = .001$). Again, the divergence between geographic regions was due to significantly higher mean relative consumption rate, averaged across plant diet, of grasshoppers from Colorado (mean = 0.080) relative to California (mean = 0.072; table 2-3). The significant diet-induced plastic increase in relative consumption rate of soft test plants for grasshoppers raised on soft-plant diets (mean = 0.086), relative to hard-plant diets (mean = 0.066), does not match the predictions of an adaptive hypothesis. Geographic variation in diet-induced norms of reaction and genetic divergence in phenotypic plasticity were apparent in the significant geographic region × diet interaction. The mean phenotypic plasticity of the Colorado populations was two to three times greater than the California populations (table 2-3). Although grasshoppers in both geographic regions expressed behavioral plasticity, the low overall relative consumption rate and low phenotypic plasticity of California grasshoppers cannot be accounted for with a simple adaptive hypothesis.

Nonadaptive geographic variation in behavioral norms of reaction could evolve through genetic drift of alleles with environment-sensitive expression. The prediction from a hypothesis of genetic drift is that there would be random divergence of mean norms of reaction among populations for samples that come from more than one effectively random-mating population and, therefore, significant population × environment interactions within geographic regions. In this preliminary analysis of variation between grasshopper populations within the geographic regions of Colorado and California, there was no evidence of population × diet interaction. More experiments are needed to test the hypothesis of adaptive behavioral plasticity and to determine whether discrepancies in the patterns of phenotypic plasticity observed in experiments with third nymphal instar and adult *M. sanguinipes* are due to experimental differences in the source populations and the types of plants used in test diets, or due to differences arising from the ontogeny of feeding behavior.

Genetic Variation for Phenotypic Plasticity of Relative Consumption Rate

Theoretical analyses of different quantitative genetic models of the evolution of adaptive phenotypic plasticity all have generated the same general conclusions (Via and Lande 1985, Via 1987, de Jong 1990a,b, Gomulkiewicz and Kirkpatrick 1992, Gavrilets and Scheiner 1993; reviewed by Scheiner 1993). The evolutionary response to natural selection of a population's mean norm of reaction is determined by additive genetic variation for norms of reaction. Although actual predictions about the trajectory of evolutionary change in a population's mean norm of reaction require estimation of trait genetic correlations or covariances across environments (Via and Lande 1985, Gomulkiewicz and Kirkpatrick 1992), measurement of genotype × environment interaction can be used to assess overall genetic variation in phenotypic plasticity.

Evolution of phenotypic plasticity entails an increase or decrease in slope of a population's mean norm of reaction, the average phenotype expressed in each of a series of environments. For a simple norm of reaction expressed in two environments (e.g., fig. 2-3), genetic variation for phenotypic plasticity is equivalent to genetic variation in the slopes of the reaction norm lines (Via 1987, de Jong 1990a). In other words, if genotypes vary in their phenotypic response to rearing environments, the lines connecting the mean phenotype expressed in each environment (genotype norms of reaction) will have different slopes. In an ANOVA, the statistical significance and magnitude of genetic variation in slopes is assessed with the genotype × environment interaction. Lack of genotype × environment interaction in a population indicates low or nonexistent genetic variation for plasticity. Details of the ANOVA approach for estimating genotype × environment interactions with full- and half-sib quantitative genetic designs are discussed in Via (1987, 1988) and Falconer (1989).

From the feeding experiments with adult *Melanoplus*, I have estimated genotype × environment interaction for relative consumption rate on hard and soft plants (table 2-2, fig. 2-3). Because these estimates are based on differences among full-sib families rather than half-sib families, the genotype × environment interaction (family × plant diet) includes nonadditive genetic and common environment sources of variation (Falconer 1989). In addition, my conclusions about genetic variation of behavioral plasticity are limited by the statistical problems inherent in analyzing small, unbalanced data sets. However, the detection, in a single experiment, of significant genotype × environment interaction for one behavioral measure but not another seems to reflect actual differences in the magnitude of genetic variation for plasticity rather than differences in the magnitude of sampling error.

For measurements of relative consumption rate in *M. sanguinipes*, I used 30 full-sib families with an average of 2.6 grasshoppers per family per plant diet. For relative consumption rate of grasshoppers feeding on a hard test plant, the family × plant diet interaction was significant ($p = .02$). However, the interaction for grasshoppers feeding on a soft test plant was not significant ($p = .93$). The significant genetic variation in phenotypic plasticity of relative consumption rate can be seen in the variation in the slopes of reaction norms for different families in figure 2-3. For measurements of relative consumption rate in *M. femurrubrum*, I used 16 full-sib families (Thompson 1992). The family × plant diet interaction was not significant for grasshoppers feeding on hard test plants ($p = .42$), but was significant for grasshoppers feeding on soft test plants ($p = .02$).

If natural populations exhibit differences in genetic variation for behavioral plasticity similar to those I have measured, evolution of adaptive behavioral norms of reaction could be constrained. If evolution of the mean norm of reaction for *M. sanguinipes* feeding on soft test plants was constrained by low genetic variation in some populations, evolved patterns of geographic variation in behavioral plasticity would not match an adaptive hypothesis. Indeed, with low genetic variation in phenotypic plasticity, the evolutionary trajectory of a population is governed mainly by evolution in the environment with the greatest natural selection or the largest proportion of the population (Via and Lande 1985). Low genotype × environment interaction or high positive genetic correlations across environments that could constrain evolution have been reported in other studies of behavioral norms of reaction (Lynch et al. 1988, Lynch 1992, Carroll and Corneli this volume).

There are theoretical reasons that low genetic variation could frequently constrain the evolution of reaction norms. In models of reaction norms expressed in a continuum of environments (infinite-dimensional traits), Gomulkiewicz and Kirkpatrick (1992) found that the optimal reaction norm could evolve in a predictable fashion only if there was additive genetic variation for all conceivable evolutionary changes in the mean reaction norm. The greater the number of environments in which a trait was expressed and subject to natural selection, the higher the probability of genetic constraints on evolution. This general result was also derived in models of labile or reversible norms of reaction expressed in temporally varying environments. Although Gomulkiewicz and Kirkpatrick (1992) set up their models with each environment ordered along a continuum, it may be useful, heuristically, to consider behavioral norms of reaction that are expressed in a continuum of *sets* of all possible temporal sequences of environments encountered by individuals.

For example, I have quantified the mean norm of reaction (rate of incision) for grasshoppers that were raised on the same plant diet (hard or soft) throughout development and for grasshoppers that were switched to the opposite plant diet two-thirds of the way through development (fig. 2-4). The order in which plant environments were encountered during ontogeny significantly affected the expression of feeding behavior. If these temporal sequences of environments occurred in nature, in conjunction with selective differences between environments, individuals in a population would be selected to express appropriate behaviors in response to different ordered combinations of environmental stimuli.

If the sequence of environments encountered influences the expression of behavioral phenotypes and fitness, the number of environmental states in which a

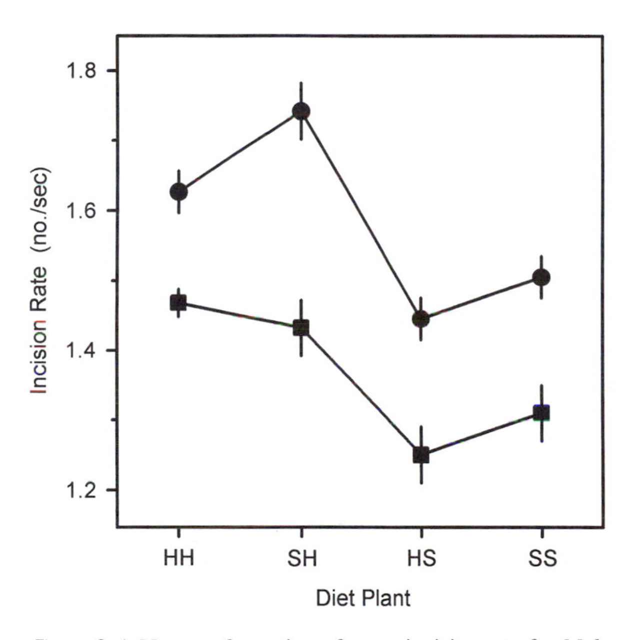

Figure 2-4 Norms of reaction of mean incision rate for *Melanoplus sanguinipes* raised on different combinations of hard and soft plant diets. The norms of reaction are measured for grasshoppers feeding on hard (buffalo grass) test plants (squares) and soft (rye grass) test plants (circles). The letters along the x-axis represent hard (H) or soft (S) plant diets in early development (first to fourth nymphal stages) and late development (fourth to adult stages), respectively. The sequence of exposure to rearing diets had a significant effect on incision rate (ANOVA, $p < .05$) for both norms of reaction.

population evolves will be substantially greater (all combinations of sequences of environments) than the actual number of separate environments. For example, grasshoppers encountering sequences of two plant environments have reaction norms expressed in four sets of environments (fig. 2-4). Because behavior is likely to be influenced by a wide variety of environmental stimuli over different time scales, it is unlikely that a population could harbor genetic variation in the sensitivity of expression of the behavior for all possible combinations of environments. This line of reasoning, in conjunction with the results of Gomulkiewicz and Kirkpatrick (1992), leads to the conclusion that lack of genetic variation for complex reaction norms may frequently constrain the evolution of adaptive behavioral plasticity.

Conclusions

There are several themes that connect the topics I have addressed in this chapter. First, the hierarchies of spatial scales of environmental variation, natural selection, and gene flow found in nature can produce a variety of different evolutionary outcomes. I think it is useful to consider the evolution of behavior in a conceptual framework that encompasses all these spatial scales. Depending on the scale of gene flow, ethologists should be able to identify distributions of environmental variables that could be predicted to lead to the evolution of behavioral phenotypic plasticity, behavioral geographic variation, or geographic variation in phenotypic plasticity.

The second theme concerns the use of comparative studies framed within spatial hierarchies to infer the influences of evolutionary processes. Careful choice of replicated populations, sampled from different-sized environmental patches and different levels of a spatial hierarchy, should allow ethologists to empirically test the predicted effects of natural selection, genetic drift, and, to a limited extent, gene flow. Even if geographic comparisons of behavioral plasticity cannot provide definitive tests of adaptive and nonadaptive hypotheses, they are useful in identifying the important patterns and hypotheses to examine with direct estimates of natural selection (e.g., Arnold and Wade 1984). Third, quantitative genetic methods for measuring heritability of behavior and genetic variation for behavioral phenotypic plasticity can be used to test preconditions for the evolution of behavior. Use of these methods will be particularly important if it is found that complex behavioral traits or norms of reaction sometimes lack sufficient genetic variation for evolutionary change.

Finally, because the evolution of behavior can be influenced by factors such as genetic drift, genetic constraints, colonization events, and past environmental change, studies of geographic variation should address a wide variety of nonadaptive hypotheses in addition to hypotheses of natural selection. With the incorporation of quantitative genetic methods and consideration of the influences of history, comparisons of geographically varying populations should continue to provide insights on behavioral evolution.

Acknowledgments Portions of this work were supported by National Science Foundation grant BSR-8907386 and funds from the University of Nevada, Las Vegas. The critical comments of J. Travis and J. Endler were particularly useful in revisions of this chapter, as were the comments of an anonymous reviewer.

References

Arnold, S. J. 1981. Behavioral variation in natural populations. I. Phenotypic, genetic, and environmental correlations between chemoreceptive responses to prey in the garter snake, *Thamnophis elegans*. Evolution 35:489–509.

Arnold, S. J. 1994. Multivariate inheritance and evolution: A review of concepts. *In* C. R. B. Boake, ed. Quantitative genetic studies of behavioral evolution, pp. 17–48. University of Chicago Press, Chicago.

Arnold, S. J., and M. J. Wade. 1984. On the measurement of natural and sexual selection: Theory. Evolution 38:709–719.

Bernays, E. 1986. Diet-induced head allometry among foliage-chewing insects and its importance for graminivores. Science 231:495–497.

Boake, C. R. B. 1994. Evaluation of applications of the theory and methods of quantitative genetics to behavioral evolution. *In* C. R. B. Boake, ed. Quantitative genetic studies of behavioral evolution, pp. 305–325. University of Chicago Press, Chicago.

Bradshaw, A. D. 1965. Evolutionary significance of phenotypic plasticity in plants. Advances in Genetics 13:115–155.

Brown, J. H., and E. J. Heski. 1990. Temporal changes in a Chihuahuan Desert rodent community. Oikos 59:290–302.

Chapco, W. 1980. Genetics of the migratory grasshopper, *Melanoplus sanguinipes*: Orange stripe and its association with tibia color and red-back genes. Annals of the Entomological Society of America 73:319–322.

de Jong, G. 1990a. Genotype by environment interaction and the genetic covariance between environments: Multilocus genetics. Genetica 81:171–177.

de Jong, G. 1990b. Quantitative genetics of reaction norms. Journal of Evolutionary Biology 3:447–468.

Dewsbury, D. A., D. L. Lanier, and A. Miglietta. 1980. A laboratory study of climbing behavior in 11 species of muroid rodents. American Midland Naturalist 103:66–72.

Dice, L. R. 1938. Variation in nine stocks of the deermouse, *Peromyscus maniculatus*, from Arizona. Occasional Papers of the Museum of Zoology 375:1–19.

Dice, L. R. 1940. Ecologic and genetic variability within species of *Peromyscus*. American Naturalist 74:212–221.

Dingle, H., T. A. Mousseau, and S. M. Scott. 1990. Altitudinal variation in life cycle syndromes of California populations of the grasshopper, *Melanoplus sanguinipes* (F.). Oecologia 84:199–206.

Endler, J. A. 1973. Gene flow and population differentiation. Science 179:243–250.

Endler, J. A. 1977. Geographic variation, speciation, and clines. Princeton University Press, Princeton, NJ.

Endler, J. 1986. Natural selection in the wild. Princeton University Press, NJ.

Falconer, D. S. 1989. Introduction to quantitative genetics, 3rd ed. Longman, New York.

Gavrilets, S., and S. M. Scheiner. 1993. The genetics of phenotypic plasticity. V. Evolution of reaction norm shape. Journal of Evolutionary Biology 6:31–48.

Gomulkiewicz, R., and M. Kirkpatrick. 1992. Quantitative genetics and the evolution of norms of reaction. Evolution 46:390–411.

Gray, L. 1981. Genetic and experiential components of feeding behavior. *In* A. C. Kamil, and T. D. Sargent, eds. Foraging behavior: Ecological, ethological, and psychological approaches, pp. 455–473. Garland Press, New York.

Gurney, A. B., and A. R. Brooks. 1959. Grasshoppers of the *mexicanus* group, genus *Melanoplus* (Orthoptera: Acrididae). Proceedings of the United States National Museum 110:1–93.

Hoffmeister, D. F. 1986. Mammals of Arizona. University of Arizona Press and Arizona Game and Fish Department, Tucson, AZ.

Holbrook, S. J. 1979. Vegetational affinities, arboreal activity, and coexistence of three species of rodents. Journal of Mammalogy 60:528–542.

Horner, E. B. 1954. Arboreal adaptations of *Peromyscus*, with special reference to use of the tail. Contributions of the Laboratory of Vertebrate Biology University of Michigan 61:1–85.

King, J. A. 1961. Swimming and reaction to electric shock in two subspecies of deer mice (*Peromyscus maniculatus*) during development. Animal Behaviour 9:142–150.

King, J. A., E. D. Price, and P. G. Weber. 1968. Behavioral comparisons within the genus *Peromyscus*. Papers of the Michigan Academy of Sciences, Arts, and Letters 53: 113–136.

King, J. A., and N. Shea. 1969. Subspecific differences in the responses of young deer mice on an elevated maze. Journal of Heredity 50:14–18.

Layne, J. N. 1970. Climbing behavior of *Peromyscus floridanus* and *Peromyscus gossypinus*. Journal of Mammalogy 51:580–591.

Lynch, C. B. 1992. Clinal variation in cold adaptation in *Mus domesticus*: Verification of predictions from laboratory populations. American Naturalist 139:1219–1236.

Lynch, C. B, D. S. Sulzbach, and M. S. Connolly. 1988. Quantitative genetic analysis of temperature regulation in *Mus domesticus*. IV. Pleiotropy and genotype-by-environment interaction. American Naturalist 132:521–537.

Maruyama, T., and M. Kimura. 1980. Genetic variability and effective population size when local extinction and recolonization of subpopulations are frequent. Proceedings of the National Acadademy of Sciences USA 77:6710–6714.

May, R. M., J. A. Endler, and R. E. McMurtrie. 1975. Gene frequency clines in the presence of selection opposed by gene flow. American Naturalist 109:659–676.

Mayr, E. 1963. Animal species and evolution. Belknap Press, Cambridge, MA.

Mitchell-Olds, T., and J. J. Rutledge. 1986. Quantitative genetics in natural plant populations: A review of the theory. American Naturalist 127:379–402.

Sandoval, C. P. 1994. The effects of the relative geographic scales of gene flow and selection on morph frequencies in the walking stick *Timema cristinae*. Evolution 48: 1866–1879.

Scheiner, S. M. 1993. Genetics and the evolution of phenotypic plasticity. Annual Review of Ecology and Systematics 24:35–68.

Schmalhausen, I. 1949. Factors of evolution: The theory of stabilizing selection. Blakiston, Philidelphia, PA.

Slatkin, M. 1973. Gene flow and selection in a cline. Genetics 75:733–756.

Slatkin, M. 1978. Spatial patterns in the distributions of polygenic characters. Journal of Theoretical Biology 70:213–228.

Sumner, F. B. 1930. Genetic and distributional studies of three subspecies of *Peromyscus*. Journal of Genetics 23:275–376.

Terman, C. R. 1968. Population dynamics. *In* J. A. King, ed. Biology of *Peromyscus*

(Rodentia), pp. 412–450. Special Publication 2. The American Society of Mammalogists.

Thompson, D. B. 1990. Different spatial scales of adaptation in the climbing behavior of *Peromyscus maniculatus*: Geographic variation, natural selection, and gene flow. Evolution 44:952–965.

Thompson, D. B. 1992. Consumption rates and the evolution of diet-induced plasticity in the head morphology of *Melanoplus femurrubrum* (Orthoptera: Acrididae). Oecologia 89:204–213.

Van Devender, T. R., and W. G. Spaulding. 1979. Development of vegetation and climate in the southwestern United States. Science 204:701–710.

Via, S. 1987. Genetic constraints on the evolution of phenotypic plasticity. *In* V. Loeschcke, ed. Genetic constraints on adaptive evolution, pp. 46–71. Springer-Verlag, Berlin.

Via, S. 1988. Estimating variance components: Reply to Groeters. Evolution 42:633–634.

Via, S., and R. Lande. 1985. Genotype–environment interaction and the evolution of phenotypic plasticity. Evolution 39:505–523.

Wade, M. J., and D. E. McCauley. 1988. Extinction and recolonization: Their effects on the genetic differentiation of local populations. Evolution 42:995–1005.

Wecker, S. C. 1963. The role of early experience in habitat selection by the deer mouse. Ecological Monographs 33:307–325.

Wells, P. V. 1983. Paleobiogeography of montane islands in the Great Basin since the last glaciopluvial. Ecological Monographs 53:341–382.

3

The Evolution of Behavioral Norms of Reaction as a Problem in Ecological Genetics

Theory, Methods, and Data

SCOTT P. CARROLL
PATRICE SHOWERS CORNELI

Behavior, like other phenotypic traits, varies as a function of genes and environment. Variation occurs at all demographic levels, within individuals over time, between individuals, and between populations and species. Whether variation is important will depend on the behavior and its context. For example, whether a bird scratches its head by extending a leg above or below the adjacent wing may not have profound fitness consequences, although species differences in this character may shed light on phylogenetic relationships (e.g., Wallace 1963; Simmons 1964). In contrast, other behaviors, such as the instantaneous decision to migrate or not, may affect fitness directly by altering the schedule of fecundity or mortality (Dingle et al. 1982). Such strategic behaviors (Maynard Smith 1982), which often depend for their expression on the assessment of local cues (Moran 1992), are complicated and important evolutionary traits. The phenotypic variability that defines them, however, has hindered our ability to treat them with formal evolutionary–genetic analyses that are central to the complete understanding of any putative adaptation.

Much of the evolutionarily important variation observed in strategic behavior probably stems from differences among individuals due to genotype–environment interactions. To illustrate this in the most general terms, consider that behavioral distinctions among individuals may be based on (1) differences in the environmental conditions they experience, (2) differences in genetic elements that code for specific tactics or predispositions, or (3) differences in the genotype–environment interaction, manifested through developmental or facultative pathways, that is, "norms of reaction" (Schmalhausen 1949). Norms of reaction are functions that describe how a genotype is translated into a phenotype by the environment. They are becoming widely employed as a paradigm in evolutionary studies of physiological and life-history traits (e.g., Dingle 1992; reviewed by Stearns 1989), but are not yet used widely in studies of behavioral traits (but see Thompson this volume). Because much of the variation that behaviorists observe within populations and species is likely the result of a complex combination of individual differences in genetic code and differences in environment, norms of reaction need to be explored as a method for understanding the sources and structure of behavioral variation.

In nature, behavioral variation occurs not just within populations, but among them as well. Comparative methods distinguish historical (e.g., phylogenetic) versus ecological contributions to phenotypes (e.g., Endler 1982), and geographic comparisons within species are useful in limiting phylogenetic variation, while taking advantage of environmental differences that may cause strategic divergence. For testing theoretical ideas about behavioral adaptation, this between-population approach (e.g., in birds: Reyer 1980, Dhont 1987, Koenig and Stacey 1990, Dunn and Robertson 1992; in fish: Kodric-Brown 1981, Mousseau and Collins 1987, Foster 1988, 1995; Houde and Endler 1990; in mammals: cf. Sherman 1989; in insects: Riechert 1986b, Carroll 1993; and most of the chapters in this volume) is similar to that taken in studies that examine conditional variation (norms of reaction) within a population as a function of resource variation (e.g., reviews of behavior by Thornhill and Alcock 1983, Lott 1991).

However, studies of geographic variation that do not also examine conditionality within populations may confound the sources of variation. Environmental differences between populations may cause behavioral differences even if the populations do not differ genetically. Or the population differences may have a genetic basis. Without making this distinction, population differences in behavior may be incorrectly assumed to have a genetic basis, while in fact neither the intra- nor interpopulation approach normally directly addresses genetic contributions to behavioral variation.

In figure 3-1, we illustrate the complexity implicit in interpreting the causes of geographic variation. To emphasize our point, we depict a one-dimensional domain with simple relationships between environmental and phenotypic values. Natural situations are typically more complex because environmental "gradients" are often nonlinear mosaics, multiple environmental factors interact, and complex phenotypes such as behavior may exhibit an array of potentially interchangeable and nonlinear (e.g., threshold) values, which may in turn influence the environmental conditions experienced.

Figure 3-1A shows the first empirical step: the observation of a difference in the mean phenotypic value of a trait between geographically distinct populations along an environmental gradient. Such an environmental axis could be a gradient in the physical environment, such as temperature variation with latitude, or a gradient in the social system of a species or in the community ecology of its habitat. Figure 3-1B–D show the predictions of three hypotheses that could explain the observation. Figure 3-1B depicts the hypothesis of genetic determination: the population difference will be maintained when the observations are made in the reciprocal environments. An analysis of variance (ANOVA) would show a significant population (genetic) effect, but nonsignificant environmental and population-by-environment interaction effects. Figure 3-1C depicts the hypothesis

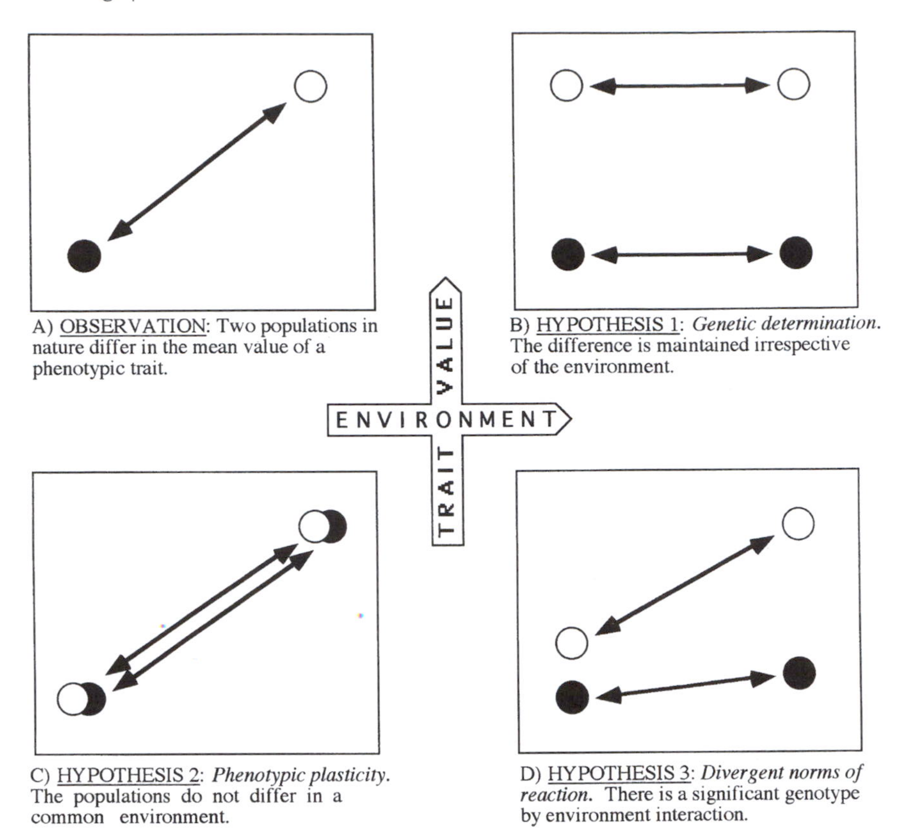

Figure 3-1 Alternative hypotheses for geographic variation in a phenotypic trait.

of species-wide phenotypic plasticity: members of the two populations show the same mean phenotypes when observed under identical conditions. In this case ANOVA would show a significant environment effect, but nonsignificant population and interaction effects. Figure 3-1D depicts the hypothesis of differentiation in norms of reaction: mean phenotypes depend on the environment in a different way in each population. The upper population exhibits greater phenotypic plasticity than does the lower. In this case, ANOVA would show significant population, environmental, and interaction terms. Note that in each case, the hypothesis is tested by observing each population at more than one point along the environmental gradient.

The norms of reaction depicted in figure 3-1 represent phenotypic mean responses. To the extent that these responses are genetically determined, the response of each population could be decomposed to show its constituent genotypic norms of reaction. Depending on intrapopulation heterogeneity, these could be tightly clustered about the mean responses or widely spread about the means. Gathering data at this level has the value of clarifying the nature and extent of differentiation both within and between populations. In addition, because plasticity and genetic polymorphism are evolutionarily interdependent, as discussed below, more fully accounting for the sources of behavioral variation is a critical aspect of understanding the evolution of behavior in general.

In this chapter, we discuss some of the reasons strategic behavioral traits, and especially behavioral plasticity, are important and unusual evolutionary traits. We present the merits of studying geographic variation in behavioral norms of reaction and review some salient literature on the evolutionary genetics of population differentiation in behavior and phenotypic plasticity. Then, using an example from our work with soapberry bugs, we present an analysis of geographic variation in the plasticity of the male mating strategy and its underlying genetic variation. We find that this intraspecific comparison presents a special opportunity for considering the effects of both environmental and genetic differences.

Behavioral Plasticity and Genetic Variation

Behavioral traits are even more complex than many physiological and life-history components of fitness (e.g., Price and Schluter 1991) because they may often be more flexible (i.e., exhibit reversible change) and more frequently revised (i.e., altered after assessment). The flexibility and responsiveness of behavior are among the traits of greatest functional importance in all of evolution because they provide a program for "adapting" to environmental changes throughout each individual's lifetime (Thoday 1953, Slobodkin and Rapoport 1974). Two general views have been developed to describe the evolutionary importance of behavioral flexibility. First, such within-generation plastic responses may promote homeostasis and thereby "buffer" the genetic effects of natural selection (Wright 1931, Sultan 1987). Alternatively, adaptive behavioral flexibility may increase the variety of habitats to which a genotype has access, ultimately enhancing a population's potential for evolutionary change (Morgan 1896, Waddington 1953, Wcislo 1989, West-Eberhard 1989). These ideas are developed below.

The power of natural selection to shape populations depends in part on three aspects of genotype–environment interaction that determine the relative fitness of individuals under diverse conditions: the capacity for adaptive plasticity inherent within genotypes, the pattern of diversity among genotypic norms of reaction within populations, and the distribution of environmental variability (Levins 1968, Sultan 1987, 1993). As a result, plasticity and genetic polymorphism will interact in evolutionary time. For example, both phenotypic plasticity (de Jong 1989) and genetic polymorphism may be maintained by spatial variation in selection (Levene 1953, Maynard Smith and Hoekstra 1980, Via and Lande 1985), while plasticity, evolved in response to spatial or especially to temporal variation (Moran 1992), may act to reduce the intensity of diversifying selection (sensu Wright 1931, Levins 1963, Slobodkin and Rapoport 1974). In other words, once an organism is sufficiently plastic, it will not experience spatial variation in selection at the level experienced by less plastic counterparts. On the other hand, in a population of plastic individuals, disparate genotypes may converge on a common adaptive phenotype, effectively shielding genetic variation. Moreover, even when genotypes differ in their responses to environment, these differences may sum to equal fitnesses across environments, again shielding them from natural selection (Haldane 1946, Via 1987, Gillespie and Turelli 1989, Barton and Turelli 1989). These hypotheses are important because the potential for evolutionary change in a population depends in part on its genetic variation (Fisher 1958).

The inherent complexity of genotype–environment interactions indicates the importance of studying genetic and environmental variation in tandem. A foundation for such an approach comes from the work of quantitative geneticists, who in recognizing the importance of environmental variation in determining phenotypic values, work to control the environmental conditions in which they conduct their studies (e.g., Falconer 1981). Until recently, the converse could not be said of ecologists, and behavioral and other variation is still most often related to environmental variation with scant consideration for genetic variation, especially at the empirical level. One useful technique is the "common garden experiment," developed by botanists, in which study subjects from disparate environments are observed in the same setting to test for genetic differences. This approach has been used to show genetically based population differentiation in the social behavior of amphipods (Strong 1973), spiders (Uetz and Cangialosi 1986), and fish (Magurran 1986, Magurran and Seghers 1990).

The common garden technique may be readily extended to permit the measurement of norms of reaction by observing populations across a range of reciprocal environmental conditions. At least four studies have used this approach to study population divergence in strategic behavior. Lynch (1992) studied the effects of temperature on nest-building behavior in mice, Riechert (1986a) examined food availability and territoriality in spiders, Dingle (1994) studied the effects of temperature on flight propensity in milkweed bugs, and Carroll and Corneli (1995) examined the effects of sex ratio on male mate-guarding decisions in the soapberry bug. All but Riechert (1986a) observed differentiation among populations in reaction norms; Riechert (1993, this volume) provides evidence that gene flow among her study populations has retarded divergence.

These studies, as well as other recent genetic studies (summarized in table 3-1), offer some preliminary answers to questions about the genetics of behavioral traits strongly tied to fitness. Most basically, behavioral variation among populations often has some genetic basis. Behavioral variation among individuals within populations has a genetic basis as well, with instances of both Mendelian and quantitative control documented (Orr and Coyne 1992). In addition, some behavior patterns are tightly correlated with other traits and may form genetically based, coadapted complexes with morphological values (e.g., male mating morphs within populations of swordtails, *Xiphophorus nigrensis* [Ryan et al. 1992], and of sponge-dwelling isopods [Shuster 1989]), life-history values (e.g., differences in age of first reproduction in differentially migratory morphs between populations of milkweed bugs, *Oncopeltus fasciatus* [Dingle 1994]), and behavioral values (e.g., mating success [Hoffmann and Cacoyianni 1989]).

Taken together, these results show that genetic differences among individuals can be important in behavioral differentiation both within and among populations.

Table 3-1 Examples of the use of quantitative genetics as a tool for studying the biology of behaviors related to fitness.

Behavior	Reference
Intraspecific crosses	
Foraging	Arnold 1981, Schemmel 1980, Hedrick and Riechert 1989
Migration	Berthold and Querner 1980
Predator avoidance	Riechert and Hedrick 1990
Courtship	Krebs 1990
Territorial and agonistic behavior	Riechert 1986a, Riechert and Maynard Smith 1989
Parent–offspring regression	
Migratory behavior	Caldwell and Hegmann 1969
Host preference	Fox 1993
Half-sib designs	
Migratory behavior	Dingle 1988, Fairbairn and Roff 1990
Dispersal	Greenwood et al. 1979, Mikasa 1990
Agonistic behavior	Riddell and Swain 1991
Antipredator behavior	Brodie 1989, Breed and Rogers 1991
Male mating strategy	This chapter
Full-sib designs	
Host preference	Via 1986, Fox 1993
Among-colony comparisons (honeybees)	
Behavioral ontogeny	Page et al. 1992
Artificial selection	
Migratory behavior	Palmer and Dingle 1989
Agonistic behavior	Ruzzante and Doyle 1991
Male mating strategy	Cade 1981
Territorial and mating success	Hoffmann and Cacoyianni 1989

Most of these studies have not formally considered behaviors as norms of reaction, but several have investigated the genetics of behavioral differences between populations.

An example of how this genetic perspective can be extended to comparisons of behavioral reaction norms is the subject of the next section.

Geographic Variation in Behavioral Plasticity in the Soapberry Bug: Environmental, Population, and Additive Genetic Effects

We are comparing the form and flexibility of male mating tactics (mate guarding versus nonguarding) among populations of soapberry bugs from two types of environments: those that exhibit spatial and temporal stability in male/female ratios (southern Florida), and those that exhibit exceedingly variable male/female ratios (Oklahoma). Our focus is to test whether (1) behavioral differences between populations have a genetic basis, (2) males from the more variable environment show a more plastic mating strategy than do those from the more constant environment, and (3) the populations differ in the amount of additive genetic variation underlying the behavioral reaction norms.

The soapberry bug, *Jadera haematoloma* (Insecta: Hemiptera: Rhopalidae), is a mainly neotropical seed predator. It follows the distribution of one of its host plants, the western soapberry tree, *Sapindus saponaria* v. *drummondii* northward into the temperate south-central United States, and the distribution of another, the balloon vine, *Cardiospermum corindum*, into subtropical Florida. Thus it forms two ecologically divergent and geographically disjunct metapopulations in the United States. We have studied aggregations at host plants in central and west-central Oklahoma and in the upper Florida Keys for several years. Bugs from the two regions are essentially identical in appearance and are interfertile, but several lines of evidence suggest that there is probably little gene flow between them (Carroll and Boyd 1992). As a result, they may evolve differentially in response to regional differences in selection (see Thompson this volume for further discussion).

The principal mating decision that adult male soapberry bugs can make is whether to guard mates after insemination or depart and search for additional matings (sensu Parker 1978). Sperm transfer is completed in 10 min or less, but males often remain in copula with females after inseminating them. In the field, pairs of marked individuals have been observed to remain together for as long as 11 days, with the female laying several clutches of eggs while the male attends (Carroll 1991).

Males in the two regions may have evolved differences in mating behavior at three levels. First, differences between bug populations in the relative costs and benefits of mate guarding versus nonguarding could select for differences in male propensity to exhibit either tactic (the mean of the reaction norm). Second, any differences in the variability of mating opportunities could select for differences in behavioral plasticity (the slope of the reaction norm). Third, any differences specifically in the spatial variability of mating opportunities could also result in the maintenance of different levels of genetic variation for the male behavioral reaction norm.

As indicated above, the regional difference with the greatest potential impact on the male mating system is in the mean and variability of sex ratio in large reproductive aggregations around the host plants. In Oklahoma, sex ratios have ranged from 0.62 to 4.71 males per female (mean = 2.60 ± 1.02 *SD*, $n = 33$ aggregations); in contrast, they are restricted much more closely to 1 : 1 in Florida (range = 0.56–1.67, mean = 1.07 ± 0.29 males/female, $n = 21$ aggregations). Sex ratio is also significantly more variable in Oklahoma (variance = 1.27) than in Florida (variance = 0.08; $F(1, 54) = 17.0$, $p < .0001$; Carroll and Corneli 1995). Thus, the populations differ in both the magnitude and variability of female availability as mates. This pattern may cause the populations to differ in the form and intensity of sexual selection on male mating behavior as well as other characters (Carroll and Salamon 1995). The primary sex ratio is 1 : 1 in both populations, but greater female than male mortality occurs during most phases of the life cycle in Oklahoma, apparently in association with environmental and developmental stresses related to ephemeral breeding opportunities in a highly variable climate (Carroll 1988, 1991).

The difference in mean sex ratio is reflected in the costs of mate searching in the two populations. In field experiments, the search time required for a male to find a mate was about three times greater in an Oklahoma aggregation (sex ratio 3 : 1) than in a Florida aggregation (sex ratio 1 : 1). Consistent with predictions based on this difference, brief (unguarded or minimally guarded) pairings were almost twice as common in Florida as in Oklahoma (Carroll 1993). This behavioral observation is of the type depicted in figure 3-1A. The regional difference in sex ratio variation also suggested that males from Oklahoma, but not Florida, might show the capacity to alter their tactical allocation as a function of the sex ratio experienced.

To test this hypothesis, we observed males from both populations in greenhouse arenas over a range of four experimental sex ratios (a "common garden" design; male:female 1 : 2, 1 : 1, 2 : 1, and 3 : 1). The results showed that the strategic differences between the populations have a genetic basis and that the population from the more variable sex ratio environment (Oklahoma) appears to be more plastic behaviorally (details below; Carroll and Corneli 1995).

Further, we incorporated a half-sib breeding design into this experiment to investigate how the plasticity of the phenotype interacts with its genetic variation at the population level (fig. 3-2, Falconer 1981). This gave us our third basic result: the Oklahoma population appears to have significant additive genetic variation for the male strategy, whereas the Florida population does not. These findings are also detailed below. Because of our small number of sire families, we followed the suggestion of Via (1986) to avoid relying on estimates of variance components and instead limited our treatment to analyses of the interaction between sire family and sex ratio within each population.

The nature of the data (sequences of states observed at discrete time intervals) and of the hypotheses suggested modeling the sequences of mating behavior as first-order Markov chains. These are stochastic processes in which the probability of an event occurring depends only on the immediately preceding event. For a male making tactical decisions, the present mating state should depend, in part, on the previously sampled one. For example, a guarding male should be more likely to stay with the same female from one observation to the next than a male who devotes more effort to searching. The latter male should switch from one female to another relatively more often. The probabilities of transitions from one behavioral state to another should differ among males and populations employing different mating strategies. By fitting the mating data to suitable probabilistic models and comparing the results of the fits, the differences should be revealed in log-likelihood ratio tests (LRT) of maximum likelihood estimators (MLEs) (Carroll and Corneli 1995). The null hypothesis is that the transition probabilities do not differ among the four sex ratios. We used first-order chains rather than more complex models because the mating state just before the current one is probably much more likely to have a significant effect on the current one than are more removed mating states.

The main results are shown in figure 3-3. Guarding behavior changed significantly as a linear function of sex ratio in the Oklahoma population (LRT = 66.8, df = 1, $p < .001$), but not in the Florida population (LRT = 1.6, df = 1, $p > .10$; Carroll and Corneli 1995). To examine additive genetic variation, we used a nes-

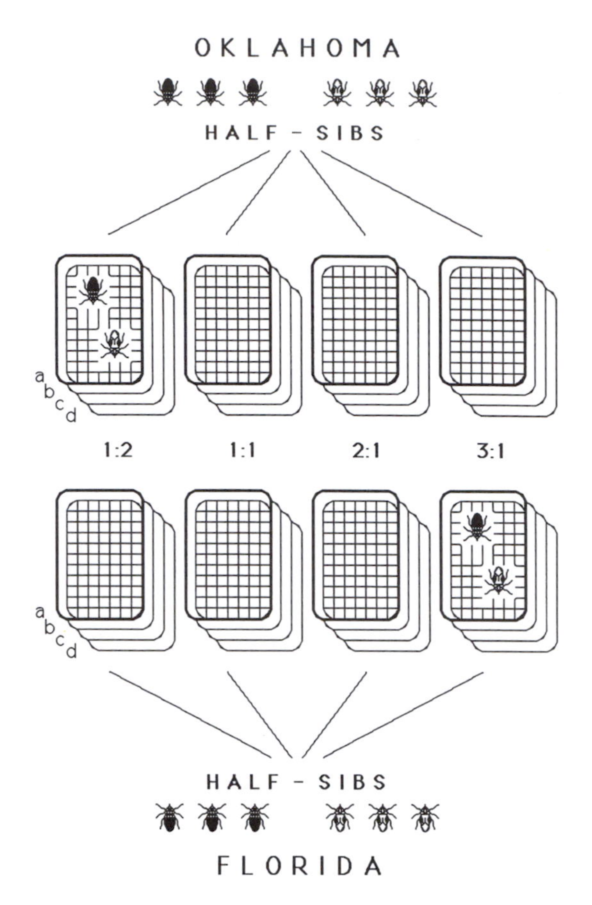

Figure 3-2 Stylized depiction of the experimental design. The grandparents of the bugs used in this experiment were collected from Boiling Springs State Park in Woodward County, Oklahoma, USA, and Plantation Key in Monroe County, Florida, USA. They were held in captivity in identical rearing cages at similar densities, where they reproduced feeding on the seeds of their native host plants. First-generation adults were paired in a half-sib mating design, with seven sires for Oklahoma and eight for Florida, each mated to three or four different females. Their offspring were similarly reared in full sib-groups. Experimental (second-generation) individuals were taken from these parents as newly molted (naive virgin) adults, measured, and given an individually identifying number on the dorsum. Members of each full-and half-sib family were distributed nearly uniformly through the sex ratio treatment replicates. Arenas were plastic boxes $33 \times 24 \times 11$ cm high. Twenty-four virgin adults were placed in each arena, in groups consisting of 8 males + 16 females, 12 males + 12 females, 16 males + 8 females, and 18 males + 6 females. Each sex ratio treatment was simultaneously replicated four times per population. Light and temperature conditions simulated those typical of reproduction in the field, and food and water were provided ad libitum. The mating status (copulating or single) of all individuals, was recorded at 3-h intervals, eight times each day, for 8 days.

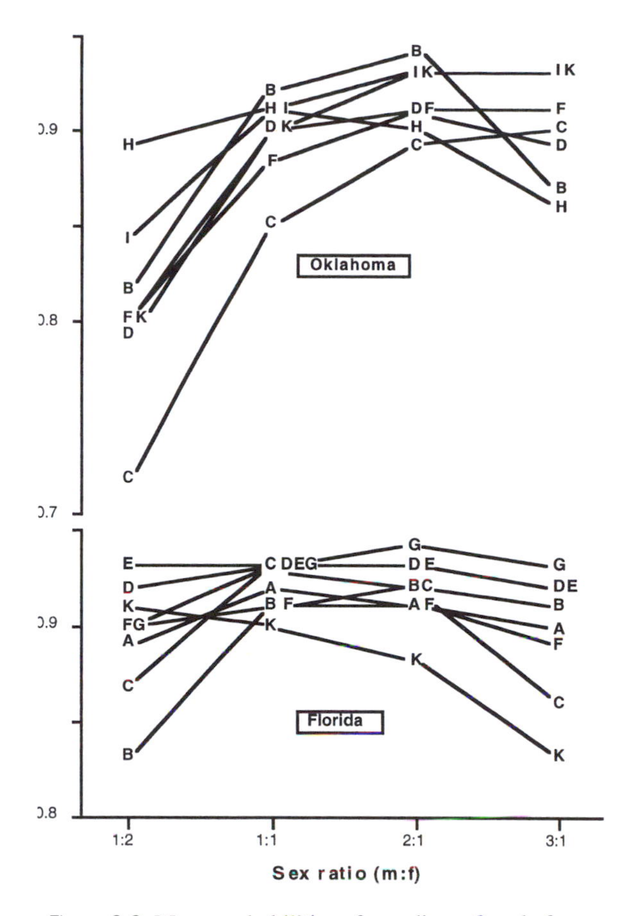

Figure 3-3 Mean probabilities of guarding a female from one observation period to the next (3-h observation interval) for each of seven half-sib families (Oklahoma, above), or eight half-sib families (Florida, below), measured at four sex ratios. Letters designate each family within a population.

ted design and a binomial model in which we fit a slope to each family's norm of reaction across sex ratio treatments (comparable to an ANCOVA). The interaction between family and ratio was significant for Oklahoma (LRT = 15.5, df = 6, $p <$ .025), indicating that slopes differ among families as a result of additive genetic variation. The results of the "ANCOVA" model analysis also suggest that slopes are significantly different from zero ($p < .05$) for each family except B and H.

In contrast, Florida families did not differ in guarding behavior (LRT = 7.5, df = 7, $p > .10$) when all families were considered simultaneously (fig. 3-3). Likewise, examination of the individual slope parameters and their standard errors for each family separately showed no change in mating behavior as a linear function of sex ratio, with the possible exception of family B ($p < .05$).

Interpretation of the Results and General Conclusions

Without the results of our common garden, norm of reaction experiment, one could readily argue that the phenotypic differences in soapberry bug mate-guarding originally observed between populations in nature by Carroll (1993) reflect an evolved, unconditional difference in mating strategy. One biologist could equally well argue that what was observed is part of the range of behavior available to all male soapberry bugs depending on the environmental conditions. Both analyses would be incomplete and potentially misleading. The populations have differentiated in both senses; it is the reaction norm that has differentiated, resulting in an inherited differential response that depends on the conditions a male experiences.

More broadly, we have found variation across environmental conditions, between populations, and perhaps as a function of additive genetic variation as well. Phenotypic variation at all of these levels results from a combination of behavioral plasticity and evolutionary response to diversifying selection. Behavioral plasticity is predicted to evolve chiefly in response to temporal variation in environmental conditions (Moran 1992), but to the extent that individuals may encounter variable conditions by moving, spatial variation may play an important role as well. Both kinds of variation are experienced by male soapberry bugs during their adult lifetimes (Carroll 1988, 1993, unpublished data).

It is interesting to consider how these findings relate to the general and often conflicting notions about the interaction of plasticity and genetic variation. In the Oklahoma population, relatively great behavioral plasticity and genetic diversity co-occur. This indicates that in the more variable environment, genetic differences among individuals in behavioral predisposition exist in spite of plasticity that potentially shields genetically different bugs from differential reproductive success. In addition, two of the seven Oklahoma half-sib families were not plastic, and one of the eight Florida families was. Thus the genetic predispositions for behavioral plasticity and nonplasticity appear to exist currently in both populations. Exploring such questions further will require larger sample sizes for each population to give greater statistical power for analyzing genetic and environmental components of variance, as well as data from more populations in diverse environments to test the generality of our results.

The co-occurrence of plasticity and genetic variation in the male mating strategy of the Oklahoma population also relates to theories addressing "evolutionary potential" in populations. Genetic variation is required for evolutionary change, and the rate of evolutionary response to selection is directly proportional to the amount of additive genetic variation present (Fisher 1958). In addition, West-Eberhard (1989), independently developing a premise originally put forth by Morgan (1896), suggested that relatively plastic organisms are more likely to encounter novel conditions that could increase the diversity of genotype by environment interaction manifested and thus increase the diversity of potential evolutionary trajectories. Because of the special power of behavioral plasticity to influence the selective environments experienced by other phenotypic traits, this argument should be especially relevant for behavioral plasticity as compared to plasticity in other fitness-related traits.

To the extent that both strategic plasticity and genetic variation for the behavioral strategy are greater in Oklahoma, that population's potential for evolutionary change may be greater than that of the more tropical, environmentally static Florida population. Oklahoma bugs, which inhabit an unstable environment at the northern edge of the species range, are probably derived from more tropical ancestors. Their local adaptations to environmental variability, including behavioral plasticity, may serve to accelerate their rate of evolutionary divergence from tropical antecedents beyond that which would be predicted from measurements of mean selection intensity and genetic variation alone.

Our study does not test whether individuals in each population exhibit optimal or evolutionarily stable tactics within and across sex ratios. Nor do our results distinguish the genetic basis of the population difference. Our focus was simply to ask whether there is plasticity and whether this allows the individual to do well in its own environment. In this sense we have taken a sufficiency rather than an optimality approach. This approach has permitted us to ask evolutionary questions about complicated fitness-associated traits that are difficult to model, and it has revealed a fascinating tactical complex within the species.

Dobzhansky (1951) argued that norms of reaction, rather than specific traits, are the targets of selection, a perspective that has continued to be explored into the present (Via et al. 1995). He was making a plea for incorporating more of the complexity of nature into scientific study, a perspective that clearly applies to behavior hypothesized to be adaptively flexible. All the current model approaches to behavioral evolution—evolutionarily stable strategies, optimality, quantitative genetics, norms of reaction—are simplistic caricatures of the true complexity of natural systems. Yet, because they are complementary and can be combined, as shown here and in our related work (Carroll 1993, Carroll and Corneli 1995), progress can be made in analyzing the evolution of strategic conditionality.

Behavioral strategies may be viewed as tool kits that organisms use to solve problems and to take advantage of the opportunities they encounter. Because the form and frequency of problems and opportunities differ among environments, geographic comparisons are valuable for testing hypotheses about the selective basis of strategic variation. When possible, it is also important to test whether population differences in behavior result from genetic diversification or from phenotypic plasticity. Only by doing so can a researcher know whether the phenomena under study are different expressions of the same strategy or different strategies altogether; whether a biologist uses adaptation as a "null hypothesis" or has specifically demonstrated adaptive differentiation, it will be valuable to know the manner in which the pattern results from environment and/or genetics. Distinguishing these sources of variation is basic to understanding both the structure of adaptation within and among populations and the processes by which new behavioral phenotypes evolve. To ignore this issue is to remain at the crossroads in the battle between those who champion the power of selection in phenotypic evolution and those who argue for the importance of "phylogenetic constraints." By instead comparing strategies as norms of reaction, in reciprocal environments, it may be possible to move beyond this superficial dichotomization of "adaptation versus constraints" and study instead the adaptive process within a genetic lineage.

Acknowledgments For fruitful discussion and valuable support, we thank D. Berrigan, E. Charnov, P. Coley, O. Cuellar, D. Davidson, H. Dingle, V. Eckhart, J. Endler, C. Fox, H. Horn, J. Loye, M. McGinley, D. Rubenstein, J. Seger, S. Sultan, S. Tavaré, and J. Thompson, Jr. Thoughtful comments by S. Riechert, J. Travis, and an anonymous reviewer improved the chapter. Research funds were provided by National Science Foundation grant BSR 8715018, National Institutes of Health training Grant 5 T32 GM0764-11 in genetics, Sigma Xi, and the Animal Behavior Society. We also benefitted from the support of National Science Foundation grant IBN 9306818.

References

Arnold, S. J. 1981. Behavioral variation in natural populations. I. Phenotypic, genetic and environmental correlations between chemoreceptive responses to prey in the garter snake, *Thamnophis elegans*. Evolution 35:489–509.

Barton, N. H., and M. Turelli. 1989. Evolutionary quantitative genetics: How little do we know? Annual Review of Genetics 23:337–370.

Berthold, P., and U. Querner. 1981. Genetic basis of migratory behavior in European warblers. Science 212:77–79.

Breed, M. D., and K. B. Rogers. 1991. The behavioral genetics of colony defense in honeybees-genetic variability for guarding behavior. Behavior Genetics 21:295–307.

Brodie, E. D. III. 1989. Genetic correlations between morphology and antipredator behaviour in natural populations of the garter snake *Thamnophis ordinoides*. Nature 342: 542–543.

Cade, W. 1981. Alternative male strategies: Genetic differences in crickets. Science 212: 563–564.

Caldwell, R. L., and J. P. Hegmann. 1969. Heritability of flight duration in the milkweed bug *Lygeaus kalmii*. Nature 223:91–92.

Carroll, S. P. 1988. Contrasts in the reproductive ecology of temperate and tropical populations of *Jadera haematoloma* (Rhopalidae), a mate-guarding hemipteran. Annals of the Entomological Society of America 81:54–63.

Carroll, S. P. 1991. The adaptive significance of mate guarding in the soapberry bug, *Jadera haematoloma* (Hemiptera: Rhopalidae). Journal of Insect Behavior 4:509–530.

Carroll, S. P. 1993. Divergence in male mating tactics between two populations of the soapberry bug: I. Mate guarding versus not guarding. Behavioral Ecology 4:156–164.

Carroll, S. P., and Boyd, C. 1992. Host race radiation in the soapberry bug: Natural history, with the history. Evolution 46:1052–1069.

Carroll, S. P., and P. S. Corneli. 1995. Divergence in male mating tactics between two populations of the soapberry bug: II. Genetic change and the evolution of a plastic reaction norm in a variable social environment. Behavioral Ecology 6:46–56.

Carroll, S. P., and M. H. Salamon. 1995. Variation in sexual selection on male body size within and between populations of the soapberry bug. Animal Behaviour 50: 1463–1474.

de Jong, G. 1989. Phenotypically plastic characters in isolated populations. *In* A. Fontdevila, ed. Evolutionary biology of transient unstable populations, pp. 125–152. Springer-Verlag, New York.

Dhont, A. D. 1987. Polygynous blue tits and monogamous great tits: Does the polygyny threshold model hold? American Naturalist 129:213–220.

Dingle, H. 1988. Quantitative genetics of life-history evolution in a migrant insect. *In* G. de Jong, ed. Population genetics and evolution, pp. 83–93. Springer-Verlag, New York.

Dingle, H. 1992. Food level reaction norms in size-selected milkweed bugs (*Oncopeltus fasciatus*). Ecological Entomology 17:121–126.

Dingle, H. 1994. Genetic analyses of animal migration. *In* C. R. B. Boake, ed. Quantitative genetic studies of behavioral variation, pp. 145–164. University of Chicago Press, Chicago.

Dingle, H., W. S. Blau, V. K. Brown, and J. P. Hegmann. 1982. Population crosses and the genetic structure of milkweed bug life histories. *In* H. Dingle and J. P. Hegmann, eds. Evolution and genetics of life histories, pp. 209–229. Springer-Verlag, New York.

Dobzhansky, T. 1951. Genetics and the origin of species. Columbia University Press, New York.

Dunn, P. O., and R. J. Robertson. 1992. Geographic variation in the importance of male parental care and mating systems in tree swallows. Behavioral Ecology 3:291–299.

Endler, J. 1982. Problems of distinguishing historical from ecological factors in biogeography. American Zoologist 22:441–452.

Fairbairn, D. J., and D. A. Roff. 1990. Genetic correlations among traits determining migratory tendency in the sand cricket, *Gryllus firmus*. Evolution 44:1787–1795.

Falconer, D. S. 1981. Introduction to quantitative genetics, 2nd ed. Longman, New York.

Fisher, R. A. 1958. The genetical theory of natural selection. Dover, New York.

Foster, S. A. 1988. Diversionary displays of paternal stickleback: defenses against cannibalistic groups. Behavioral Ecology and Sociobiology 22:335–340.

Foster, S. A. 1995. Understanding the evolution of behavior in threespine stickleback: The value of geographic variation. Behaviour 132:1107–1129.

Fox, C. W. 1993. A quantitative genetic analysis of oviposition preference and larval performance on two hosts in the bruchid beetle (*Callosobruchus maculatus*). Evolution 47:166–175.

Gillespie, J. H., and M. Turelli. 1989. Genotype–environment interactions and the maintenance of polygenic variation. Genetics 121:129–138.

Greenwood, P. J., P. H. Harvey, and C. M. Perrins. 1979. The role of dispersal in the great tit (*Parus major*): The causes, consequences, and heritability of natal dispersal. Journal of Animal Ecology 48:123–142.

Haldane, J. B. S. 1946. The interaction of nature and nurture. Annals of Eugenics 13: 197–205.

Hedrick, A. V., and S. E. Riechert. 1989. Genetically-based variation between two spider populations in foraging behavior. Oecologia 80:533–539.

Hoffmann, A. A., and Z. Cacoyianni. 1989. Selection for territoriality in *Drosophila melanogaster*: Correlated responses in mating success and other fitness components. Animal Behaviour 38:23–34.

Houde, A. E., and J. A. Endler. 1990. Correlated evolution of female mating preference and male color patterns in the guppy *Poecilia reticulata*. Science 248:1405–1408.

Kodric-Brown, A. 1981. Variable breeding systems in pupfishes (Genus *Cyprinodon*): Adaptations to changing environments. *In* R. J. Naiman, and D. L. Soltz, eds. Fishes in North American deserts, pp. 347–363. John Wiley and Sons, New York.

Koenig, W. D., and P. B. Stacey. 1990. Acorn woodpeckers: Group-living and food storage under contrasting ecological conditions. *In* P. B. Stacey, and W. D. Koenig, eds. Cooperative breeding in birds, pp. 413–453. Cambridge University Press, Cambridge.

Krebs, R. A. 1990. Courtship behavior and control of reproductive isolation in *Drosophila mojavensis*—genetic analysis of population hybrids. Behavioral Genetics 20:535–543.

Levene, H. 1953. Genetic equilibrium when more than one ecological niche is available. American Naturalist 87:331–333.

Levins, R. 1963. Theory of fitness in a heterogeneous environment. II. Developmental flexibility and niche selection. American Naturalist 97:75–90.

Levins, R. 1968. Evolution in a changing environment. Princeton University Press, Princeton, NJ.

Lott, D. F. 1991. Intraspecific variation in the social systems of wild vertebrates. Cambridge University Press, Cambridge.

Lynch, C. B. 1992. Clinal variation in cold adaptation in *Mus domesticus*: Verification of predictions from laboratory populations. American Naturalist 139:1219–1236.

Magurran, A. E. 1986. Population differences in minnow anti-predator behavior. *In* R. J. Blanchard, P. F. Brain, D. C. Blanchard, and S. Parmigiani, eds. Ethoexperimental approaches to the study of behavior, pp. 192–199. Kluwer Academic, Dordrecht.

Magurran, A. E., and B. Seghers. 1990. Population differences in the schooling behavior of newborn guppies, *Poecilia reticulata.* Ethology 84:334–342.

Maynard Smith, J. 1982. Evolution and the theory of games. Cambridge University Press, Cambridge.

Maynard Smith, J., and R. Hoekstra. 1980. Polymorphism in a varied environment: How robust are the models? Genetics Research 35:45–57.

Mikasa, K. 1990. Genetic study on emigration behavior of *Drosophila melanogaster* in a natural population. Japanese Journal of Genetics 65:299–307.

Moran, N. A. 1992. The evolutionary maintenance of alternative phenotypes. American Naturalist 139:971–989.

Morgan, C. L. 1896. On modification and variation. Science 4:733–740.

Mousseau, T. A., and N. C. Collins. 1987. Polygyny and nest site abundance in the slimy sculpin (*Cottus cognatus*). Canadian Journal of Zoology 65:2827–2829.

Orr, H. A., and J. A. Coyne. 1992. The genetics of adaptation: A reassessment. American Naturalist 140:725–742.

Page, R. E. Jr., G. E. Robinson, D. S. Britton, and M. K. Fondrik. 1992. Genetic variability for rates of behavioral development in worker honeybees (*Apis mellifera* L.). Behavioral Ecology 3:173–180.

Palmer, J. O., and H. Dingle. 1989. Responses to selection on flight behavior in a migratory population of the milkweed bugs (*Oncopeltus fasciatus*). Evolution 43:1805–1808.

Parker, G. A. 1978. Searching for mates. *In* J. R. Krebs, and N. B. Davies, eds. Behavioural ecology: An evolutionary approach, pp. 214–244. Blackwell, Oxford.

Price, T., and D. Schluter. 1991. On the low heritability of life-history traits. Evolution 45:853–861.

Reyer, H. 1980. Flexible helper structure as an ecological adaptation in the pied kingfisher (*Ceryle rudis rudis* L.). Behavioral Ecology and Sociobiology 6:219–227.

Riddell, B. E., and D. P. Swain. 1991. Competition between hatchery and wild coho salmon (*Oncorhynchus kisutch*)—genetic variation for agonistic behavior in newly emerged wild fry. Aquaculture 98:161–172.

Riechert, S. E. 1986a. Between population variation in spider territorial behavior: Hybrid-pure population line comparisons. *In* M. D. Huettel, ed. Evolutionary genetics of invertebrate behavior, pp. 33–42. Plenum, New York.

Riechert, S. E. 1986b. Spider fights as a test of evolutionary game theory. American Scientist 74:604–610.

Riechert, S. E. 1993. Investigation of potential gene flow limitation of adaptation in a desert spider population. Behavioral Ecology and Sociobiology 32:355–363.

Riechert, S. E., and A. V. Hedrick. 1990. Levels of predation and genetically based antipredator behaviour in the spider, *Agelenopsis aperta*. Animal Behaviour 40:1738–1745.

Riechert, S. E., and J. Maynard Smith. 1989. Genetic analyses of two behavioural traits linked to individual fitness in the desert spider *Agelenopsis aperta*. Animal Behaviour 37:624–637.

Ruzzante, D. E., and R. W. Doyle. 1991. Rapid behavioral changes in medaka (*Oryzias latipes*) caused by selection for competitive and non-competitive growth. Evolution 45:1936–1946.

Ryan, M. J., Pease, C. M., and M. R. Morris. 1992. A genetic polymorphism in the swordtail *Xiphophorus nigrensis*: Testing the prediction of equal fitnesses. American Naturalist 139:21–31.

Schemmel, C. 1980. Studies on the genetics of feeding behaviour in the cave fish *Astyanax mexicanus*. An example of apparent monofactoral inheritance by polygenes. Zeitschrift für Tierpsychologie 53:9–22.

Schmalhausen, I. I. 1949. Factors of evolution: The theory of stabilizing selection. Blakiston, Philadelphia, PA.

Sherman, P. W. 1989. Mate guarding as paternity assurance in Idaho ground squirrels. Nature 338:418–420.

Shuster, S. M. 1989. Male alternative reproductive strategies in a marine isopod crustacean *Paracerceis sculpta*—the use of genetic markers to measure differences in fertilization success among alpha-males, beta-males, and gamma-males. Evolution 43:1683–1698.

Simmons, K. E. L. 1964. Feather maintenance. *In* A. L. Thompson, ed. A new dictionary of birds, pp. 278–286. MacGraw-Hill, New York.

Slobodkin, L. B., and A. Rapoport. 1974. An optimal strategy of evolution. Quarterly Review of Biology 49:181–200.

Stearns, S. C. 1989. The evolutionary significance of phenotypic plasticity. Bioscience 39: 436–445.

Strong, D. R., Jr. 1973. Amphipod amplexus, the significance of ecotypic variation. Ecology 54:1383–1388.

Sultan, S. E. 1987. Evolutionary implications of phenotypic plasticity in plants. Evolutionary Biology 21:127–178.

Sultan, S. E. 1993. Phenotypic plasticity in *Polygonum persicaria*. I. Diversity and uniformity in genotypic norms of reaction. Evolution 47:1009–1031.

Thoday, J. M. 1953. Components of fitness. Symposium of the Society for Experimental Biology 7:96–113.

Thornhill, R., and J. Alcock. 1983. The evolution of insect mating systems. Harvard University Press, Cambridge, MA.

Uetz, G. W., and K. R. Cangialosi. 1986. Genetic differences in social behavior and spacing in populations of *Metepeira spinipes*, a communal-territorial orb weaver (Araneae, Araneidae). Journal of Arachnology 14:159–173.

Via, S. 1986. Genetic covariance between oviposition preference and larval performance in an insect herbivore. Evolution 40:778–785

Via, S. 1987. Genetic constraints on the evolution of phenotypic plasticity. *In* V. Loeschcke, ed. Genetic constraints on adaptive evolution, pp. 47–71. Springer-Verlag, Berlin.

Via, S., and R. Lande. 1985. Genotype–environment interaction and the evolution of phenotypic plasticity. Evolution 39:505–522.

Via, S., R. Gomulkiewicz, G. de Jong, S. M. Scheiner, C. D. Schlichting, and P. H. van Tienderen. 1995. Adaptive phenotypic plasticity—concensus and controversy. Trends in Ecology and Evolution 10:212–217.

Waddington, C. H. 1953. Genetic assimilation of an acquired character. Evolution 7:118–126.

Wallace, G. J. 1963. An introduction to ornithology. MacMillan, New York.

Wcislo, W. T. 1989. Behavioral environments and evolutionary change. Annual Review of Ecology and Systematics 20:137–169.

West-Eberhard, M. J. 1989. Phenotypic plasticity and the origins of diversity. Annual Review of Ecology and Systematics 20:249–278.

Wright, S. 1931. Evolution in mendelian populations. Genetics 16:97–159.

4

Geographic Variations on Methodological Themes in Comparative Ethology

A Natricine Snake Perspective

GORDON M. BURGHARDT
JAMES M. SCHWARTZ

The most distinctive and characteristic emphasis of early ethology was also what set it off from other post-Darwinian studies of animal behavior. This was the view that behavior varied among species in the same way as did morphological characters and that behavioral differences were as much a product of the evolutionary drama as were the characters that could be measured in museum collections (Tinbergen 1960, Lorenz 1981, Burghardt 1985, Burghardt and Gittleman 1990, Gittleman and Decker 1994). The logical extensions of this view were that behavioral phenotypes could be used in reconstructing phylogenetic histories, that the evolution of behavioral phenotypes could be studied in the same way as the evolution of other classes of traits, and that many of the behavioral differences among taxa reflected underlying genetic differentiation at both the species (Hinde and Tinbergen 1958) and population (Foster and Cameron 1996) levels. Behavior may also initiate evolutionary changes in other attributes of organisms (Mayr 1960, 1965, Wcislo 1989, Gittleman et al. 1996).

Although the role of genes in behavioral determination remained controversial for years (see Gottlieb 1992, de Queiroz and Wimberger 1993 for current critiques), many behavior patterns have proven heritable (Mousseau and Roff 1987; papers in Boake 1994b). Indeed, some complex, "species-typical" behavior patterns are performed normally without opportunity for learning (Lorenz 1965). Such behavior patterns can be expressed early or late in development (Lorenz 1981). At the other extreme, many complex behavioral phenotypes are learned with only slight, if any, genetically based predisposition to perform particular behavior patterns. Between these extremes is a diversity of interactions between genes and environment, including imprinting and complex developmental trajectories produced by interactions between neural development and experience.

Many of the currently interesting and controversial questions in the nature–

nurture debate do not center around species-typical behavior patterns. Instead, they concern the nature of genetic differences among individuals and populations in the performance of particular behavior patterns and in the ability to modify their performance with experience. Thus the problem must be conceptualized as one in which the interactions of specific genetic constitutions with specific environmental contexts need to be evaluated.

For those interested in the evolution and causation of behavior, the very complexity of the interactions between genes and environment, as expressed in complex developmental trajectories, is fascinating. However, the same features can be extremely frustrating for those interested in establishing that behavioral differences among individuals, populations, or species are heritable, or in evaluating, for example, genetic correlations among phenotypic traits including behavior. Unfortunately, regardless of goal, the complexity of the interactions and the pitfalls of common garden or heritability experiments are often unanticipated or can render data difficult to interpret (for discussion, see Arnold 1994, Boake 1994a, Cheverud and Moore 1994).

Our first purpose in this chapter is to describe research designed to evaluate the extent of genetic differentiation underlying geographic differences in the behavior of a well-studied genus of natricine snake, the garter snakes, *Thamnophis*. Garter snakes exhibit substantial variation both within and among species in certain aspects of behavior, particularly foraging and predator-avoidance behavior. We provide an overview of geographic variation in behavior in this group and then describe differences in selected behavioral, physiological, and morphological traits between two populations of common garter snakes (*Thamnophis sirtalis*). We also present evidence of underlying genetic differentiation and trait correlations as revealed by testing neonates from wild-caught females.

We conclude with a discussion of issues in methodology and interpretation in the garter snake studies. Some of the issues we address are relevant to all research on the genetic bases of behavior; others are relevant primarily to live-bearing animals that do not provide parental care. The latter features are characteristic of the natricine snakes. Highly precocial neonatal *Thamnophis*, born into large litters, have provided extensive quantitative genetic data on phenotype development and expression (reviewed in Brodie and Garland 1993). These positive attributes are somewhat offset by the confounding effects of multiple paternity and maternal effects (Schwartz et al. 1989). Our discussion should provide methodological insights of value to those interested in exploring evolutionary patterns through geographic comparison within species.

Geographic Variation in Garter Snake Behavior

The considerable inter- and intraspecific variation in the ubiquitous North American genus of garter snakes, *Thamnophis*, poses a long-recognized challenge for students of evolution (Cope 1891, Ruthven 1908). The genus, encompassing approximately 30 species (de Queiroz and Lawson 1994), includes both cosmopolitan and geographically restricted species (Stebbins 1985, Conant and Collins

1991). Some are foraging specialists, while others are generalists. There is considerable interspecific variation in body size, body shape, coloration, markings, habitat use, food habits, reproductive characteristics, and behavior (Fitch 1965, 1985; Rossman et al. 1996). Although less well studied, population differences in behavioral and perceptual characteristics have also been convincingly documented in several species (table 4-1). Geographic variation in size (King 1988, 1989), color patterns (Rossman et al. 1996), home range (Macartney et al. 1988), resistance to toxic prey (Brodie and Brodie 1991), and reproductive characteristics (e.g., Fitch 1985, Gregory and Larsen 1993, 1996) have also been demonstrated. Similar geographic variation has been described in other snakes as well (e.g., Gove and Burghardt 1975, Hasegawa and Moriguchi 1989, King and Lawson 1995).

Foraging and antipredator behavior display especially high levels of variation at both inter- and intraspecific levels and offer excellent opportunities for the study of population differentiation in relation to ecological and geographical variation. In contrast, elements of social behavior other than courtship pheromone recognition (Ford 1986, Mason 1992) and tendency to aggregate (Gillingham 1987) appear to vary little within and among species of garter snakes. This impression may, however, reflect the lack of detailed study of these often seasonal and subtle responses, rather than an absence of variation (cf. Ford 1986).

Among North American snakes, the common garter snake, *Thamnophis sirtalis*, is particularly well studied in the field and in captivity. It is the most widespread snake species in North America (fig. 4-1), and its mating behavior, reproductive biology, movements, genetics, physiology, and food habits have been well characterized (see reviews in Carpenter 1952, Fitch 1965, Seigel et al. 1987, Ernst and Barbour 1989, Seigel and Collins 1993, Rossman et al., 1996). Although more subspecies have been named, 12 are recognized currently on the basis of size, scalation, and coloration (Fitch 1980).

In comparison to most garter snakes, *T. sirtalis* is a prey and habitat generalist. One reflection of this is the substantial variation in diet across populations (Fitch 1965, Kephart 1982). Dix (1968) suggested that genetic differentiation contributed to geographic variation in food preference based on a demonstration that newborn *T. sirtalis* from Florida preferred fish to earthworms, whereas those in New England preferred worms to fish. Similarly, Burghardt (1970) demonstrated that neonates from three populations in the Midwest (Wisconsin, Iowa, Illinois) differed in response to chemicals derived from a diverse array of potential prey items, with Wisconsin neonates, for example, responding more to fish chemicals than to worm chemicals. Although based on small samples, these early results supported the hypothesis of genetic differentiation and illustrated the importance of chemical cues in prey choice and recognition by garter snakes (Halpern 1992).

A second well-studied species that is also a widespread prey generalist is the western terrestrial garter snake, *Thamnophis elegans*, of the American West. Like *T. sirtalis*, this species displays considerable geographic variation in dietary preferences (Arnold 1977, Kephart 1982), and the differences are similarly mediated by chemosensory cues (Arnold 1981a,b, 1992). Neonate inland snakes responded more strongly to fish and coastal snakes responded more strongly to slugs, reflect-

Table 4-1 Selected studies on geographical variation in behavior and diet in *Thamnophis*.

Traits	Species	Localities	Age at testing	Study
Feeding Behavior				
Prey choice				
Laboratory	*T. sirtalis*	Florida, Massachusetts	Neonates	Dix 1968
Field	*T. sirtalis*	Inland and central California	Adults	Arnold 1977, 1981b, 1992
	T. elegans	Central Washington		
	T. sirtalis	West Coast, USA	Adults	Kephart 1982
	T. elegans			
	T. sirtalis	Vancouver Island	Adults	Gregory 1984
				Gregory and Nelson 1991
	T. melanogaster	Lakes, ponds, streams in Mexico	Adults	Macias Garcia and Drummond 1990
				Drummond, personal communication
Chemosensory prey preference	*T. sirtalis*	Illinois, Iowa, Wisconsin	Neonates	Burghardt 1970
	T. elegans	Inland and coastal California	Neonates	Arnold 1981b, 1992
Foraging and prey capture	*T. elegans*	Inland and coastal California	Neonates	Drummond and Burghardt 1982
	T. melanogaster	Mexico	Neonates	Macias Garcia and Drummond 1990
Antipredator Behavior				
Defensive strikes	*T. sirtalis*	Michigan, Wisconsin	Neonates	Herzog and Schwartz 1990
Defensive strikes; aggression score	*T. elegans*	Arizona, central and northern California, New Mexico, Oregon	Adults	de Queiroz 1992
Covariance structure	*T. ordinoides*	Washington	Neonates	Brodie 1993b

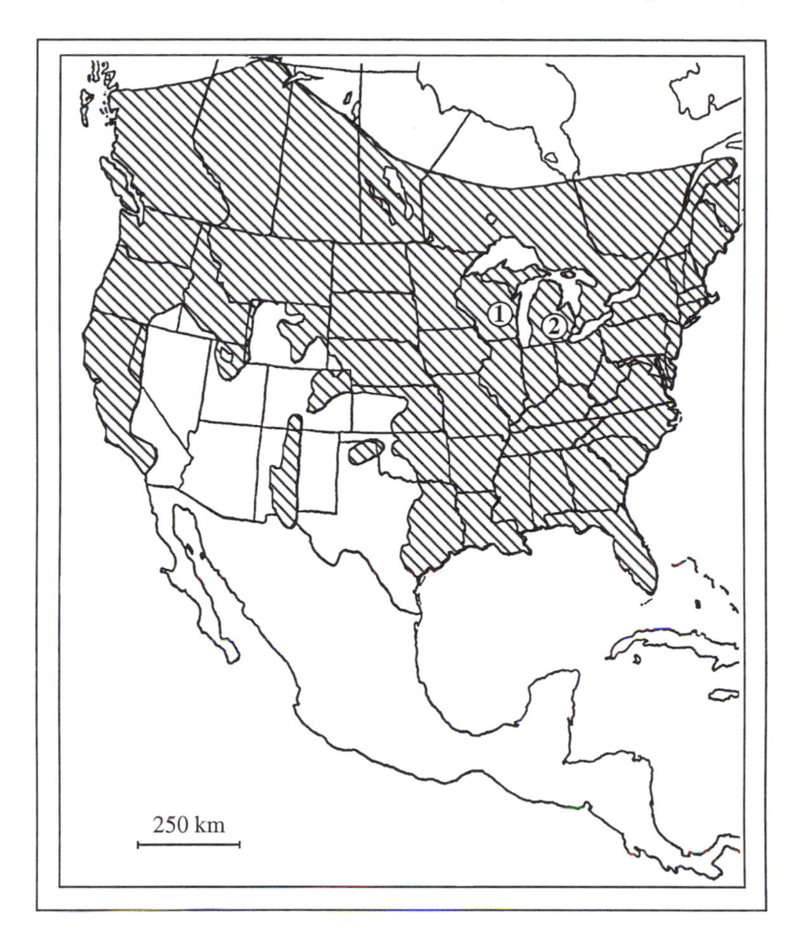

Figure 4-1 Geographic distribution of *Thamnophis sirtalis* showing the location of the populations studied in Wisconsin (1) and Michigan (2). Based on Fitch (1965).

ing regional food habits and availability. Inland snakes nearly always rejected slugs, whereas neonate snakes from both areas responded to fish. The inland snakes were more accomplished foragers on fish than were neonates from coastal areas (Drummond and Burghardt 1983). Although geographic differences in diet and prey preference tend to reflect prey availability, Arnold (1992) has recently documented dietary and neonatal chemosensory prey preference differences in three sympatric populations of *T. sirtalis* and *T. elegans* on the West Coast of the United States. Even where prey availability was the same, diet and chemical responses differed somewhat across species, with *T. elegans* exhibiting somewhat broader responses.

Although anecdotal reports of the tendency of snakes to bite when captured are common (e.g., Ernst and Barbour 1989), documentation of regional variation is slight. Antipredator behavior should be sensitive to local conditions (see Coss

1991, this volume; Magurran this volume), and geographic variation does occur in garter snake antipredator behavior (Herzog and Schwartz 1990, Brodie, 1993b).

In conclusion, the pattern that emerges from the study of geographic differentiation in the behavior of garter snakes is one of substantial variation in at least two classes of behavior: dietary preference and antipredator behavior. This variation is mirrored in the variation in morphological characteristics across geographical regions and in the relatively large number of recognized subspecies (Rossman et al. 1996). In the next section we demonstrate population differences in a greater array of behavioral and physiological phenotypes than have been examined previously, and we examine the genetic bases of the behavioral differences between two populations.

A Case Study: Two Populations of *Thamnophis sirtalis*

The study described below (Schwartz 1989) was carried out to compare multiple litters from two populations on opposite sides of Lake Michigan. The distance between these populations is not great relative to the wide distribution of *T. sirtalis* in North America (fig. 4-1), and both are considered the Eastern garter snake subspecies (*T. sirtalis sirtalis*). A major goal of the research was to estimate the frequency of multiple paternity in wild populations and to determine the extent to which multiple paternity would influence heritability estimates made assuming a single father for all sibs. Multiple paternity was common and did affect quantitative genetic measures, but we do not review that aspect of the study here (Schwartz et al. 1989).

The results we describe are based on studies of the neonates born to 24 females from a population of *T. sirtalis* from Wayne County, Michigan, USA, and to 5 females from the Wolf River Drainage, Wisconsin, USA (fig. 4-1). The females gave birth to 452 neonates (4–40 per litter), which were immediately separated from the mother. Data were collected on a number of behavioral, morphological, and physiological traits for each neonate in a set schedule over the snakes' first 21 days of life. The heritability estimates we provide are based on full-sibling calculations, although these are underestimates because multiple paternity occurred in many of these litters.

Population Differences Indicated by Neonatal Tests

Morphological differences between the populations were immediately evident. The Michigan neonates were significantly longer and heavier than neonates from the Wisconsin population. Heritability estimates are not given due to possible maternal diet effects on offspring size. Because all mothers ate well in captivity, the population difference should not have been affected, however. The number of ventral and subcaudal scales, measures commonly used in snake systematics (and closely linked to number of vertebrae), were both greater in Wisconsin animals than in Michigan animals, in spite of the shorter length of the Wisconsin animals (table 4-2).

Table 4-2 Population differences between two populations of *Thamnophis sirtalis* for several traits.

Contrast	$(p)n$	Mean ± SE	h^2 ± SE
Snout Vent Length (cm)	(<.0001)		
Michigan	348	15.1 ± 0.56	
Wisconsin	60	14.7 ± 1.60	
Mass (g)	(<.0001)		
Michigan	348	1.50 ± 0.02	
Wisconsin	60	1.23 ± 0.04	
Ventrals	(<.0001)		
Michigan	352	150.1 ± 0.24	0.63 ± 0.15*
Wisconsin	40	154.6 ± 0.63	0.90 ± 0.40*
Subcaudals	(<.0001)		
Michigan	352	65.8 ± 0.38	0.57 ± 0.14*
Wisconsin	40	70.6 ± 0.80	0.33 ± 0.32
Mean Litter Nearest-Neighbor Distance (mm)	(ns)		
Michigan	315	83.8 ± 6.0	0.26 ± 0.11*
Wisconsin	47	97.1 ± 12.9	0.06 ± 0.17
Mean Litter Time under Cover (%)	(.05)		
Michigan	315	58.0 ± 0.05	
Wisconsin	47	76.6 ± 0.06	
Mean Worm TFAS	(ns)		
Michigan	350	33.9 ± 1.39	0.25 ± 0.10*
Wisconsin	52	33.7 ± 4.79	0.35 ± 0.28*
Mean Fish TFAS	(<.02)		
Michigan	350	19.1 ± 1.82	0.24 ± 0.10*
Wisconsin	52	44.1 ± 4.44	0.52 ± 0.33*
Defensive Behavior (total score)	(<.0001)		
Michigan	349	5.8 ± 0.7	0.67 ± 0.15*
Wisconsin	52	25.9 ± 3.4	0.30 ± 0.27*
CTmin (°C)	(.06)		
Michigan	355	7.2 ± 0.08	0.56 ± 0.07*
Wisconsin	44	6.6 ± 0.23	0.78 ± 0.20*

For each population comparison the significant probability value (p) and sample size (n) are given.
*h^2 differs significantly from 0.0 at $p < .05$.
TFAS, tongue flick–attack score; CTmin, critical thermal minima.

Neonates from the two populations also differed in their use of cover. When observed over a 5-day period in a circular arena containing six identical objects that offered cover (Lyman-Henley and Burghardt 1994), neonates from the Michigan population were more often found in the open than were their Wisconsin counterparts (table 4-2). However, after controlling for litter sizes, Michigan snakes found under cover were in significantly larger groups than were Wisconsin animals. Michigan snakes appeared to have smaller nearest-neighbor distances than Wisconsin snakes (table 4-2), although the difference was not significant.

A standardized antipredator behavioral test (Herzog and Burghardt 1986) followed the aggregation observations (table 4-2; fig. 4-2). These tests revealed that Wisconsin neonates were more likely to strike than were Michigan neonates (see also Herzog and Schwartz 1990). The observation that the less aggressively defensive Michigan animals were more likely to be found at a distance from cover suggests different levels or types of predation in the two regions, but this possibility remains to be explored.

The two populations also differed in their responses to aqueous extracts prepared from surface substances of two prey types commonly eaten by eastern garter snakes: earthworms and fish. Employing standardized aqueous stimulus preparation and testing methods (Burghardt et al. 1988, Lyman-Henley and Burghardt 1995), chemical prey preferences were measured for earthworm extract (*Lumbricus terrestris*), fathead minnow extract (*Pimephales promelas*), and deionized water controls. Each neonate was tested twice with each prey type. Both attack data and a combined measure of tongue flick rate and attack latency, termed the "tongue flick–attack score" (TFAS), were analyzed (Burghardt et al. 1988, Cooper and Burghardt 1990).

When both TFAS and attack frequency were analyzed by stimulus, clear differences between populations emerged (table 4-2, fig. 4-3). Wisconsin neonates attacked fish chemicals more than they did worm chemicals, replicating earlier results for Wisconsin *T. sirtalis* (Burghardt 1970). Michigan animals directed a slightly greater frequency of attacks at earthworm than fish chemicals, but this difference was only significant using the TFAS measure of response intensity.

Finally, we tested for differences in critical thermal minima (CTmin) between the populations using 20-day-old animals. The CTmin is defined as the lowest body temperature at which the animal maintains its ability to right itself. Latitude is inversely related to CTmin in *Thamnophis* (Doughty 1994), suggesting the

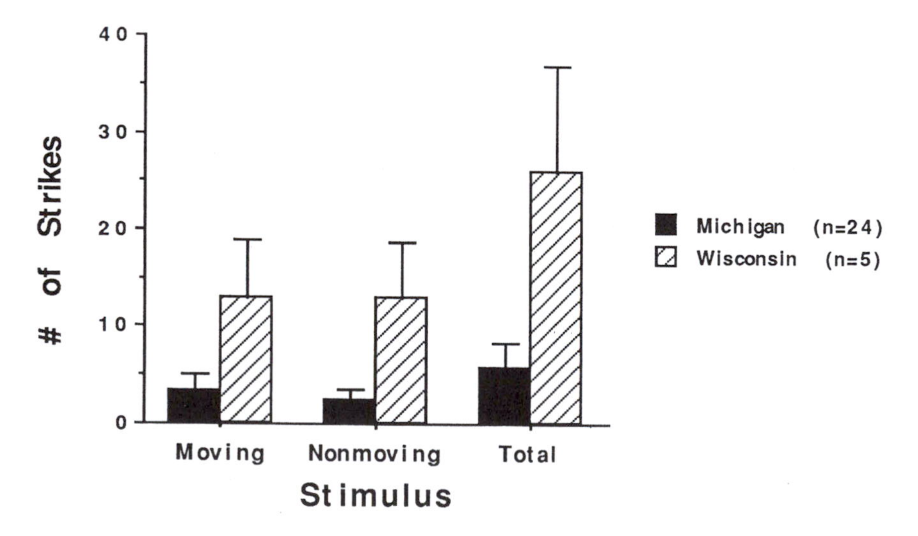

Figure 4-2 Population differences (±SE) in antipredator striking behavior of neonatal *Thamnophis sirtalis* directed toward moving and nonmoving stimuli.

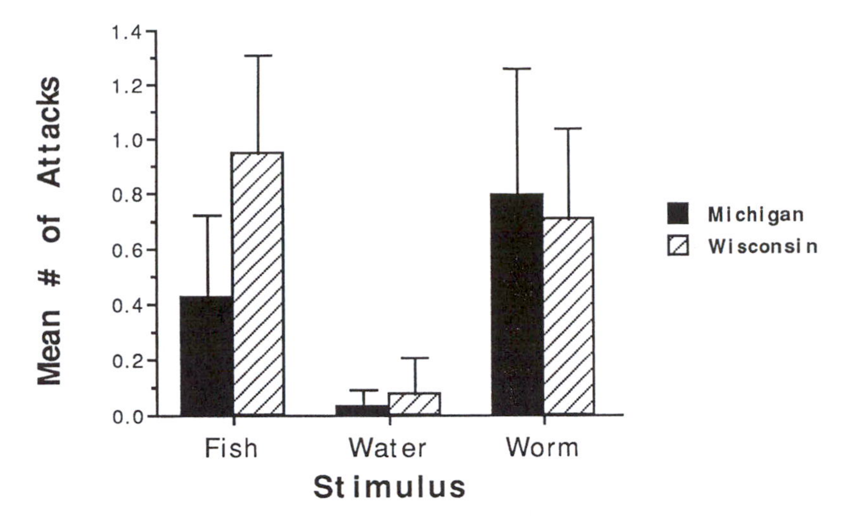

Figure 4-3 Population differences in the mean number of attacks (±SE) of neonatal *Thamnophis sirtalis* elicited by water controls and fish and worm prey chemicals.

possibility of population differentiation of this physiological trait. Body temperature was recorded by inserting a thermocouple into the cloaca of the animal and placing it into a cold (–4°C) chamber (see Doughty 1994 for details). Despite their smaller body size and resulting more rapid heat loss, the Wisconsin snakes exhibited a lower CTmin than the Michigan snakes (table 4-2). The Wisconsin animals came from an area that averages 3°C cooler throughout the year than the Michigan population and is exposed to an average of 21 fewer frost-free days. If, as Kitchell (1969) hypothesized, different behavior and temperature relationships of snakes are correlated with their distributions, selection may act on these traits. The large amount of heritable variation in CTmin (table 4-2) suggests that these populations could respond rapidly to selection. Sexton et al. (1992) have shown that *T. sirtalis* at higher latitudes have different denning and cool-weather activities; these may also have a genetic component.

Thus, as we anticipated, many differences were documented between the offspring of Michigan and Wisconsin snakes in size, scalation, antipredator behavior, prey chemical responses, cover use, and critical thermal minima. Because only two populations were sampled, each at a single time, firm conclusions about cause cannot be reached, but the differences are intriguing and merit further study. In the next section we describe genetic and phenotypic correlations among the traits that could have influenced evolutionary potentials in these populations.

Genetic and Phenotypic Correlations among Traits

Arnold (1981a,b, 1992) demonstrated that chemical responses of *T. elegans* and *T. sirtalis* to different prey are strongly enough correlated genetically that local adaptation to one of these prey can lead to associated changes in responses to chemicals from the other. Similarly, the degree of skin striping in *Thamnophis ordinoides* is related to defensive behavior (Brodie 1989, 1993b). Thus, examination of correlations not only between behavioral traits, but also among behavioral, morphological, and life-history traits, is essential if we are to understand constraints on evolutionary change or if we are to predict the directions of microevolutionary change under specific selective regimes (reviewed in Arnold 1994).

Because the patterns of correlation are often complex and unexpected, a number of kinds of traits should be examined simultaneously. For example, Arnold and Bennett (1988) studied five morphological and four locomotor performance variables (in *Thamnophis radix*) and found that single morphological and performance trait correlations explained insignificant amounts of variation. However, canonical correlation of the effect of all five morphological measures on all four performance measures explained a significant 24% of the variation. Performance was best predicted by a compound measure of morphology consisting of large body length, few body abnormalities, and few tail vertebrae.

We also discovered correlations among traits within both *T. sirtalis* populations, as well as some intriguing differences between the populations. The genetic and phenotypic correlations among traits within the Wisconsin and Michigan populations were, in many cases, very similar (table 4-3). Mantel tests (Dietz 1983, Smouse and Long 1992) involving all permutations (40,320) showed that the matrix correlation between the genotypic and phenotypic correlation matrices for the Michigan snakes was significantly greater than zero (matrix $r = .86$, $p < .0001$), as was that for the Wisconsin snakes (matrix $r = .70$, $p = .0007$). The observed correlation of 0.86 in the Michigan population was the highest correlation of the 40,320 permutations. Only 28 correlations of the permutations were greater than or equal to the observed correlation in the Wisconsin population.

The matrix correlation between the Michigan and Wisconsin phenotypic correlation matrices was also significantly different from zero (matrix $r = .80$, $p < .0001$), as was the correlation between genetic correlation matrices (matrix $r = .64$, $p = .002$). Interestingly, the genetic structure seems to differ between the populations even more than the resulting phenotypes.

Although fewer of the individual correlations were significant in the Wisconsin population than in the Michigan population (an outcome that is to be expected given the smaller sample sizes), some interesting correlation patterns emerged. For example, genetic correlations between antipredator and total chemical test response results were highly negative in both populations. In contrast, CTmin showed a negative genetic correlation with number of ventral scales in the Michigan population and a positive correlation in the Wisconsin population.

These results suggest that genetic correlations may influence responses to selection across morphological, physiological, and behavioral traits in garter snakes and that the patterns of the constraints may vary among populations. Unfortunately, the large number of possible contrasts and the nonindependence of parameter estimates within the matrices restrict our ability to make statistical statements about individual values within the correlation matrices. These kinds of problems, and others, are likely to be encountered by those conducting research on the

Table 4-3 Genetic and phenotypic correlations among traits in neonate *Thamnophis sirtalis* from Michigan and Wisconsin

	Michigan							
	Ventral	Subcaudal	SVL	TL	Mass	Antipredator	CTmin	Chemical
Ventral		**0.45**[a]	**0.46**[a]	**0.42**[b]	**0.54**[a]	–0.12	**–0.62**[a]	0.36
Subcaudal	**0.44**[a]		–0.24	**0.37**[c]	0.18	–0.17	**–0.41**[b]	0.33
SVL	**0.28**[a]	–0.04		**0.64**[a]	**0.75**[a]	0.11	**–0.46**[a]	0.14
TL	**0.36**[a]	**0.56**[a]	**0.46**[a]		**0.78**[a]	–0.08	**–0.80**[a]	0.21
Mass	**0.28**[a]	0.08	**0.74**[a]	**0.45**[a]		0.02	**–0.43**[a]	0.21
Antipredator	–0.06	**–0.13**[c]	0.04	–0.10	0.04		0.29	**–0.65**[a]
CTmin	**–0.17**[b]	**–0.13**[c]	**–0.25**[a]	**–0.28**[a]	**–0.28**[a]	**0.12**[c]		–0.20
Chemical	0.02	0.00	**0.17**[b]	0.01	**0.14**[c]	**–0.12**[c]	**–0.13**[c]	
	Wisconsin							
	Ventral	Subcaudal	SVL	TL	Mass	Antipredator	CTmin	Chemical
Ventral		**0.99**[a]	**–0.41**[c]	–0.29	**0.55**[d]	0.36	**0.50**[c]	0.54
Subcaudal	**0.71**[a]		–0.11	–0.08	–0.27	–0.18	–0.06	**1.18**[a]
SVL	0.07	**–0.38**[c]		**1.02**[a]	**1.00**[a]	–0.29	**–1.01**[a]	0.72
TL	0.26	**0.62**[a]	**0.67**[a]		**1.02**[a]	**–1.17**[a]	**–1.21**[a]	**0.87**[a]
Mass	–0.12	0.02	**0.90**[a]	**0.58**[a]		–0.43	**–1.00**[a]	**0.50**[c]
Antipredator	–0.08	0.03	0.09	0.12	0.04		0.83[a]	**–0.47**
CTmin	0.25	0.12	**–0.38**[c]	–0.19	**–0.52**[d]	0.27		–0.55
Chemical	0.23	0.20	0.30	0.24	**0.36**[c]	–0.18	**–0.41**[b]	

Genetic correlations are shown above the diagonal, phenotypic correlations below. Traits include the number of ventral and subcaudal scales, snout vent length (SVL), tail length (TL), mass, total antipredator score, CTmin in temperature, and the total tongue-flick attack score to fish and worm prey extracts and water controls.
[a]P < 0.0001, [b]P < 0.01, [c]P<0.05, [d]P < 0.001.

evolution of geographic variation in behavior. In the next section we address methodological issues relevant to such efforts.

Methodological Issues

There are three main categories of methodological problems that plague studies of geographical variation in behavior. These include experimental design and measurement issues, the need to take into account developmental changes in behavior, and the effects of psychological processes. This discussion presents the major issues in general terms, and then offers examples from the research we presented here and from other research on natricine snakes. Those studying other taxa will probably encounter many of these problems and may find unexpected twists in their expression. We hope the discussion will help others avoid some of these potential pitfalls. The cited references provide additional detail.

Experimental Design

There are two general approaches to evaluating genetic differences underlying population differences in behavior. In the first, to the maximum extent possible, all individuals are exposed to the same environment or sets of environments, and behavioral measurements are made at the same developmental stage according to a set routine. We used such a common garden design. However, conditions of captive rearing such as diet, habitat structure, temperature, social experience, handling, and even previous ambient odors in housing can dramatically affect the behavior of individuals from the same species (e.g., Burghardt and Layne 1995). If geographic factors interact with experiential factors such as these, a better approach may be to examine heritable differences in reaction norms (for discussion, see Carroll and Corneli this volume, Thompson this volume). This method provides insights into the nature of the interaction, particularly when specific environmental differences are suspected to have influenced the evolution of the behavioral traits of interest, but it is not necessary for the demonstration of heritable population differences. With either method, every effort must be made to ensure uniformity of appropriate environmental variables across all individuals, especially when experiential effects are as well documented as they are for feeding and antipredator behavior in *Thamnophis* (e.g., Herzog 1990, Bowers 1992, Burghardt 1992, 1993, Lyman-Henley and Burghardt 1995). The following methodological issues are relevant to both approaches.

Observer Reliability

Different observers may record the same event differently, or the way the same observer records behavioral events may change over time. The importance of such interobserver and intraobserver reliability and methods for assessing and controlling it are widely recognized (e.g., Martin and Bateson 1993), but distressingly infrequently used. The more one knows about the system and the more hypothesis-driven the study, the more critical this issue becomes. Double-blind testing, in which the experimenter is unaware of the hypotheses being evaluated, the stimulus presented, prior history of individuals, and the population of origin, is all too rare. Independent analysis of video-recorded trials is a recommended solution, but it is not always practical.

In field studies, the double-blind system may be nearly impossible to use, but it typically can be used in studies of captive animals. Because the double blind method can increase confidence in experimental results considerably, it should be employed wherever possible. Similarly, observer reliability should be routinely assessed in behavioral studies.

Sample Size

The sample sizes needed to detect genetic differences between populations will depend on trait heritabilities, the distribution of individuals among families (family sizes), and the magnitude of the differences among populations. Litters of varying size may capture more or less of the variation within and among populations. There is no clear answer except to stress that larger sample sizes will increase power to detect differences (see Arnold 1994, and references therein), and power to detect differences will generally increase with heritability and the magnitude of the difference between populations. Because heritabilities of many behavioral traits appear to be low (Mousseau and Roff 1987), the value of research designed to detect population differences in the genetic bases of behavior should be carefully considered before embarking on labor-intensive research in which sample sizes are restricted by housing, logistics, or restrictions on acquisition of study subjects.

Additionally, as is evident from the results presented here, sample sizes necessary to evaluate the correlation structure between a large number of characters may be large. Again, given the labor-intensive nature of rearing studies designed to evaluate behavioral heritabilities, careful planning and goal assessment is necessary. Arnold (1994) provides a lucid discussion of the relative merits of studies in which sample sizes are marginal, but the research is integrated with other goals, versus those designed to estimate genetic parameters with greater certainty.

Test Order

Because prior experience can have dramatic effects on the expression of behavior, the order of stimulus presentation can affect the results of research in which multiple behavioral traits or experimental conditions are measured in the same individuals. Given the time-intensive nature of rearing studies, measurement of multiple traits and repeated testing are typically desirable and, in studies of character correlations and trait stability, necessary. Repeated measures can, however, lead to habituation effects. These need to be assessed (Burghardt 1969, Herzog et al. 1989, Arnold 1992) and studied in their own right as important learning phenomena of potential biological significance (Bowers 1992).

Most experimenters realize that order of stimulus presentation should be balanced. If, for example, habituation occurs over trials across stimuli, then testing stimulus *A* first will lead to greater response than stimulus *B* even if they have equal intrinsic salience. The opposite effect, termed "sensitization," can also occur. Order effects may have contributed to the finding that *T. radix* exhibited greater antipredator behavior at low than at high temperatures (Arnold and Bennett 1984), whereas this was not found in a study of *T. sirtalis* when the temperatures were presented in a balanced order (Schieffelin and de Queiroz 1991). Conclusions about species differences are thus rendered suspect.

When individuals or populations are being compared, the best approach is to test all animals identically to eliminate individual variation induced by presentation order. This method is not without pitfalls, however. Consider a scenario in which stimulus *B* is always presented after stimulus *A*. If an animal responds strongly to *A*, the response to *B* may be enhanced even if *B* is a blank or control. In this case, a population that is more responsive to *A* than is a second population may also appear to differ in response to *B*, even if there is no real difference in the population responses to *B*. As an example, several species of *Thamnophis* exhibited greater tongue-flick rates to distilled water controls if the preceding prey extract elicited an attack than if it did not, and the rate was even higher if the preceding two extracts had elicited attacks (Burghardt 1969). Such differences may prove fairly inconsequential when species with widely different food habits are compared, but may be troublesome in population comparisons because diets are often more similar between populations than between species.

The procedure used in comparing the responses of Wisconsin and Michigan snakes to fish and worm stimuli was to combine identical stimulus sequence and counterbalancing stimulus order. Each individual was tested on seven stimuli in the following order: water, worm, fish, water, fish, worm, water. Thus, in each replicate the order varied with the fish stimulus presented both before and after the worm stimulus, separated by an aqueous control. When responses by test position rather than pooled stimulus scores (fig. 4-3) were analyzed (fig. 4-4), the second worm stimulus elicited a greater response than the first. Fortunately, both populations responded in the same manner. Nonetheless, position in the test sequence had an effect. Furthermore, whereas the three water controls elicited similar TFAS scores from Wisconsin snakes, the second and third water controls elicited higher scores from the Michigan snakes (Kruskal-Wallis test, $\chi^2 = 6.54$, $df = 2$, $p = .038$). Thus, caution must be exercised not only in comparing stimuli within populations (Arnold 1992), but also in comparisons among populations.

Trait Repeatability

As these results suggest, individuals do not respond in the same way to the same stimuli on all occasions (e.g., Lessels and Boag 1987, Boake 1989). Recent experience, as reflected in presentation order, is only one cause of response variation. Other causes of temporal trends in response are developmental changes in the individual and motivational changes, each of which is addressed below.

When there is no temporal trend in repeated measures of a behavior (as assessed in a pilot study), multiple measures can be averaged, and the repeatability

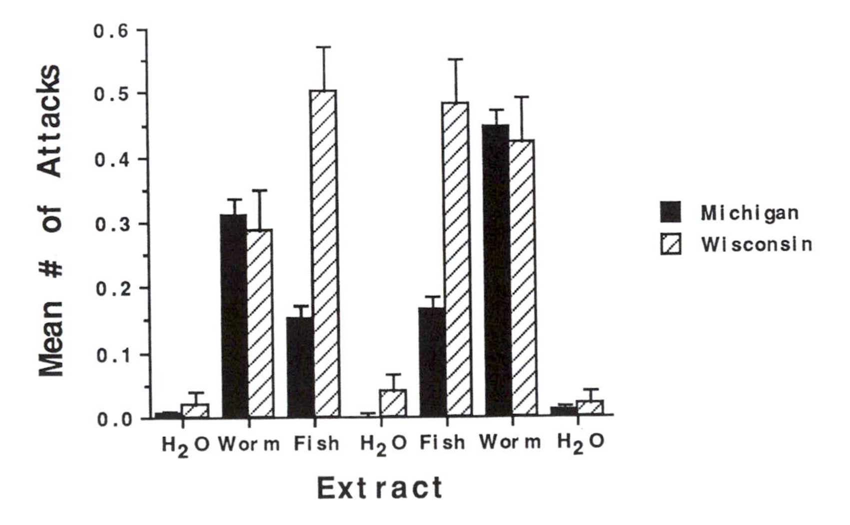

Figure 4-4 Population differences in mean number of attacks (±SE) of neonatal *Thamnophis sirtalis* elicited by chemical stimuli in order of presentation (left to right).

of these averages increases with the number of measures. Thus, when there is no evidence of a temporal trend in response values within individuals, multiple measures should be used in quantitative genetic studies of behavior (Boake 1989, Arnold 1994). Use of individual averages, when appropriate, will increase the ability to detect population differences, as well as differences between individuals within a population.

Traditionally two or three replicate behavior measurements are taken; these may accomplish statistical needs to reduce variability in heritability calculations (Lessels and Boag 1987, Boake 1989, Jayne and Bennett 1990a, Arnold 1994). They are typically not sufficient to establish that individuals actually differ from each other to a statistically significant degree. Thus, in one study, seven replicates of each prey stimulus (14 of each prey class) were employed to identify snakes that preferred fish versus earthworm stimulus (Burghardt 1975). Without knowing the consistency of intra- and interindividual differences, inferences concerning the possible role of selection are suspect (see also Arnold 1994).

Developmental Effects

Repeated measures of behavior, when separated in time, can provide quite different behavioral scores because behavioral phenotypes change during the life of an animal. Thus, in addition to repeatability studies over short intervals to reduce variance, measurements at other ages are necessary if we want to generalize across age classes. Repeated testing over months was needed to show that individual differences in antipredator behavior in 1-day-old *Thamnophis melanogaster* reflect individual differences at over 1 year of age (Herzog and Burghardt 1988). Brodie (1993a) has demonstrated that *T. ordinoides* shows stability over years in antipredator behavior in the field. Locomotor performance in *T. sirtalis* is somewhat stable over years (Jayne and Bennett 1990b). Nevertheless, many behavior patterns do change with age, and although some of these changes may largely reflect genetically mediated changes in the animal's internal environment (maturation), others may be environmentally induced or products of genotype–environment interactions. For convenience we divide our discussion of developmental effects into effects before birth or hatching and those occurring afterward.

Prenatal Development

Testing precocial species at birth or hatching is a common means of assessing behavioral characteristics within and across populations and species. The assumption is often that the behavior of newborns directly reflects underlying genotypes because newborns have not had the opportunity to modify their behavior as a consequence of learning. The problem with this assumption is that the prenatal environment has been shown to influence neonate behavior (see Cheverud and Moore 1994 for discussion). Typically called "maternal effects," these influences can be the products of differences in maternal provisioning or health during pregnancy. However, similar influences can be the products of differences in care provided by fathers or other relatives before hatching in species with external fertilization (Cheverud and Moore 1994). When the genotypes of kin affect the prenatal environment, their influence can be difficult to distinguish from that of genes borne by the neonate, and experiments must be designed specifically to detect the effects (Wilham 1972, Falconer 1989, Cheverud and Moore 1994).

Maternal or kin effects have rarely been examined in behavioral research (Cheverud and Moore 1994). However, given that maternal effects on other aspects of phenotype have been discovered in a diversity of taxa, they are to be expected (Cheverud and Moore 1994). Geographic differences in maternal effects on behavior are also to be expected in many instances, but they have only been examined (and demonstrated) in a single study of diapause in the striped ground cricket (Mousseau 1991).

The apparent ubiquity of maternal effects on other aspects of phenotype suggests that the typical method of removing pregnant female snakes from the wild and testing captive-borne neonates under standard laboratory conditions may confound maternal effects and offspring genotype. Although maternal effects have rarely been examined in natricine snakes, egg incubation temperature does influence locomotor, defensive, and chemosensory behavior in three genera of American oviparous colubrid snakes (Berger 1989, 1990, 1991). Maternal gestation temperature has been shown to influence other aspects of the phenotype including the percentage of live births and deformities, but not sex ratio, in *T. sirtalis* (Burghardt and Layne 1995) or patterns of scalation (Fox 1948, Osgood 1978, Peterson et al. 1993) in other natricine snakes. In a related natricine, *Storeria dekayi*, the time wild-caught females are held in captivity before giving birth can also affect offspring size and condition (King 1994). In captive, pregnant *T. elegans*, similar variation was found, but the effect differed across four populations (Farr and Gregory 1991).

Genetic, environmental, and interaction effects have all been demonstrated in thermal and life-history characteristics of frog populations (*Rana*) at different altitude and temperature regimes (Berven et al., 1979, Berven 1982). Although rearing animals from two different populations under identical conditions can identify genetic contributions, rearing populations in alternative environments (e.g., montane animals in lowlands, lowland animals in montane areas) was shown to be critical in identifying environmental and interaction contributions. Such studies are needed with garter snakes as well.

Because temperature is so important to the behavior of poikilotherms, it is useful to explore a bit more deeply how population differences may be traced both proximately and evolutionarily to climatic differences. Suppose that the adult females from Michigan and Wisconsin experienced, and physiologically adapted to, different temperatures that affected their thermal preference. Even in the laboratory, females from these populations could select portions of their cages with different thermal properties and the embryos could be exposed to different temperatures and thus show a nongenetic difference. Surprisingly, however, preferred body temperatures in the field do not differ significantly either geographically or interspecifically in *Thamnophis* (Rosen 1991). A recent study also showed that in *Thamnophis elegans* maternal temperature does not affect offspring thermal preference (Arnold et al. 1995). This would suggest that the CTmin difference reported here is not biased by maternal effects.

Population differences may also be affected by the prior nutritional history of the mother, as the ova and embryos were provisioned during the time in the field before capture. The quality and quantity of prey can show considerable year-to-year and geographic variation (Kephart 1982). There is no evidence for specific behavioral effects of maternal diet on offspring prey preferences based on experimental (Burghardt 1971) or mating (Arnold 1981b) studies. However, nutritional conditions could affect offspring size, condition, endurance, or other factors that indirectly influence behavior in one population more than another (Ford and Seigel 1989).

The probable role of maternal influences on behavioral phenotypes must thus be taken into account in future research on the natricine snakes and on other taxa in which kin effects are likely. The influences on behavior may be complex and unanticipated. For example, in rats, the position of young in the uterus may affect sexually dimorphic behavior (vom Saal 1981). Experimental designs should attempt to account for such effects where appropriate and should minimize variation to which parents are exposed. Ideally, second-generation, laboratory-reared animals should be used as parents, and rearing conditions should be as uniform as possible spatially and temporally. Use of second-generation, laboratory-reared animals should reduce or eliminate influences of environmental differences between geographic sites of origin, although the cost and difficulty of laboratory rearing may preclude this approach for some taxa. Additional generations of laboratory rearing are not desirable in studies of geographic variation because of the possibility that laboratory selection will alter the genetic composition of the populations, thereby obscuring the population differences of interest.

Postnatal Development

Many aspects of behavior change with age. When this occurs, a behavioral analysis must account for the age of the animal, and the trait value must be considered an age-specific attribute. Although more complex, genetic analyses of developmental changes in behavior can be conducted using appropriate experimental designs. Recent studies suggest not only that closely related species may differ in the ontogenetic trajectories of their behavior (e.g., Herzog et al. 1992), but that populations of a single species can also evolve differences in typical patterns of behavioral development (West and King 1988, Magurran this volume).

The propensity of *T. sirtalis* neonates to feed on earthworms switches to a propensity for feeding on amphibians as individuals grow in some, but not in all, populations (e.g., Fitch 1965, Greenwell et al. 1984). In the related natricine, *Nerodia erythogaster*, there is a change from fish to frog dietary preference with size in the field that has been replicated using chemosensory tests on laboratory populations (Mushinsky and Lotz 1980, Mushinsky et al. 1982). This shift appears maturational in that it occurred in snakes reared on both all-fish and all-amphibian diets. Such changes may also occur in antipredator behavior. In *T.*

radix there is an ontogenetic increase in fleeing behavior in a standardized anti-predator testing situation between days 1 and 21, but the closely related *T. butleri* shows no such change (Herzog et. al. 1992). These two allopatric species were once considered the same species, and to date no molecular differences between them have been reported (de Queiroz and Lawson 1994). Interestingly, unpublished recent data from our laboratory suggest that *T. butleri* litters from an isolated population in southeastern Wisconsin, the only population with a range abutting if not overlapping slightly with *T. radix*, show an ontogenetic shift.

In short, the role of different maturational rates and their possible mechanisms must be accounted for in experimental designs. Differences between mother and offspring behavior may reflect maturational change (Brodie and Garland 1993) rather than genetic differences. If the ontogenetic element is not recognized, incorrect assessment of trait heritabilities will result (Riska et al. 1989). Indeed, as Endler (1986, p. 113) has argued, "Developmental changes can give misleading or even false evidence for natural selection."

Psychological Issues

Psychological issues generally encompass those dealing with cognition (perception, decision making, and learning), motivation, and emotion (Burghardt 1991). We briefly mention each of these in turn.

Cognition and Learning

How animals process and use information can differ among individuals and, potentially, among populations. This has been shown elegantly in the work of King and West (1987) on cowbird courtship and song. The sensitivity of both predatory and antipredator behavior of snakes to context and prior experience has been well documented (Burghardt 1993, Burghardt and Layne 1995, Herzog 1990), as it has in several species of fish (review in Magurran this volume). The methodological consequences of habituation processes has already been discussed in the section on test order. Researchers should also be encouraged to study population differences in information processing and use as important avenues for the expression of genotypic differences among populations.

Motivation

Motivation is a major topic in psychology (see Mook 1987 for a review) that can be examined using geographic comparisons. Experimenters are generally careful about testing foraging responses at controlled intervals after feedings. Testing neonates is a bit different, especially in animals such as snakes that may go many days, weeks, or even months before ingesting their first meal. Testing animals at the same age does not guarantee similar results if motivational levels differ.

Consider neonates who have never eaten but who are being tested on prey stimuli. As shown here, animals from different populations of *T. sirtalis* differ in neonatal size; other populations show much more extreme variation (Gregory and

Larsen 1993, 1996). Larger snakes with more yolk reserves can typically survive longer without food than smaller individuals. Thus, animals from different populations tested at the same age may differ in their readiness to respond to prey cues due to differential hunger. This has been documented in our laboratory for two closely related species of the natricine genus *Regina* that differ almost threefold in their mass at birth (R. M. Waters, unpublished data). When populations differ considerably in body size, some allometric procedures might well be incorporated into testing feeding responses. Similar, but more subtle, motivational differences may arise in testing other behavior systems.

Subject Reactivity

Subject reactivity, usually expressed as how flighty or adaptable animals are to captivity, may also complicate the design of experiments involving population comparisons. This is certainly true of wild-caught adult animals brought into the laboratory (see also Boinski this volume). Wild-caught adult green iguanas would never become popular captive companions. Even with neonates, we have found that wild-born habu pit vipers (*Trimeresurus flavoviridis*) adapt to captivity with greater difficulty than do young from eggs laid in captivity by wild-caught mothers. Differences among populations in defensive temperament, as shown here for Wisconsin and Michigan laboratory-born *T. sirtalis*, may affect their responsivity, and even suitability, for tests of the same type applied to conspecifics from other populations. Thus, differences among populations on one dimension may reflect differences in a behavior system not being studied.

Conclusions

Modern studies of geographic variation use techniques for assessing the nature of differences at finer levels of behavioral and genetic analysis than ever imagined by those comparing species in the early decades of ethology. Indeed, classical ethologists probably would have been astonished at the extent of geographic variation in behavior within species. Garter snakes are prime examples of species that display variation in behavior across populations, and like other such species, they offer a promising opportunity for exploring the evolution of behavioral phenotypes.

If, however, research on geographic variation in behavior is to meet the promise it holds, rigorous experimental design is essential. We hope that the issues we have raised here will help others avoid potential difficulties and aid in designing experiments appropriate to the questions they wish to address. We believe that the opportunity to examine the genetic bases of behavior and the causes of evolutionary change in behavior is without parallel in other kinds of study, and we hope that this chapter will encourage other scientists to take the approach of using geographic variation as a source of evolutionary insight.

We conclude by noting that, although human behavior is not the subject of this volume, one test of our accomplishments with research on geographic variation in nonhuman animals will come as the growing effort to reevaluate variation in human behavior using genetic and evolutionary theory is played out (Darwin, 1871, Herrnstein and Murray, 1994, Wilson, 1994, Rushton, 1995, Whitney, 1995). Those of us studying nonhuman animals can play an important role, conceptually and methodologically, in understanding the evolutionary history of our own species. Thus, studies of geographic and population variation in the behavior of nonhuman animals should meet standards that would address scientific critiques of the often controversial studies of humans.

Acknowledgments These studies were supported in part by National Science Foundation research grants awarded to G.M.B. and the Science Alliance at the University of Tennessee. We thank Gary McCracken and the many students and laboratory personnel who assisted in various aspects of the laboratory work. Lyle Konigsberg performed the Mantel tests. We also thank Paul Andreadis, Christine Boake, Hugh Drummond, John Gittleman, Constantino Macias Garcia, Emília Martins, Susan Riechert, the editors, and anonymous reviewers for their comments and suggestions.

References

Arnold, S. J. 1977. Polymorphism and geographic variation in the feeding behavior of the garter snake, *Thamnophis elegans*. Science 197:676–678.

Arnold, S. J. 1981a. Behavioral variation in natural populations. II. The inheritance of a feeding response in crosses between geographic races of the garter snake, *Thamnophis elegans*. Evolution 35:510–513.

Arnold, S. J. 1981b. The microevolution of feeding behavior. *In* A. Kamil and T. Sargent, eds. Foraging behavior: Ecological, ethological, and psychological approaches, pp. 409–453. Garland STPM, New York.

Arnold, S. J. 1992. Behavioral variation in natural populations. VI. Prey responses by two species of garter snakes in three regions of sympatry. Animal Behaviour 44:705–719.

Arnold, S. J. 1994. Multivariate inheritance and evolution: A review. *In* C. R. B. Boake, ed. Quantitative genetic studies of behavioral evolution, pp. 17–48. University of Chicago Press, Chicago.

Arnold, S. J., and A. F. Bennett. 1984. Behavioral variation in natural populations. III. Antipredator displays in the garter snake *Thamnophis radix*. Animal Behaviour 32: 1108–1118.

Arnold, S. J., and A. F. Bennett. 1988. Behavioral variation in natural populations. V. Morphological correlates of locomotion in the garter snake (*Thamnophis radix*). Biological Journal of the Linnean Society 34:175–190.

Arnold, S. J., Peterson, C. R., and J. Gladstone. 1995. Behavioral variation in natural populations. VII. Maternal body temperature does not affect juvenile thermoregulation in a garter snake. Animal Behaviour 50:623–633.

Berger, J. 1989. Incubation temperature has long-term effects on behavior of young pine snakes (*Pituophis melanoleucus*). Behavioural Ecology and Sociobiology 24:201–207.

Berger, J. 1990. Effects of incubation temperature on behavior of young black racers (*Coluber constrictor*) and king snakes (*Lampropeltis getulus*). Journal of Herpetology 24: 158–163.

Berger, J. 1991. Responses to prey chemical cues by hatchling pine snakes (*Pituophis melanoleucus*): Effects of incubation temperature and experience. Journal of Chemical Ecology 17:1069–1078.

Berven, K. A. 1982. The genetic basis of altitudinal variation in the wood frog *Rana sylvatica*. I. An experimental analysis of life history traits. Evolution 36:962–983.

Berven, K. A., D. E. Gill, and S. J. Smith-Gill. 1979. Countergradient selection in the green frog, *Rana clamitans*. Evolution 33:609–623.

Boake, C. R. B. 1989. Repeatability: Its role in evolutionary studies of mating behavior. Evolutionary Ecology 3:173–182.

Boake, C. R. B. 1994a. Evaluation of applications of the theory and methods of quantitative genetics to behavioral evolution. *In* C. R. B. Boake, ed. Quantitative genetic studies of behavioral evolution, pp. 305–325. University of Chicago Press, Chicago.

Boake, C. R. B. (ed.). 1994b. Quantitative genetic studies of behavioral evolution. University of Chicago Press, Chicago.

Bowers, B. B. 1992. Habituation of antipredator behaviors and responses to chemical prey extracts in four species of garter snakes, *Thamnophis* (PhD dissertation). University of Tennessee, Knoxville.

Brodie, E. D., III. 1989. Genetic correlations between morphology and behaviour in natural populations of the garter snake *Thamnophis ordinoides*. Nature 342:542–543.

Brodie, E. D., III. 1993a. Consistency of individual differences in anti-predator behaviour and color pattern in the garter snake, *Thamnophis ordinoides*. Animal Behaviour 45: 851–861.

Brodie, E. D., III. 1993b. Homogeneity of the genetic variance-covariance matrix for anti-predator traits in two populations. Evolution 47:844–854.

Brodie, E. D., III, and E. D. Brodie, Jr. 1991. Evolutionary response of predators to dangerous prey: Reduction of toxicity of newts and resistance of garter snakes in island populations. Evolution 45:221–224.

Brodie, E. D., III, and T. Garland, Jr. 1993. Quantitative genetics of snake populations. *In* R. A. Seigel and J. T. Collins, eds. Snakes: Ecology and behavior, pp. 315–362. McGraw-Hill, New York.

Burghardt, G. M. 1969. Comparative prey attack studies in naive snakes of the genus *Thamnophis*. Behaviour 33:77–114.

Burghardt, G. M. 1970. Intraspecific geographical variation in chemical food cue preferences of newborn garter snakes (*Thamnophis sirtalis*). Behaviour 36:246–257.

Burghardt, G. M. 1971. Chemical-cue preferences of newborn snakes: Influence of prenatal maternal experience. Science 171:921–923.

Burghardt, G. M. 1975. Chemical prey preference polymorphism in newborn garter snakes, *Thamnophis sirtalis*. Behaviour 52:202–225.

Burghardt, G. M., ed. 1985. Foundations of comparative ethology. Van Nostrand Reinhold, New York.

Burghardt, G. M. 1991. Cognitive ethology and critical anthropomorphism: A snake with two heads and hognose snakes that play dead. *In* C. A. Ristau, ed. Cognitive ethology: The minds of other animals, pp. 53–90. Erlbaum, San Francisco.

Burghardt, G. M. 1992. Prior exposure to prey cues influences chemical prey preference and prey choice in neonatal garter snakes. Animal Behaviour 44:787–789.

Burghardt, G. M. 1993. The comparative imperative: Genetics and ontogeny of chemoreceptive prey responses in natricine snakes. Brain Behavior and Evolution 41:138–146.

Burghardt, G. M., and J. L. Gittleman. 1990. Comparative and phylogenetic analyses: New wine, old bottles. *In* M. Bekoff and D. Jamieson, eds. Interpretation and explanation in the study of behavior, vol. 2. Comparative perspectives, pp. 192–225. Westview Press, Boulder, CO.

Burghardt, G. M., S. E. Goss, and F. M. Schell. 1988. Comparison of earthworm- and fish-derived chemicals eliciting prey attack by garter snakes (*Thamnophis*). Journal of Chemical Ecology 14:855–881.

Burghardt, G. M., and D. G. Layne. 1995. Effects of early experience and rearing condition. *In* C. Warwick, F. L. Frye, and J. B. Murphy, eds. Health and welfare of captive reptiles, pp. 165–185. Chapman & Hall, London.

Carpenter, C. C. 1952. Comparative ecology of the common garter snake (*Thamnophis s. sirtalis*), the ribbon snake (*Thamnophis s. sauritus*), and Butler's garter snake (*Thamnophis butleri*) in mixed populations. Ecological Monographs 22:235–258.

Cheverud, J. M., and A. J. Moore. 1994. Quantitative genetics and the role of the environment provided by relatives in behavioral evolution. *In* C. R. B. Boake, ed. Quantitative genetic studies of behavioral evolution, pp. 67–100. University of Chicago Press, Chicago.

Conant, R., and J. T. Collins. 1991. A field guide to reptiles and amphibians. Eastern and Central North America, 3rd ed. Houghton Mifflin, Boston.

Cooper, W. E. Jr., and G. M. Burghardt, 1990. A comparative analysis of scoring methods for chemical discrimination of prey by squamate reptiles. Journal of Chemical Ecology 16:45–65.

Cope, E. D. 1891. A critical review of the characters and variations of the snakes of North America. Proceedings of the United States National Museum 14(882):589–694.

Coss, R. G. 1991. Evolutionary persistence of memory like processes. Concepts in Neuroscience 2(2):129–168.

Darwin, C. 1871. The descent of man and selection in relation to sex. John Murray, London.

de Queiroz, A. 1992. The evolutionary lability of behavior (PhD dissertation). Cornell University, Ithaca, New York.

de Queiroz, A., and R. Lawson. 1994. Phylogenetic relationships of the garter snakes based on DNA sequence and allozyme variation. Biological Journal of the Linnean Society 53:209–229.

de Queiroz, A., and P. H. Wimberger. 1993. The usefulness of behavior for phylogeny estimation: levels of homoplasy in behavioral and morphological characters. Evolution 47:46–60.

Deitz, E. J. 1983. Permutation tests for association between two distance matrices. Systematic Zoology 32:21–61.

Dix, M. W. 1968. Snake food preference: Innate intraspecific geographic variation. Science 1159:1478–1479.

Doughty, P. 1994. Critical thermal minima of garter snakes (*Thamnophis*) depends on species and body size. Copeia 1994:537–540.

Drummond H., and G. M. Burghardt. 1983. Geographic variation in the foraging behavior of the garter snake, *Thamnophis elegans*. Behavioral Ecology and Sociobiology 12: 43–48.

Endler, J. A. 1986. Natural selection in the wild. Princeton University Press, Princeton, NJ.

Ernst, C. H., and R. W. Barbour. 1989. Snakes of eastern North America. George Mason University Press, Fairfax, VA.

Falconer, D. S. 1989. An introduction to quantitative genetics, 3rd ed. Wiley, New York.

Farr, D. H., and Gregory, P. T. 1991. Sources of variation in estimating litter characterisitcs of the garter snake, *Thamnophis elegans*. Journal of Herpetology 25:261–268.

Fitch, H. S. 1965. An ecological study of the garter snake, *Thamnophis sirtalis*. University of Kansas Publications of the Museum of Natural History 15:493–564.

Fitch, H. S. 1980. *Thamnophis sirtalis*. Catalog of American Amphibians and Reptiles 270:1–4.

Fitch, H. S. 1985. Variation in clutch and litter size in New World reptiles. University of Kansas Museum of Natural History Miscellaneous Publications 76:1–76.

Ford, N. B. 1986. The role of pheromone trails in the sociobiology of snakes. *In* D. Duvall, D. Müller-Schwarze, and R. M. Silverstein, eds. Chemical signals in vertebrates, vol. 4. Ecology, evolution and comparative biology, pp. 261–278. Plenum Press, New York.

Ford, N., and R. A. Seigel. 1989. Phenotypic plasticity in reproductive traits: evidence from a viviparous snake. Ecology 70:1768–1774.

Foster, S. A., and S. A. Cameron. 1996. Geographic variation in behavior: A phylogenetic framework. *In* E. P. Martins, ed. Phylogenies and the comparative method in animal behavior, pp. 138–165. Oxford University Press, Oxford.

Fox, W. W. 1948. Effect of temperature on development of scutellation in the garter snake, *Thamnophis elegans atratus*. Copeia 1948:252–262.

Gillingham, J. C. 1987. Social behavior. *In* R. A. Seigel, J. T. Collins, and S. S. Novak, eds. Snakes: Ecology and evolutionary biology, pp. 184–209. McGraw-Hill, New York.

Gittleman, J. L., C. G. Anderson, M. Kot, and H.-K. Luh. 1996. Phylogenetic data and rates of evolution: A comparison of behavioral, morphological and life history traits. *In* E. P. Martins, ed. Phylogenies and the comparative method in animal behavior, pp. 166–205. Oxford University Press, Oxford.

Gittleman, J. L., and D. M. Decker. 1994. The phylogeny of behaviour. *In* P. J. B. Slater and T. R. Halliday, eds. Behaviour and evolution, pp. 80–105. Cambridge University Press, Cambridge.

Gottlieb, G. 1992. Individual development and evolution. Oxford University Press, New York.

Gove, D., and G. M. Burghardt. 1975. Responses of ecologically dissimilar populations of the water snake, *Natrix s. sipedon*, to chemical cues from prey. Journal of Chemical Ecology 1:25–40.

Greenwell, M. G., M. Hall, and O. J. Sexton. 1984. Phenotypic basis for a feeding change in an insular population of garter snakes. Developmental Psychobiology 17:457–463.

Gregory, P. T. 1984. Habitat, diet, and composition of assemblages of garter snakes (*Thamnophis*) at eight sites on Vancouver Island. Canadian Journal of Zoology 62: 2013–2022.

Gregory, P. T., and K. W. Larsen. 1993. Geographic variation in reproductive characteristics among Canadian populations of the common garter snake (*Thamnophis sirtalis*). Copeia 1993:946–958.

Gregory, P. T., and K. W. Larsen. 1996. Are there any meaningful correlates of geographic life history variation in the garter snake, *Thamnophis sirtalis*? Copeia 1996:183–189.

Gregory, P. T., and K. J. Nelson. 1991. Predation on fish and intersite variation in the diet of common garter snakes, *Thamnophis sirtalis*, on Vancouver Island. Canadian Journal of Zoology 69:988–994.

Halpern, M. 1992. Nasal chemical senses in reptiles: Structure and function. Biology of the Reptilia 18:423–523.

Hasegawa, M., and H. Moriguchi. 1989. Geographic variation in food habits, body size and life history traits of the snakes on the Izu Islands. *In* M. Matui, T. Hikida, and R. C. Goris, eds. Current herpetology in East Asia, pp. 414–432. Herpetological Society of Japan, Tokyo.

Herrnstein, R. J., and C. Murray. 1994. The bell curve. Free Press, New York.

Herzog, H. A., Jr. 1990. Experiential modification of defensive behaviors in garter snakes, *Thamnophis sirtalis*. Journal of Comparative Psychology 104:334–339.

Herzog, H. A. Jr., B. B. Bowers, and G. M. Burghardt. 1989. Development of antipredator responses in snakes: IV. Interspecific and intraspecific differences in habituation of defensive behavior in neonate garter snakes. Developmental Psychobiology 22:489–508.

Herzog, H. A. Jr., B. B. Bowers, and G. M. Burghardt. 1992. Development of antipredator responses in snakes: V. Species differences in ontogenetic trajectories. Developmental Psychobiology 25:199–211.

Herzog, H. A. Jr., and G. M. Burghardt. 1986. The development of antipredator responses in snakes: I. Defensive and open-field behaviors in newborns and adults of three species of garter snakes (*Thamnophis melanogaster, T. sirtalis, T. butleri*). Journal of Comparative Psychology 100:372-379.

Herzog, H. A. Jr., and G. M. Burghardt. 1988. The development of antipredator responses in snakes: III. Stability of individual and litter differences over the first year of life. Ethology 77:250–258.

Herzog, H. A. Jr., and J. M. Schwartz. 1990. Geographical variation in the anti-predator behavior of neonate garter snakes, *Thamnophis sirtalis*. Animal Behaviour 40:597–601.

Hinde, R., and N. Tinbergen. 1958. The comparative study of species specific behavior. *In* A. Roe and G. G. Simpson, eds. Behavior and evolution, pp. 251–268. Yale University Press, New Haven, CT.

Jayne, B. C., and A. F. Bennett. 1990a. Scaling of speed and endurance in garter snakes: A comparison of cross-sectional and longitudinal allometries. Journal of Zoology 220: 257–277.

Jayne, B. C., and A. F. Bennett. 1990b. Selection on locomotor performance capacity in a natural population of garter snakes. Evolution 44:1204–1229.

Kephart, D. G. 1982. Microgeographic variation in the diets of garter snakes. Oecologia 52:287–291.

King, A. P., and M. J. West. 1987. The experience of experience: An exogenetic program for social competence. Perspectives in Ethology 7:153–182.

King, R. B. 1988. Polymorphic populations of the garter snake *Thamnophis sirtalis* near Lake Erie. Herpetologica 44:451–458.

King, R. B. 1989. Body size variation among Island and mainland snake populations. Herpetologica 45:84–88.

King, R. B. 1994. Determinants of offspring number and size in the brown snake, *Storeria dekayi*. Journal of Herpetology 27:175–185.

King, R. B., and R. Lawson. 1995. Color-pattern variation in Lake Erie water snakes: The role of gene flow. Evolution 49:885–896.

Kitchell, J. F. 1969. Thermophilic and thermophobic responses of snakes in a thermal gradient. Copeia 1969:189–191.

Lessells, C. M., and P. T. Boag. 1987. Unrepeatable repeatabilities: A common mistake. Auk 104:116–121.

Lorenz, K. 1965. Evolution and modification of behavior. University of Chicago Press, Chicago.

Lorenz, K. 1981. The foundations of ethology. Springer-Verlag, New York.
Lyman-Henley, L., and G. M. Burghardt. 1994. Opposites attract: Effects of social and dietary experience on snake aggregation behaviour. Animal Behaviour 47:980–982.
Lyman-Henley, L., and G. M. Burghardt. 1995. Diet, litter, and sex effects on chemical prey preference, growth, and site selection in two sympatric species of *Thamnophis*. Herpetological Monographs 9:140–160.
Macartney, J. M., P. T. Gregory, and K. W. Larsen. 1988. A tabular survey of data on movements and home ranges of snakes. Journal of Herpetology 22:61–73.
Macias Garcia, C., and H. Drummond. 1990. Population differences in fish-capturing ability of the Mexican aquatic garter snake (*Thamnophis melanogaster*). Journal of Herpetology 24:412–416.
Martin, P., and P. Bateson. 1993. Measuring behaviour, 2nd ed. Cambridge University Press, Cambridge.
Mason, R. T. 1992. Reptilian pheromones. Biology of Reptiles 18:114–228.
Mayr, E. 1960. The emergence of evolutionary novelties. *In* S. Tax, ed. The evolution of life, pp. 349–380. University of Chicago Press, Chicago.
Mayr, E. 1965. Animal species and evolution. Harvard University Press, Cambridge, MA.
Mook, D. G. 1987. Motivation: The organization of action. W. W. Norton, New York.
Mousseau, T. A. 1991. Geographic variation in maternal-age effects on diapause in a cricket. Evolution 45:1053–1059.
Mousseau, T. A., and D. A. Roff. 1987. Natural selection and the heritability of fitness components. Heredity 59:181–197.
Mushinsky, H. R., J. J. Hebrard, and D. S. Vodopich. 1982. Ontogeny of water snake foraging ecology. Ecology 63:1624–1629.
Mushinsky, H. R., and K. H. Lotz. 1980. Responses of two sympatric water snakes to the extracts of commonly ingested prey species: Ontogenetic and ecological considerations. Journal of Chemical Ecology 6:523–535.
Osgood, D. W. 1978. Effects of temperature on meristic characters in *Nerodia fasciata*. Copeia 1978:33–47.
Peterson, C. R., A. R. Gibson, and M. E. Dorcas. 1993. Snake thermal ecology: The causes and consequences of body temperature variation. *In* R. A. Seigel and J. T. Collins, eds. Snakes: Ecology and behavior, pp. 241–314. McGraw-Hill, New York.
Riska, B., T. Trout, and M. Turelli. 1989. Laboratory estimates of heritabilities and genetic correlations in nature. Genetics 123:865–871.
Rosen, P. C. 1991. Comparative field study of thermal preferenda in garter snakes (*Thamnophis*). Journal of Herpetology 25:301–312.
Rossman, D., R. Seigel, and N. Ford. 1996. Garter snakes: Systematics and ecology. University of Oklahoma Press, Norman.
Rushton, P. 1995. Race, evolution, and behavior. Transaction, Brunswick, NJ.
Ruthven, A. G. 1908. Variations and genetic relationships of the garter snakes. Bulletin of the United States National Museum no. 61.
Schieffelin, C. D., and A. de Queiroz, A. 1991. Temperature and defense in the common garter snake: Warm snakes are more aggressive than cold snakes. Herpetologica 47: 230–237.
Schwartz, J. M. 1989. Multiple paternity and offspring variability in wild populations of the garter snake, *Thamnophis sirtalis* (PhD dissertation). University of Tennessee, Knoxville.
Schwartz, J. M., G. F. McCracken, and G. M. Burghardt. 1989. Multiple paternity in wild populations of the garter snake, *Thamnophis sirtalis*. Behavioral Ecology and Sociobiology 25:269–273.

Seigel, R. A., J. T. Collins, and S. S. Novak. 1987. Snakes: Ecology and evolutionary biology. Macmillan, New York.
Seigel, R. A., and J. T. Collins. 1993. Snakes: Ecology and behavior. McGraw-Hill, New York.
Sexton, O. J., P. Jacobson, and J. E. Bramble. 1992. Geographic variation in some activities associated with hibernation in neartic pitvipers. *In* J. A. Campbell and E. D. Brodie, Jr., eds. Biology of the pitvipers, pp. 337–345. Selva Press, Tyler, TX.
Smouse, P. E., and J. C. Long. 1992. Matrix correlation analysis in anthropology and genetics. Yearbook of Physical Anthropology 35:187–213.
Stebbins, R. C. 1985. A field guide to western reptiles and amphibians, 2nd ed. Houghton Mifflin, Boston.
Tinbergen, N. 1960. Behaviour, systematics, and natural selection. *In* S. Tax, ed. Evolution after Darwin, vol. 1: The evolution of life, pp. 595–613. University of Chicago Press, Chicago.
vom Saal, F. S. 1981. Variations in phenotype due to random intrauterine positioning of male and female fetuses in rodents. Journal of Reproduction and Fertility 62:633–650.
Wcislo, W. T. 1989. Behavioral environments and evolutionary change. Annual Review of Ecology and Systematics 20:137–169.
West, M. J., and A. P. King. 1988. Female visual display affects the development of male song in the cowbird. Nature 334:244–246.
Whitney, G. 1995. Ideology and censorship in behavior genetics. Mankind Quarterly 35: 327–342.
Wilham, R. L. 1972. The role of maternal effects in animal breeding: III. Biometrical aspects of maternal effects in animals. Journal of Animal Science 35:1288–1293.
Wilson, D. S. 1994. Adaptive genetic variation and human evolutionary psychology. Ethology and Sociobiology 15:219–235.

5

Geographic Variation in Behavior of a Primate Taxon

Stress Responses as a Proximate Mechanism in the Evolution of Social Behavior

SUE BOINSKI

> We have the common observation that species behave differently—indeed it is the "special flavor" of each species' behavior that provides the pleasure of comparative studies. Part of the difference lies in the different signals used, or the different frequency of similar signals in each species. But there are also differences in "tone" that make one feel justified in describing species with such adjectives as quarrelsome or relaxed; differences in the amount and predictability of subordinancy observed. . . .
> Thelma Rowell (1979, p. 13)

Temperament is a complex behavioral trait that describes characteristic patterns of response to environmental, particularly social, conditions and perturbations. Disparities in the tendency to approach or avoid novelty or readiness to engage in aggressive interactions have been documented in comparisons between species (Christian 1970), subspecies (Gonzalez et al. 1981), populations within species (Champoux et al. 1994), inbred lines of laboratory animals (Scott and Fuller 1965), domesticated versus wild populations (Price 1984), and individuals within a species (Benus et al. 1992). Differences in physiological stress response systems (Selye 1937) are commonly identified as an important proximate mechanism underlying these temperament differences (Huntingford and Turner 1987, Kagan et al. 1988).

Social systems of animals are perceived as emerging from relationships between individuals (Hinde 1983). Individual interactions, in turn, are hypothesized to reflect individual behavioral strategies which maximize inclusive fitness (Silk 1987). Selection on a physiological system, which can dramatically affect the pattern and outcomes of individual interactions, could produce evolutionary change in social organization and social behavior. Many workers explicitly suggest that temperament differences among primate species are adaptive in many instances, yet admit that the specific ecological and social selection pressures to which the neuroendocrine system is responding are often unclear (Thierry 1985, Clarke et al. 1988, Richard et al. 1989). Species-level comparisons have not offered many testable comparative models, probably because of confounding effects such as large phylogentic distances or uncertain phylogeny, inadequate knowledge of ecological and social conditions in the wild, drift, and convergent evolution.

In short, little progress has been made toward understanding the evolution of stress-response patterns in primates. In this chapter I suggest that comparisons of geographically and genetically separated primate populations or subspecies may be an alternative and more successful approach to addressing the evolution of stress responses and the disparate social behaviors that result. Population and geographic comparisons are likely to be profitable for three reasons: (1) comparisons are less likely to be confounded by phylogenetic disparities (Arnold 1992), (2) the factors imposing different selective regimes among localities can perhaps be more readily identified, (3) hypothesis testing may be facilitated because populations suitable for testing a model will be easier to identify than new species.

In this chapter I first summarize the physiological mechanisms underlying reactivity to stress-inducing situations. This is followed by examples of inter- and intraspecific differences in reactivity among primates. I then review the opportunities for natural selection to operate on the stress-response systems, and ultimately social behavior. Next I present a model based on the Neotropical squirrel monkey, genus *Saimiri*, on the proximate and ultimate factors underlying genetic differentiation of social behavior. Laboratory and field data for this taxon suggest that geographic variation in many aspects of social behavior is often a consequence of geographic variation in environmental stressors such as social competition and predation risk. In squirrel monkeys, geographic variation in the distribution of food patches and the pattern of competition it engenders appears to be a powerful variable in explaining geographic differences in social behavior. An important mechanism linking population differences in social behavior with these ecological factors is the adaptation of the physiological systems regulating and responding to direct and indirect food competition.

Stress Reactivity

A major axis of behavioral variation in primates, and in mammals in general, is described by the distribution of differences in emotional and behavioral reactivity to environmental challenges. Responses to uncertainty, novelty, or fear-provoking situations range from shyness and timidity to extroversion and boldness. Species

(Biben 1983) and individual (Fox 1972, Armitage 1982) differences in reactivity are well recognized. Breeding studies with rats (Broadhurst 1975), dogs (Scott and Fuller 1965), and other taxa have shown that individual differences in reactivity have a significant genetic component.

Selye (1946) provided the initial model of the Generalized Adaptive Syndrome or "fight or flight response"—the physiological mechanisms by which an organism is quickly mobilized to cope with external stressors. A major component of the stress response is the rapid activation of several physiological systems. These include the autonomic nervous system, particularly the sympathetic system, and the hypothalamic–pituitary–adrenal axis (Brain 1979, Gunnar 1988). The activated systems permit rapid mobilization of resources, including blood sugar, for quick action. Physiological activities not immediately required for survival are suspended. Components of the stress response include decreases in sensitivity to pain and heart rate variability and increases in blood pressure, heart rate, respiration rate, and blood allocation to muscle tissue (Munck et al. 1984, Silverman 1989).

A stress response is triggered by a cognitive interpretation of an external threat followed by the emotional arousal of the limbic system (Wiepkema and Schouten 1990). The hypothalamus is the nexus between the neural limbic system and the endocrine system. From the hypothalamus, the adrenal medulla is stimulated by sympathetic nerve pathways. The stimulated adrenal medulla releases the catecholamines epinephrine and norepinephrine, which affect body organs and have a negative feedback on the autonomic nervous system. Corticotropin-releasing factor secreted from the pituitary acts on the pituitary to release adrenocorticotrophic hormone (ACTH). In turn, the adrenal cortex is stimulated by ACTH and releases cortisol and other glucocorticoids, which affect body organs and have a negative feedback effect on the pituitary gland, the hypothalamus, and other brain regions. Because of the multiple feedback loops, a stress response is usually a temporary deviation from a homeostatic baseline, with a rapid onset followed by a gradual return to baseline.

Although catecholamine, ACTH, and cortisol release have long been regarded as the classic endocrine indices of the stress response (Brain 1979, Levine et al. 1989), physiological responses to stress are not restricted to the hypothalamus, pituitary, and adrenal glands and the sympathetic–adrenomedullary systems (Huntingford and Turner 1987). Catecholamines and cortisol and other glucocorticoids have been shown to affect the production of a panoply of other hormones, including gonadal hormones, growth hormone, thyroxines, and insulin. The magnitude of the change in the titers of these hormones from their baseline levels is often correlated with the magnitude and category of the external stressor (Hennessy and Levine 1977, Alberts et al. 1992, Hennessey 1997). Similarly, the overt behavioral reactivity component of the stress response frequently corresponds to the change in endocrine levels and the magnitude of the stressor that instigated the stress response (Fokkema et al. 1988, Kagan et al. 1988). However, in some circumstances behavioral and physiological responses to stress may not covary in intensity, particularly when the external stressor is severe (Coe et al. 1982).

Species Differences in Reactivity

More than 50 years ago three species of apes were described as differing in personality. Yerkes and Yerkes (1929) found gorillas (*Gorilla gorilla*) to be diffident and shy, chimpanzees (*Pan troglodytes*) outgoing, expressive, and impulsive, and orangutans (*Pongo pygmaeus*) withdrawn and brooding. Since then, the number of studies addressing the diverse reactivity patterns of primates, as well as their physiological concomitants, have rapidly increased in number and sophistication. Differences in physiological responsiveness, as determined at least in part by hypothalamic–pituitary–adrenal and sympathetic–adrenomedullary systems, are explicitly recognized as a major source of temperament differences among primate species (Mason 1976, 1978; Clarke et al. 1988, Mendoza and Mason 1989, Clarke and Boinski 1995). Furthermore, species differences in representative or modal social organization are seen as emergent phenomena resulting from interactions among individuals with disparate reactivity patterns or stress-response characteristics. Evolution of the phenotypic expression of social organization thus could result from selection on the physiological systems underlying behavioral reactivity to stress.

Comparisons of titi monkeys (*Callicebus moloch*) and squirrel monkeys (*Saimiri sciureus*), two Neotropical primates of relatively distant phylogenetic affinity, provide strong evidence of the evolutionary interplay between temperament, physiological characteristics of the stress response, and social and ecological specializations (Cubicciotti et al. 1986). Titis form close, monogamous pair bonds, are territorial, and have a significant folivorous component to their diet (Kinzey 1981). In contrast, squirrel monkeys live in large, nonterritorial, multimale and multifemale groups and are largely insectivorous (Boinski 1987b). Thus, the titis can be regarded as adapted to a niche less demanding of rapid sustained and vigorous activity than are squirrel monkeys. Concomitantly, titis have a lower heart rate and plasma cortisol response to experimentally imposed stressors than do squirrel monkeys. Even within familiar, non-stressful situations, squirrel monkeys have a higher heart rate and higher cortisol levels than titis. Stress-inducing situations increase this disparity (Cubicciotti et al. 1986, Mendoza and Mason 1989).

The Old World monkey genus *Macaca* presents a tantalizing opportunity for comparative studies. It comprises at least 12 species distributed across Asia with relict populations in Morocco and Algeria (Melnick and Pearl 1987). Although intraspecific variation in behavior is great (Melnick and Pearl 1987), there appear to be consistent species differences in temperament and reactivity that could be due to diet and habitat specializations in the wild (Caldecott 1986, Richard et al. 1989).

Laboratory studies of macaques support the hypothesis that strong interrelationships exist between temperament, the physiological systems affecting the expression of temperament, and ecological and social adaptations. A comparison across four macaque species (rhesus [*Macaca mulatta*], cynomolgus [*M. fascicularis*], bonnet [*M. radiata*], and lion-tailed [*M. silenus*]) demonstrates robust patterns in tendencies to approach, interact, and manipulate novel objects and situations (Clarke and Mason 1988, Clarke et al. 1988, Clarke and Lindburg 1993). For example, compared to the passive and diffident cynomolgus macaques, lion-tailed macaques are bold, adaptable, vigilant, manipulative, and exploratory. Clarke and Lindburg (1993) suggest that the greater "environmental curiosity" of lion-tailed macaques is consistent with their diverse, omnivorous diet relative to the frugivorous cynomolgus. Other studies with captive rhesus, bonnet, and cynomolgus macaques simultaneously measured behavioral, cardiac, and adrenocortical responses to stressful situations (Clarke et al. 1988, 1994). Across all test conditions the cynomolgus exhibit the highest adrenocortical and heart rate response and rhesus the lowest. Rhesus also had the lowest indices of depression and highest levels of motor activity.

Intraspecific Variation in Reactivity

Within species of primates, robust, heritable, and developmentally stable individual differences are present in the response patterns of the hypothalamic–pituitary–adrenal axis and the sympathetic nervous system to environmental changes and disturbances. About 20% of rhesus individuals are easily disrupted and stressed (Suomi 1991). They typically have higher baseline levels of catecholamine turnover, particularly norepinephrine, and higher heart rates than do the other 80%. In contrast, the majority exhibit rapid adaptation to environmental perturbations such as social separation. These individual differences can be linked with divergent social outcomes. Juvenile rhesus macaques with high reactivity to experimentally induced stressors are less likely to attain high dominance status as adults than low-reactive juveniles (Golub et al. 1979).

Breeding studies demonstrate that these behavioral and physiological response traits of rhesus macaques have a significant genetic component (Suomi 1981). Individual differences in these traits can first be detected at 1 month of age and persist into adulthood (Suomi 1987). Robust behavioral reactivity and attention and orientation differences also exist between neonate rhesus macaques of Chinese origin and Chinese-Indian hybrids (Champoux et al. 1994). Sex differences can exaggerate individual and population differences in response to stress even further. Male rhesus, and males of other primate species, are typically more aggressive (Nagel and Kummer 1974, Mitchell 1979) and apparently are less reactive to novel stimuli and opportunities to explore (Buirski et al. 1978, Stevenson-Hinde and Zunz 1978, Bolig et al. 1992).

Individual differences in reactivity have also been documented in humans and baboons. The distribution and development of reactivity profiles in humans is strikingly reminiscent of that documented in rhesus macaques (Kagan et al. 1988). Fifteen percent of white American children can be assigned to one of two extreme categories on the basis of behavioral and physiological response to novelty. When confronted with a wide variety of challenges, such as the presence of an unfamiliar adult, some children are shy and restrained, while others are outgoing and spontaneous. Comparable personality contrasts in adults are labeled introversion and extroversion. In a longitudinal study of 400 children from age 2 to 7, the temperaments of the extremely shy and the extremely outgoing minority remained stable, whereas the developmental trajectory of the intermediate majority could not be predicted. In addition, numerous indices of physiological reactivity were correlated with temperament. Shy, highly reactive children exhibited high and minimally variable heart rates, greater pupillary dilation, and high norepinephrine and cortisol levels (Kagan et al. 1988).

Dominance status, behavioral "style," and the turmoil of disruptions in a troop's social structure add an additional level of complexity to any discussion of individual variation in reactivity to stress (Sapolsky 1990, Alberts et al. 1992). Long-term field studies of the relationship between dominance and the physiology of stress in savannah baboons (*Papio cynocephalus*) determined that the testosterone responses of subordinate and dominant males differ after encounters with the same pyschological stressor, capture stress (Sapolsky and Ray 1989, Sapolsky 1990). Testosterone titers of dominant males quickly rose and remained elevated for about an hour after capture stress, whereas testosterone titers of subordinate males dropped precipitously and remained low. Furthermore, within the stress-inducing situation of capture, dominant males usually exhibited the largest cardiovascular stress responses and the fastest return to normal levels (Sapolsky and Share 1994).

Dominant male baboons can be distinguished by a low-reactive versus a high-reactive style (Sapolsky and Ray 1989, Sapolsky 1990). Dominant males who consistently exhibit a low-key response to psychological stressors and are also able to calibrate their response to different intensities of stress (e.g., the mere presence of an adult male versus a threat from that male) also had low basal concentrations of cortisol. In contrast, dominant males who were highly reactive to identical stressors had basal cortisol levels comparable to those of subordinate adult males. A non-reactive style might facilitate attainment of high rank in baboons and other primates (Rasmussen and Suomi 1989, Sapolsky 1990).

Substrates for Natural Selection

Individual variation and a genetic basis for the expression of physiological and behavioral traits are well established (e.g., Suomi 1991, Plomin et al. 1994), and there are many potential selection opportunities within the neuroendocrine system. These include genes and gene complexes for each hormone that affect receptor-site density and sensitivity, physiological responsiveness to the agonists and synergists specific to each hormone, levels of hormone production, and modification of the molecular structure of the hormones (Gaunt and Leathem 1967, Chrousos et al. 1982, Reynolds et al. 1997). Artificial selection on behavior typically results in a pattern of persistent, albeit small, increases in the distinction between separated lines over generations (Broadhurst 1975; see also Gottlieb 1992), indicating that many genes, each accounting for a small amount of genetic variance, are involved (Plomin et al. 1994). A recent report on the genetic basis of reactivity in mice, however, found that three loci on three different murine chromosomes account for a large proportion of the genetic variance in reactivity (Flint et al.

1995). Flint and his coauthors (1995) explicitly suggest that the expression of reactivity among other species, including humans, has a similarly simple genetic basis.

A Model: Geographic Comparisons of Squirrel Monkey Populations

Squirrel monkeys (genus *Saimiri*) are unique among Neotropical mammals. They are nearly ubiquitous in the Amazonian basin with extensions north into the Guyanas and south into Paraguay. There also is a relict Central American population restricted to the Pacific lowlands of Costa Rica (fig. 5-1, Costello et al. 1993). Two population groups, named Roman and Gothic, have been distinguished on the basis of pelage coloration and other characters (MacLean 1964, Hershkovitz 1984, Thorington 1985). The Central American and northern South American populations comprise the Gothic type (pointed-arch brow pattern), whereas populations in Peru, Bolivia, and southwestern Brazil are of the Roman type (rounded brow pattern).

The most recent revision of this genus (Costello et al. 1993) concluded that designation of *Saimiri* populations as distinct species was not warranted, particularly within South America. Central American squirrel monkeys appear to have accrued small dental, biochemical, and genetic differences (although only a single individual provided the biochemical and genetic data to represent the Central

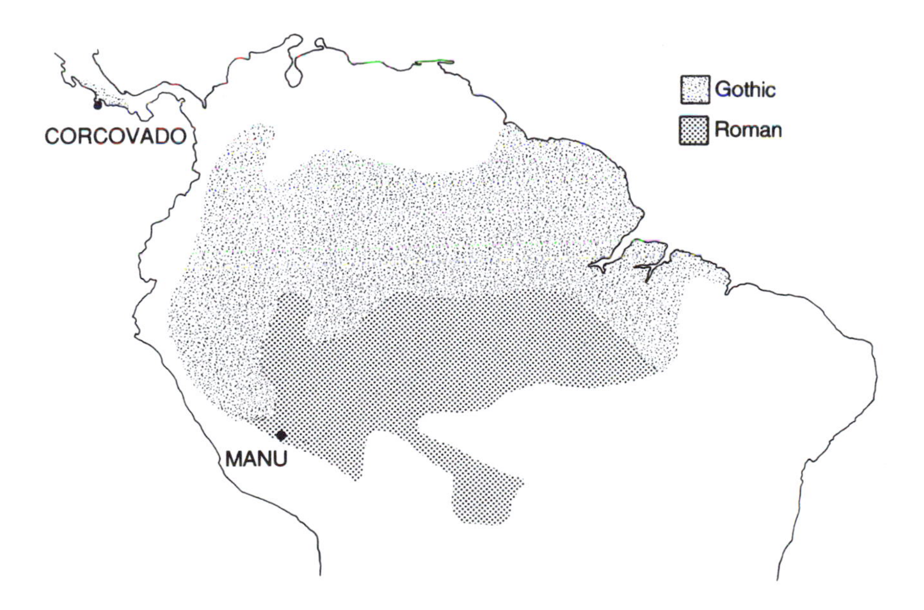

Figure 5-1 Map of the distribution squirrel monkeys in the Neotropics, including Roman versus Gothic populations and the location of the field sites in Costa Rica and Peru.

American population). The Central American population clearly has a closer genetic, and probably phylogenetic, relationship with other Gothic type squirrel monkeys in northern South America than with Roman type squirrel monkeys (Costello et al. 1993). For the purposes of this chapter, I emphasize the simplistic, although widely recognized, Gothic (*S. oerstedi*) versus Roman (*S. sciureus*) dichotomy. All squirrel monkey populations can be subsumed under the species name *S. sciureus*, but *S. oerstedi* remains a useful moniker for the Central American population because it is disjunct, exhibits some distinctions from *S. sciureus*, and is threatened with extinction (Boinski and Sirot 1997, Boinski et al. 1998).

Field Studies

Two detailed ecological and behavioral studies of Roman squirrel monkeys in Parque Nacional del Manu, Peru (Terborgh 1983, Mitchell 1990), and a comparable study of Gothic squirrel monkeys in Parque Nacional Corcovado, Costa Rica (Boinski 1987a,b,c), have contributed the bulk of the available field data. At both study sites squirrel monkeys are surprisingly similar in ecology and demography (table 5-1, Mitchell et al. 1991, Janson and Boinski 1992). Squirrel monkeys in Peru and Costa Rica live in large troops, eat the same types of fruits and insects, harvest these foods in the same ways, and spend nearly identical amounts of time engaged in these activities. The only notable difference in their ecology is that the fruit patches available to Peruvian squirrel monkeys are much larger and denser (higher quality) than those available in Costa Rican squirrel monkey habitat.

Both squirrel monkey populations form large cohesive groups, probably more as an antipredator adaptation against the documented risk of predation by raptors, felines, and snakes, than for potential foraging benefits associated with large

Table 5-1 Charactersistics of the Costa Rican (Gothic, *Saimiri oerstedi*) and Peruvian (Roman, *S. sciureus*) squirrel monkey populations (from data in Mitchell et al. 1991, Boinski 1987a,b,c).

Characteristic	Costa Rica	Peru
Body weight	<600 g, 750 g	751 g, 992 g
Group size	35–65	45–75
Habitat	Usually in second growth, but also primary forest	Successional, primary forest
Home range	200 ha	250–500 ha
Density	0.36/ha	0.6/ha
Predation[a]	>0.013 attack/h, 5 successful	0.911 attack/h, 5 successful
Between-troop territoriality or conflict	Never	Never
Fruit patch size	Uniform, small	Variable, larger
Feeding party size	Estimated 3–4, often only 1	17–18

[a]Data shown are rates of attack per observer hour with a squirrel monkey troop (including troops other than the main study troop at each site): >40 attacks seen in Costa Rica in over 3000 h of observation, 24 attacks seen in Peru in 2100 h of observation.

Table 5-2 Behavioral comparisons between squirrel monkeys in Costa Rica and Peru (data from Mitchell 1990, Mitchell et al. 1991, Boinski, submitted).

Characteristic	Costa Rica	Peru
Resource-based aggression within groups	0.004 event/h	0.286 event/h
Aggression Context		
Insect items	100%	5%
Fruit	0%	95%
Female dominance hierarchies	Not detected	Present, linear, stable
Female coalitions	Not detected	Present, kin based
Dominant sex	Sexes egalitarian aside from breeding season when males dominant	Females dominant over males
Between-sex interactions	Few aggressive or affiliative interactions aside from breeding period	Females persistently harass males
Within-troop male-male interactions	Affiliative	Aggressive and clear dominance hierarchy apparent
Troop residence	Females migrate, males natal or usurp reproductive males in another troop	Females natal, males migrate

group sizes (Boinski 1988). Densities are comparable and home range sizes at both sites are extremely large for New World primates, especially for a primate species weighing less than 1 kg. Similarly, squirrel monkeys at both sites eat a wide variety of plant and animal material, including arthropods, small (<1 cm diameter), berrylike fruit, nectar, and small vertebrates. Eighty percent of identifiable arthropods ingested at both sites were Orthoptera, especially grasshoppers and katydids, and Lepidoptera caterpillars (Janson and Boinski 1992).

The allocation of time to major activities differs little between sites. In Costa Rica squirrel monkeys spend 45–65% of their time traveling and searching for insects, 11% of their time feeding, and 5–10% of their time resting, while the respective figures for Peruvian squirrel monkeys are 50–75% searching for food, 11% feeding, and 11% resting (Terborgh 1983, Boinski 1987b, 1988). Direct between-troop food competition has never been observed at either site; squirrel monkey troops are not territorial, and overlap in the ranges of adjacent troops is extensive (Boinski 1987b, Mitchell 1990). Moreover, the same sex differences in foraging behavior are described in Costa Rica and Peru; females search for and eat foods at much greater frequencies than males (Boinski 1988, Mitchell 1990).

Given these broad ecological similarities, the differences in social behavior between the two populations are almost astounding (table 5-2, Mitchell et al. 1991). In fact, to my knowledge, no other pair of primate populations in either the New or Old World are as morphologically similar and as phylogenetically close, but have such divergent social organizations. Female squirrel monkeys in Peru form a dominance hierarchy, and stable, kin-based alliances within the troop.

These females also engage in defense of fruit resources and support other members of their alliance in resource defense. In over 3 years of observation of 51 known females, no adult female ever changed groups, and 8 juvenile females matured and bred in their natal troops (Mitchell 1990). Costa Rican females differed in showing no evidence of a hierarchy and no long-term affiliative bonds among individual females. Furthermore, female troop membership is labile (Boinski 1987a, Boinski and Mitchell 1994). Within one 11-month period, at least seven adult females left or entered a single troop in Costa Rica.

These marked differences in social organization almost certainly follow from the contrasting levels of within-group food competition in the two squirrel monkey populations. There is negligible direct within-group food competition in Costa Rica, whereas it is approximately 70 times more frequent in Peru. The majority of direct conflicts in Peru over food occur in fruit trees of <25 m crown diameter. Coalitions among individuals in aggressive food conflicts form in approximately one of every four encounters. During direct competition for fruit among Peruvian females in the smallest fruit patch category (<5 m), coalitions formed in only 25 of 136 (18.4%) of aggressive interactions. However, in fruiting trees of 5–25 m crown diameter, 11 of 22 (50%) aggressive interactions involved the formation of coalitions ($\chi = 9.88$, df = 1, $p < .002$). All food-related aggression in Costa Rica (only 12 instances observed in more than 3000 h) involves small vertebrate or invertebrate food items. Troop members in Costa Rica also never form coalitions of individuals cooperating in the aggressive acquisition or defense of these prey items.

In turn, the most plausible (and only) ecological factor that can be identified as underlying these population differences in direct food competition is site variation in available fruit patch sizes. The size of harvested fruit patches, as estimated by crown diameter of the fruiting tree, was usually quite small (<5 m diameter) in Costa Rica, and much larger in Peru (fig. 5-2, Kolmogorov-Smirnov statistic

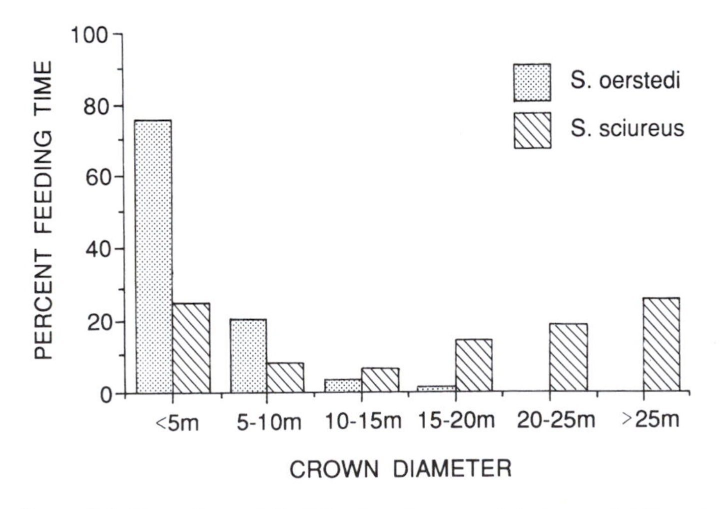

Figure 5-2 Percentage of fruit-feeding-time spent in trees of different crown sizes for *S. oerstedi* in Costa Rica and *S. sciureus* in Manu, Peru.

using the actual percentage points as data, $D_{max} = 0.621$, $p < .01$; the critical value of D_{max} for equal sample sizes of $n = m = 25$ at the $p = .01$ level of significance is 0.480; population sample sizes >25). The number of individuals simultaneously foraging in a fruit tree in Costa Rica seldom exceeded three or four, and was often one, whereas the mean fruit foraging party size was four to five times greater in Peru.

The disparity in the sizes of harvested fruit patches is unlikely to be due to differing fruit preferences; at both sites squirrel monkeys appeared to harvest all fruit species that were small, soft, and berrylike and can thus be easily handled and ingested (Janson and Boinski 1992). Similarly, habitat destruction in Costa Rica, although pervasive, is not a factor underlying site differences in the distribution of fruit patches harvested. The Costa Rican site, Corcovado, is a near-pristine tropical wet forest, and the second-growth areas the squirrel monkeys exploited had somewhat larger and denser patches of fruit than undisturbed primary forest (Boinski 1987b, unpublished data).

The bigeography of shrub and tree taxa in the Neotropics (Gentry 1988) ultimately accounts for the between-site difference in sizes of fruit patches harvested. In Central America, the predominant plant families that bear small, berrylike fruits (Melastomataceae, Rubiaceae, Piperaceae, and Verbenaceae) usually have a growth form of small trees and shrubs and thus a small patch diameter. These shrubs and small trees also tend to ripen only a few fruits per plant each day. In contrast, in Peru suitable fruits are most commonly available from a different set of plant taxa, from families that tend to have trees with larger crowns and fruits within each crown that ripen almost simultaneously (Moraceae, Myrsinaceae, Eleocarpaceae, Leguminosae, Sapotaceae).

Male social behavior also differs between Costa Rica and Peru. Outside of the 2-month long breeding season in Costa Rica, male and female dominance relationships are egalitarian, although males are aggressive and dominant within the breeding season (Boinski 1987c). Costa Rican males maintain residence in their natal troop until they either assume reproductive positions in their natal troop or, rarely, usurp the reproductive cohort in another troop in cooperation with other males in their birth cohort (Boinski and Mitchell 1994). Between-male aggression within a Costa Rican troop has never been observed, and males are never relegated to the troop periphery (Boinski 1994).

Females in Peru, however, are always dominant to males (Mitchell 1990). Peruvian males migrate from their natal troop at 4 to 5 years of age. When these Peruvian males are older, and if they gain residence in a mixed-sex troop, they are all relegated to the troop periphery by the harassment of the females, possibly because of the females' perception that males pose a threat to their offspring (Mitchell 1990). Adult males in Peruvian troops have numerous aggressive interactions with other males, and dominance hierarchies are clear; males also spend much time in vigilance and aggressive interactions against nonresident males attempting to enter the troop (Mitchell 1994).

Direct food competition does not appear to be an important factor affecting the social behavior and residence patterns of males in either population. Indirectly, however, the level of direct competition for fruit resources among females is thought to affect male residence patterns and thus patterns of male–male aggression (Boinski and Mitchell 1994). In Peru, high levels of direct food competition likely make female coalitions and female philopatry advantageous. Inbreeding avoidance (Hoogland 1982) is the strongest explanation for the consistent emigration of Peruvian males from their natal troops. The presence of unrelated adult males in mixed-sex troops of Peruvian monkeys results in high levels of aggressive competition among males for mates.

Alliances, matrilines, and philopatry are not, in contrast, advantageous to female Costa Rican squirrel monkeys in a selection regime of low direct food competition (Mitchell et al. 1991, Boinski and Mitchell 1994). Once females started moving between troops, benefits to migrating males from inbreeding avoidance were reduced. In addition to male–male cooperation in sexual interactions, males remaining resident in the natal troop could obtain enhanced reproductive benefits from antipredator vigilance serving to protect their offspring and kin.

In summary, many aspects of female social behavior in Costa Rica and Peru are consistent with the hypothesis that the distribution and size of resource patches affect the relative value of aggression and coalition formation among female primates (van Schaik 1989). Indirectly, male social interactions will also be affected by female social behavior and dispersal patterns (Boinski 1994, Boinski and Mitchell 1994, Mitchell 1994). High levels of direct competition for fruit among female Peruvian squirrel monkeys are associated with a female dominance hierarchy, within-group stable kin alliances, and female philopatry. Migration to another troop probably would be costly in terms of food competition for a Peruvian female if female kin are important in acquiring access to fruit patches. In Costa Rica, a near absence of within-group food competition is associated with female dispersal and a lack of female bonding. Male Costa Rican squirrel monkeys, like their female counterparts, rarely participate in any aggressive interaction outside a brief, annual breeding season. Although food competition also appears to be an unimportant factor in the daily existence of Peruvian male squirrel monkeys (males spend significantly less time foraging than females in Costa Rica and Peru), Peruvian males must cope with frequent threats (often numerous times daily) and the persistent possibility of threats from females and resident and intruding males.

Laboratory Studies

Studies addressing the physiological response for coping with an external stressor in squirrel monkeys have emphasized changes in plasma cortisol titers. Fluctuations in cortisol levels appear to be the most reliable index of the emotional component of stress compared to other hormones, such as ACTH or the catecholamines (Stanton et al. 1985). Unlike the situation with many other primates, ACTH stimulation of the adrenal cortex does not appear to significantly stimulate cortisol production in squirrel monkeys (Mendoza et al. 1991). Instead, the sympathetic nervous system is apparently responsible for the stimulation of the adrenal cortex, a conclusion consistent with the high level of sympathetic activity (also indicated by high heart rates in squirrel monkeys), unimpeded by significant parasympathetic feedback inhibition (Mendoza and Moberg 1985, Mendoza and Mason 1989).

Plasma cortisol titers of squirrel monkeys even in neutral, baseline conditions are high (about 150 ng/ml) relative to other primates, in part because of allometry and phylogeny (Coe et al. 1992). In primates, baseline cortisol titers are negatively associated with body size. Thus, the relatively small New World monkeys generally have higher titers than Old World monkeys. Most prosimians and Old World monkeys and apes range between 20 and 60 ng/ml plasma cortisol in baseline conditions. Despite high baseline levels, squirrel monkey plasma cortisol titers can rise still higher, particularly after exposure to a stressor. For example, levels up to 350 ng/ml have been recorded after an hour of physical restraint (Cubicciotti et al. 1986). Such high induced cortisol levels prepare the animal to deal with stressful or arousing situations by promoting glucogenesis and other metabolic processes that permit energy reserves to be metabolized immediately (Cubicciotti et al. 1986, Mendoza and Mason 1989). In contrast, titi monkeys, a New World monkey about the same size as squirrel monkeys, have much lower baseline cortisol levels and a disproportionately reduced cortisol surge after exposure to comparable stressors. Concomitantly, titi monkeys evince a different pattern and pace in their interaction with the environment, seeming almost lethargic in comparison to squirrel monkeys (Cubicciotti et al. 1986).

Stress reactivity in squirrel monkeys is strongly affected by the presence or absence of other squirrel monkeys. Squirrel monkeys housed by themselves or two per cage exhibited a dramatic cortisol and behavioral response to a snake model (Coe et al. 1982). Those housed in groups of six also respond to the model with a pronounced behavioral avoidance, but not with a cortisol response (Vogt et al. 1981). Saltzman et al. (1991) compared plasma cortisol levels, heart rate, and behavioral parameters in females when they were housed singly and when placed in a cage with another female. The females' baseline plasma cortisol titers quickly dropped an average 22% when housed with another female, and maintained this lower level for the duration of the pair housing. When paired females were separated, several days passed before basal cortisol levels again rose (Mendoza et al. 1989).

Mendoza et al. (1991) suggested that the delayed cortisol response to separation in female squirrel monkeys is consistent with the wide dispersion of foraging of squirrel monkey troops (Boinski 1988). If there were no lag (or selectivity) in reaction to the absence of a female in close proximity, females would be in a near permanent state of stress reactivity, a situation unlikely to enhance foraging efficiency. Field studies of adult female squirrel monkey vocal behavior also provide little evidence that separation over moderate distances (<10 m) induces stress (Boinski 1991, Boinski and Mitchell 1992, 1997).

Robust, genetically based behavioral differences are well established among South American Roman and Gothic squirrel monkey morphotypes in captivity (potentially these morphotypes could have several within-morphotype populations of origin) (cf. Gonzalez et al. 1981, Costello et al. 1993). These behavioral differences, including stereotyped displays (MacLean 1964) and vocalizations (Winter 1969, Boinski and Mitchell 1992), reflect geographic separation and, presumably, regional genetic adaptation. An extensive series of captive studies is available on social behavior and physiological reactivity for Gothic (northern South America and Central America) versus Roman (Peruvian and Bolivian) morphotypes. Roman squirrel monkeys have a sexually segregated social organization in which males and females remain spatially separated outside of the breeding season (Mendoza et al. 1978, Gonzalez et al. 1981). Male and female dominance hierarchies are also segregated; all females are dominant to all males. Aggression by females toward males underlies the spatial segregation. Among captive Gothic squirrel monkeys, in contrast, social groups are integrated throughout the year, and a single linear dominance hierarchy includes both males and females. Overall, Roman females are more frequently aggressive than Gothic females. In both populations individual differences in temperament are documented for both males and females, but morphotypic differences exceed individual differences (Martau et al. 1985).

Only one behavioral study based, in part, on captive Gothic squirrel monkeys from Central America has been published (Boinski and Newman 1988). This report, together with my extensive unpublished observations of this colony (15 animals) of wild-caught and captive-born *S.oerstedi*, strongly suggest that the distinctive behavior patterns of wild *S. oerstedi* are maintained in captivity. The clear inference is that the expression of these behaviors, including negligible evidence of dominance hierarchies, social aggression, and female–female affiliative bonds, have a significant genetic component in squirrel monkeys from Central America.

Behavioral differences among the South American Gothic and Roman morphotypes are concordant with differences in their physiological reactivity. The more aggressive Roman squirrel monkeys have greater baseline cortisol levels and a more pronounced cortisol response to stress-inducing situations. Although they are larger than Gothic squirrel monkeys, and lower cortisol levels might therefore be expected (Coe et al. 1992), Roman squirrel monkeys have higher basal cortisol levels than Gothic squirrel monkeys (Gothic: female 65 ng/ml, male 140 ng/ml; Roman: female 110 ng/ml, male 200 ng/ml; Coe et al. 1985). Following formation of new social groups in captivity, the plasma cortisol levels of females in Gothic groups were unchanged from baseline levels, but those of Roman females were elevated for up to 9 weeks, only gradually decreasing over that period (Gonzalez et al. 1981). Males also exhibit notable geographic variation in the interaction between dominance and stress response. Among Gothic males, the cortisol levels following group formation were greater in subordinate than in dominant males, but in Roman groups the relationship was reversed: dominant males had higher cortisol levels than subordinates (Coe et al. 1983).

Discussion

Squirrel Monkey Comparison

In 1981 Gonzalez et al. speculated that the behavioral and physiological differences that had already been documented between captive squirrel monkey morphotypes reflected "varying ecological demands." The recently available data from the two wild populations of squirrel monkeys provide intriguing support for this hypothesis (table 5-3). Together, the physiological, behavioral, and ecological data suggest that a large component of the geographic variation in the social behavior and temperament of adult Roman and Gothic squirrel monkeys can be explained by different "settings" of the stress-response system (sensu Mendoza and Mason 1989). Roman squirrel monkey morphotypes have a much more reactive stress-response system than do Gothic morphotypes. This is consistent with the much greater levels of aggression in the social and ecological environment of the Roman population of squirrel monkeys in Peru compared to the Gothic population of squirrel monkeys in Costa Rica.

Plausibly, the Roman squirrel monkeys in Peru benefit from a greater reactivity to stress as individuals and thus are physiologically better prepared to cope with a more stressful environment. Field data further suggest that the distribution of fruit resources available to squirrel monkeys is an important factor that may engender greater levels of stress when fruit sources are clumped and thus more readily defended. Susceptibility to predation (which might make vigilance and rapid responses to threat advantageous) is only a weak explanation for the geographic differences in reactivity, as squirrel monkeys in both Costa Rica and Peru are subject to intense predation pressure.

Peruvian squirrel monkeys, particularly females, live in an environment where aggression enhances food procurement. Peruvian females exhibit and receive more social aggression than do Costa Rican females. Peruvian male squirrel monkeys that are fortunate enough to attain residence in mixed-sex troops are continu-

Table 5-3 Summary of characteristics of Roman and Gothic squirrel monkeys from field and captive studies.

Roman	Gothic
Captive	
Higher female and male cortisol titers	Lower female and male cortisol titers
Longer and more sustained adrenal response to stress	Shorter and reduced adrenal response to stress
Females more active than males in initiating agonistic interactions	Little female aggression
Females dominant to males	Males dominant to females
Dominant males have higher cortisol titers than subordinates	Dominant males have lower titers than subordinates
Females harass males into peripheral positions	Males spatially integrated into troop
Field	
Peru	Costa Rica
Frequent direct food competition	Rare food competition
Females dominant to males	Males dominant to females
Female dominance hierarchies	No female hierarchy detected
Much male–male aggression	Little male-male aggression
Frequent female aggression	Rare aggression by females

ally badgered and threatened by females, and even by juveniles and infants (Boinski, personal observation). Furthermore, Peruvian males must maintain near-constant vigilance to deter, often vigorously, persistent attempts by other males to enter their troop. One reason dominant male Roman squirrel monkeys have greater cortisol levels than subordinate males may be the dominant males' greater need for social vigilance (Gonzalez et al. 1981). In contrast, Costa Rican males are exposed to negligible within-troop aggression and to only sporadic aggression from nonresident males. This is a lower stress situation which is compatible with the laboratory profile of Gothic male morphotypes.

Testable Model

I have provided evidence of an association between levels of aggression and reactivity in a comparison of two geographically disparate morphotypes of squirrel monkeys. These differences seem to be associated with differences in ecological conditions and social structure. Although suggestive, the study is still only correlative and cannot be taken as a demonstration of causation.

Nevertheless, the associations make it possible to predict that among an array of geographically separated squirrel monkey populations in the wild, within-population baseline cortisol titers, the extent of the cortisol upsurge after exposure to an environmental stressors, and overall behavioral and physiological reactivity will be positively associated with the amount of aggression commonly encountered by individuals within that population. The premise is that higher "settings" of the physiological response system provide more effective buffers to the deleterious effects of environmental stress. Furthermore, relative levels of physiological reactivity are not expected to be tightly linked between the sexes within any population. Among female squirrel monkeys, physiological and behavioral reactivity are expected to be positively associated with the extent of within-troop direct competition for fruit. Male squirrel monkey reactivity, however, will be more directly related to the level of male mate competition.

A weakness in the model that cannot be overlooked is that the sites where field studies were conducted and the sites where the source animals for captive groups were obtained were not identical. Ideally, future field studies at other geographic localities should simultaneously obtain data on the behavior, ecology, and physiology of the study population. Physiological data, especially cortisol titers, can be collected in the field from fecal samples (Gross 1991, 1992) or from temporarily captured squirrel monkeys. This model also presumes that squirrel monkey populations have evolved under similar predation pressures and exhibit roughly equivalent antipredator vigilance and reactivity.

Population and Geographic Comparisons

In general, geographic comparisons are likely to be more useful than species comparisons in identifying factors important in the evolution of stress reactivity in primates. There is an immense benefit of scale from between-population geographic studies. First, not only are phylogenetic histories and distances often more precisely known, but similarities due to convergent evolution are reduced (Arnold 1992). Second, ecological differences that might account for significant modifications in selective regimes can be detected (Foster et al. 1992). There are, however, unlikely to be so many differences as to weaken the power of comparison. Third, hypotheses that are generated a posteriori from a set of data are best tested using independent data collected for that purpose. Particularly in primates where the number of species are few compared to many other taxa, finding "fresh" species that are appropriate to test models generated from a species-level comparison will often be difficult, if not impossible. Within-species geographic comparisons, in contrast, may offer numerous, if not a continuum, of populations and sites suitable for hypothesis testing.

There are, however, limits to the utility of geographic or population comparisons, as numerous sources of within-species behavioral variation exist. Much within-species variation in behavior can be explained as nongenetic responses to local conditions. Behavioral changes in primates have been associated with habitat and temporal differences in food distribution and availability (Whitten 1983, Barrett et al. 1992), population density (Chivers 1969), sex ratio and group composition (Datta 1989), predation risk (van Schaik and van Noordwijk 1985, Boinski 1987a), sympatric food competitors (Kinzey and Robinson 1983), and the formation of mixed-species troops (Gautier-Hion et al. 1983). Primates have large brains with a disproportionately large cortex (Jerison 1982) and a long developmental period characterized by plasticity (Mason 1965). There is also a heavy reliance on learning, particularly learning in a social environment (Pereira and Altmann 1985, Visalberghi and Fragaszy 1990). Cultural tranmission of specific behavior may be indicated in some species, particularly chimpanzees (McGrew 1992). Primate social behavior is often complex and is affected by social status, kin relationships, long-term affiliations (Smuts 1985), and social manipulations (Byrne and Whiten 1988).

Within primates, genetically based geographic variation in behavior is most likely to be detected in taxa with extensive within-population variation in morphology. The baboon, *Papio hamadryas*, a taxon in which subspecies are well differentiated genetically (Williams-Blangero et al. 1990), is a prime example. Kummer (1971) documented a mosaic of anubis (*P. h. anubis*) and hamadryas (*P. h. hamadryas*) baboon behavior in hybrid animals on a geographic boundary. Hybrid males were unsuccessful when attempting to herd females in the one-male groups characteristic of pure hamadryas male baboons. Other examples of genetically differentiated, although poorly studied primate taxa, include the 7 species or subspecies of macaques on the South Pacific Island of Sulawesi (*Macaca nigra* spp., Fooden 1980), the 6 subspecies of the Malagaszy prosimian *Lemur fulvus* (Tattersall 1993), and the South American tamarin, *Saguinus fuscicollis*, with 10 subspecies (Rylands et al. 1993).

Pertinenence of Captive and Field Studies

The predictive model clearly suggests the value of examining the covariation of specific ecological contexts in the wild with geographic patterns in stress responses. There are also many opportunities for fertile interactions between field and captive studies to explore this evolutionary interplay. Data from captive studies that examine stress responses after exposure to novel surroundings and objects (Fragaszy 1980, Clarke and Lindburg 1993), establishment of dominance hierarchies (Mendoza et al. 1979), solitary versus group housing (Saltzman et al. 1991), high- versus low-effort foraging conditions (Champoux et al. 1993), and the effects of capture (Suomi et al. 1989) might be particularly analogous to situations in which natural selection acts in the wild.

Not all the physiological and behavioral stress responses induced by captive experiments are likely to be usefully incorporated into or used to generate predictive models for other species. Two examples are stress provoked among captive nonhuman primates raised apart from others in peer groups (Chamove et al. 1973) and stress provoked by the formation of social groups from previously unfamiliar individuals (Gust et al. 1991). Another consideration in the use of data from captive studies is that the intensity of response usually varies with the intensity of the stressor (Hennessy et al. 1982), although the general pattern of reactivity exhibited by taxa are considered to be stable across many different types of stressful situations (Mendoza and Mason 1989). Moreover, in extreme situations, such as those many laboratory studies appear to represent, all individuals probably exhibit extreme reactions (Suomi 1983).

Highly reactive individuals, be they human or nonhuman primates, that exhibit sustained stress responses are usually considered to be at a disadvantage. This is a result, in part, of captive studies using nonhuman primates to model the etiology of human disorders. From the perspective of humans in modern, complex societies, the amelioration of stress responses is nearly always considered advantageous. Just as aggression is often deprecated in human societies but is often adaptive among wild populations (de Waal 1992), high reactivity profiles may be advantageous in a natural context. Several plausible scenarios can be suggested. A highly reactive and cautious monkey might, for example, be less susceptible to predation (Wilson et al. 1994). A timid, diffident male might not provoke ejection from a troop as quickly as a more boisterous, aggressive male. Perhaps no advantage or disadvantage accrues to a highly reactive individual born to a high-ranking matriline, but the same individual would experience a reduced probability of survivorship if born into a low rank within the same troop. Similarly, Clark and Ehlinger (1987) point out that among humans, a shy and timid child could be quite functional in a small, buffered, and predictable society, whereas the same child would likely have difficulty coping in a high-paced, confrontational society where outgoing, intrepid personalities are esteemed.

Many potentially important sources of psychosocial stress in the wild have seldom been addressed in experimental studies of physiological and behavioral reactivity, including food shortages and high population density (van Schaik and van Noordwijk 1983, Judge and de Waal 1993), the need for antipredator vigilance (Caine 1984, Boinski 1987a), intertroop territorial encounters (Wright 1978), mating (Bercovitch 1988) and reproductive competition (Wasser 1983), and direct and indirect food competition (Mitchell et al. 1991). These ecological factors and the social interactions they may instigate are likely to be influential in the evolution of stress-response systems. A primate within its natural habitat must cope with a number of disparate sources of stress-inducing situations in its everyday existence. Some stressors might be mild, such as an uncomfortable encounter with a novel urticaceous plant, while others are severe, such as a successful predation attempt on another troop member. A reasonable working hypothesis is that geographic variation in stress-response profiles and temperament reflect, at least in part, differential weighting of these factors in the selective regimes in which the taxa evolved.

Acknowledgments This chapter was begun at the Laboratory of Comparative Ethology, National Institute of Child Health and Human Development, National Instititutes of Health, and completed at the Department of Anthropology and Division of Comparative Medicine, University of Florida. Comments and suggestions by A. Susan Clarke, Dee Higley, Kathlyn Rasmussen, Gary Steck, Steve Suomi, two anonymous reviewers, and the editors improved the text.

References

Alberts, S. C., R. M. Sapolsky, and J. Altmann. 1992. Behavioral, endocrine, and immunological correlates of immigration by an aggressive male into a natural primate group. Hormones and Behavior 26:167–178.

Armitage, K. 1982. Social dynamics of juvenile marmots: Role of kinship and individual variability. Behavioral Ecology and Sociobiology 11:33–36.

Arnold, S. J. 1992. Behavioral variation in natural populations. VI. Prey responses by two species of garter snakes in three regions of sympatry. Animal Behaviour 44:705–719.

Barrett, L., R. I. M. Dunbar, and P. Dunbar. 1992. Environmental influences on play behavior in immature gelada baboons. Animal Behaviour 44:111–116.

Benus, R. F., J. M. Koolhaas, and G. A. Van Oorthmerssen. 1992. Individual strategies of aggressive and nonaggressive male mice in encounters with trained aggressive residents. Animal Behaviour 43:531–540.

Bercovitch, F. B. 1988. Coalitions, cooperation and reproductive tactics among adult male baboons. Animal Behaviour 36:1198–1209.

Biben, M. 1983. Comparative ontogeny of social behavior in three South American canids, the maned wolf, crab-eating fox and bush dog: Implications for sociality. Animal Behaviour 31:814–826.

Boinski S. 1987a. Birth synchrony in squirrel monkeys (*Saimiri oerstedi*): A strategy to reduce neonatal predation. Behavioral Ecology and Sociobiology 21:393–400.

Boinski S. 1987b. Habitat use by squirrel monkeys (*Saimiri oerstedi*) in Costa Rica. Folia Primatologica 49:151–167.

Boinski S. 1987c. Mating patterns in squirrel monkeys (*Saimiri oerstedi*): Implications for seasonal sexual dimorphism. Behavioral Ecology and Sociobiology 21:13–21.

Boinski S. 1988. Sex differences in the foraging behavior of squirrel monkeys in a seasonal habitat. Behavioral Ecology and Sociobiology 23:177–167.

Boinski S. 1991. The coordination of spatial position: A field study of the vocal behaviour of adult female squirrel monkeys. Animal Behaviour 41:89–102.

Boinski S. 1994. Affiliation patterns among male Costa Rican squirrel monkeys. Behaviour 130:191–209.

Boinski, S., K. Jack, C. LaMarsh, and J. A. Coltrane. 1998. Squirrel monkeys in Costa Rica: Drifting to extinction. Oryx 38:45–58.

Boinski, S., and C. L. Mitchell. 1992. Ecological and social factors affecting the vocal behavior of adult female squirrel monkeys. Ethology 92:316–330.

Boinski, S., and C. L. Mitchell. 1994. Male residence and association patterns in Costa Rican squirrel monkeys. American Journal of Primatology 34:157–169.

Boinski, S., and C. L. Mitchell. 1997. Chuck vocalizations of wild female squirrel monkeys (*Saimiri sciureus*) contain information on caller identity and foraging activity. International Journal of Primatology 18:975–993.

Boinski, S., and J. D. Newman. 1988. Preliminary observations on squirrel monkey (*Saimiri oerstedi*) vocalizations in Costa Rica. American Journal of Primatology 14:329–343.

Boinski, S., and L. Sirot. 1997. The uncertain conservation status of squirrel monkeys in Costa Rica, *Saimiri oerstedi oerstedi* and *S. o. citrinellus*. Folia primatologica 68: 181–193.

Bolig, R., C. S. Price, P. L. O'Neill, and S. J. Suomi. 1992. Subjective assessment of reactivity level and personality traits of rhesus monkeys. International Journal of Primatology 13:287–306.

Brain, P. F. 1979. Effects of the hormones of the pituitary-adrenal axis on behaviour. *In* K. Brown and S. J. Cooper, eds. Chemical influences on behaviour, pp. 331–372. Academic Press, London.

Broadhurst, P. L. 1975. The Maudsley reactive and nonreactive strains of rats: A survey. Behavior Genetics 5:299–316.

Buirski, P., R. Plutchik, and H. Kellerman. 1978. Sex differences, dominance, and personality in chimpanzees. Animal Behaviour 26:123–129.

Byrne, R. W., and A. Whiten (eds). 1988. Machiavellian intelligence: Social expertise and the evolution of intellect in monkeys, apes, and humans. Clarendon Press, Oxford.

Caine, N. 1984. Visual scanning by tamarins: A description of the behavior and tests of two derived hypotheses. Folia Primatologia 43:59–67.

Caldecott, J. O. 1986. Mating patterns, societies and the ecogeography of macaques. Animal Behaviour 34:208–220.

Chamove, A. S., L. A. Rosenblum, and H. F. Harlow. 1973. Monkeys (*Macaca mulatta*) raised only with peers: A pilot study. Animal Behaviour 21:316–325.

Champoux, M., S. J. Suomi, and M. L. Schneider. 1994. Temperament differences between captive Indian and Chinese-Indian hybrid rhesus macaque neonates. Laboratory Animal Science 44:351–357.

Champoux, M., D. Zanker, and S. Levine. 1993. Food search demand effort effects on behavior and cortisol in adult female squirrel monkeys. Physiology and Behavior 54: 1091–1097.

Chivers, D. J. 1969. On the daily behavior and spacing of howling monkey groups. Folia Primatologia 10:48–102.

Christian, J. J. 1970. Social subordination, population density, and mammalian evolution. Science 168:84–90.

Chrousos, G. P., D. Renquist, D. Brandon, C. Eil, M. Pugeat, R. Vigersky, G. B. Cutler, D. L. Loriaux, and M. B. Lipsett. 1982. Glucocorticoid hormone resistance during primate evolution: Receptor-mediated mechanisms. Proceedings of the National Academy of Sciences USA 79:2036–2040.

Clark, A. B., and T. J. Ehlinger. 1987. Pattern and adaptation in individual behavioral differences. *In* P. P. G. Bateson and P. H. Klopfer, eds. Perspectives in ethology, vol. 7, pp. 1–47. Plenum Press, New York.

Clarke A. S., and S. Boinski. 1995. Temperament in nonhuman primates. American Journal of Primatology 37:103–126.

Clarke, A. S., and D. G. Lindburg. 1993. Behavioral contrasts between male cynomolgus and lion-tailed macaques. American Journal of Primatology 29:49–59.

Clarke, A. S., and W. A. Mason. 1988. Differences among three macaque species in responsiveness to an observer. International Journal of Primatology 9:347–364.

Clarke, A. S., W. A. Mason, and G. P. Moberg. 1988. Differential behavioral and adrenocortical responses to stress among three macaque species. American Journal of Primatology 14:37–52.

Clarke, A. S., W. A. Mason, and S. P. Mendoza. 1994. Heart rate patterns under stress in three species of macaques. American Journal of Primatology 33:133–148.

Coe, C. L., D. Franklin, E. R. Smith, and S. Levine. 1982. Hormonal responses accompanying fear and agitation in the squirrel monkey. Physiological Behavior 29:1051–1057.

Coe, C. L., A. S. Savage, and L. J. Bromley. 1992. Phylogenetic influences on hormone levels across the primate order. American Journal of Primatology 28:81–100.

Coe, C. L., E. R. Smith, D. Franklin, and S. Levine. 1983. Varying influence of social status on hormone levels in male squirrel monkeys. *In* A. S. Kling and H. D. Stecklis, eds. Hormones, drugs, and social behavior, pp. 7–32. Spectrum, New York.

Coe, C. L., E. R. Smith, and S. Levine. 1985. The endocrine system of the squirrel monkey. *In* L. A. Rosenblum and C. L. Coe, eds. Handbook of squirrel monkey research, pp. 191–218. Plenum Press, New York.

Costello, R. K., C. Dickinson, A. L. Rosenberger, S. Boinski, and F. S. Szalay. 1993. Squirrel monkey (genus *Saimiri*) taxonomy: A multidisciplinary study of the biology of species. *In* B. Kimbel and L. Martin, eds. Species, species concepts, and primate evolution, pp. 177–237. Plenum Press, New York.

Cubicciotti, D. D., S. P. Mendoza, W. A. Mason, and E. N. Sassenrath. 1986. Differences between *Saimiri sciureus* and *Callicebus moloch* in physiological responsiveness: Implications for behavior. Journal of Comparative Psychology 100:385–391.

Datta, S. 1989. Demographic influences on dominance structure among female primates. *In* V. Standon and R. A. Foley, eds. Comparative socioecology: The behavioral ecology of humans and other mammals, pp. 265–284. Blackwell, Oxford.

de Waal, F. B. M. 1992. Aggression as a well-integrated part of primate social relationships: A critique of the Seville statement on violence. *In* J. Silverberg and J. P. Gray, eds. Aggression and peacefulness in humans and other primates, pp. 37–56. Oxford University Press, Oxford.

Flint, J., R. Corley, J. C. DeFries, D. W. Fulker, J. A. Gray, S. Miller, and A. C. Collins. 1995. A simple genetic basis for a complex psychological trait in laboratory mice. Science 269:1432–1435.

Fokkema, D. S., K. Smit, J. van der Gugten, and J. M. Koolhaas. 1988. A coherent pattern among social behavior, blood pressure, corticosterone and catecholamine measures in individual male rats. Physiology and Behavior 42:485–489.

Fooden, J. 1980. Classification and distribution of the living macaques. *In* D. Lindburg, ed. The macaques: Studies in ecology, behavior, and evolution, pp. 1–9. Van Nostrand Reinhold, New York.

Foster, S. A., J. A. Baker, and M. A. Bell. 1992. Phenotypic integration of life history and morphology: An example from three-spined stickleback, *Gasterosteus aculeatus* L. Journal of Fish Biology 41(suppl. B):21–35.

Fox, M. W. 1972. Socio-ecological implications of individual differences in wolf litters: A developmental and evolutionary perspective. Behaviour 46:298–313.

Fragaszy, D. M. 1980. Titi and squirrel monkeys in a novel environment. *In* J. Erwin, T. L. Maple, and G. Mitchell, eds. Captivity and behavior: Primates in breeding colonies, laboratories and zoos, pp. 172–216. Van Nostrand Reinhold, New York.

Gaunt, R., and J. H. Leathem. 1967. Comparative endocrinology: Hormones and receptor response. Federation Proceedings 26:1192–1196.

Gautier-Hion, A., R. Quris, and J.-P. Gautier. 1983. Monospecific vs. polyspecific life: A comparative study of foraging and antipredatory tactics in a community of *Cercopithecus* monkeys. Behavioral Ecology and Sociobiology 12:325–335.

Gentry, A. H. 1988. Floristic similarities and differences between southern Central America and upper and central Amazonia. *In* A. H. Gentry, ed. Four neotropical rainforests, pp. 141–157. Yale University Press, New Haven, CT.

Golub, M. S., E. N. Sassenrath, and G. P. Goo. 1979. Plasma cortisol levels and dominance in peer groups of rhesus monkey weanlings. Hormones and Behavior 12:50–59.

Gonzalez, C. A., M. B. Hennessy, and S. Levine. 1981. Subspecies differences in hormonal and behavioral responses after group formation in squirrel monkeys. American Journal of Primatology 1:439–452.

Gottlieb, G. 1992. Individual development and evolution. Oxford University Press, Oxford.

Gross, T. 1991. Development and use of fecal steroid analyses in captive exotic mammals and birds. *In* Society for Theriogenology, Proceedings of Annual Meeting, 1991, pp. 66–74.

Gross, T. 1992. Development and use of fecal steroid analyses in several captive carnivore species. *In* Proceedings, 1st International Symposium on Fecal Steroid Monitoring in Zoo Animals, 1992, pp. 46–70, Minneapolis.

Gunnar, M. R. 1988. Psychobiological studies of stress and coping: An introduction. Child Development 58:1403–1407.

Gust, D. A., T. P. Gordon, M. E. Wilson, A. Ahmed-Anari, A. R. Brodie, and H. M. McClure. 1991. Formation of a new social group of unfamiliar female rhesus monkeys affects the immune and pituitary adrenocortical systems. Brain Behavior and Immunity 5:296–307.

Hennessy, M. B. 1997. Hypothalamic-pituitary-adrenal responses to brief social separation. Neuroscience and Biobehavioral Reviews 21:11–29.

Hennessy, M. B., and S. Levine. 1977. Effects of various habituation procedures on pituitary-adrenal responsiveness in the mouse. Physiology and Behavior 18:799–802.

Hennessy, M. B., S. P. Mendoza, and J. N. Kaplan. 1982. Behavior and plasma cortisol following brief peer separation in juvenile squirrel monkeys. American Journal of Primatology 3:143–151.

Hershkovitz, P. 1984. Taxonomy of squirrel monkeys genus *Saimiri* (Cebidae, Platyrrhini): A preliminary report with a description of a hitherto unnamed form. American Journal of Primatology 6:257–312.

Hinde, R. A. (ed). 1983. Primate social relationships: An integrated approach. Blackwell, Oxford.

Hoogland, J. L. 1982. Prarie dogs avoid extreme inbreeding. Science 215:1639–1641.

Huntingford, F., and Turner, A. 1987. Animal conflict. Chapman and Hall, London.

Janson, C. H., and S. Boinski. 1992. Morphological and behavioral adaptations for foraging in generalist primates: The case of the cebines. American Journal of Physical Anthropology 88:483–498.

Jerison, H. J. 1982. Allometry, brain size, cortical surface, and convolutedness. *In* E. Armstrong and D. Falk, eds. Primate brain evolution, pp. 77–84. Plenum Press, New York.

Judge, P. G., and F. B. M. de Waal. 1993. Conflict avoidance among rhesus monkeys: Coping with short-term crowding. Animal Behaviour 46:221–232.

Kagan, J., J. S. Reznick, J. S., and N. Snidman. 1988. Biological basis of childhood shyness. Science 240:167–171.

Kinzey, W. G. 1981. The titi monkeys, genus *Callicebus*. *In* A. F. Coimbra-Filho and R. A. Mittermeier, eds. Ecology and behavior in neotropical primates, vol. 1, pp. 241–276. Academia Brasileira de Ciencias, Rio de Janeiro.

Kinzey, W. G., and J. G. Robinson. 1983. Intergroup loud calls, range size, and spacing in *Callicebus torquatus*. American Journal of Physical Anthropology 60:539–544.

Kummer, H. 1971. Primate societies. Aldine, Chicago.

Levine, S., S. Weiner, and C. L. Coe. 1989. The psychoneuroendocrinology of stress: A psychobiological perspective. *In* F. R. Brush and S. Levine, eds. Psychoendocrinology, pp. 121–145. Academic Press, New York.

MacLean, P. D. 1964. Mirror display in the squirrel monkey. Science 146:950–952.

Martau, P. A., N. G. Caine, and D. K. Candland. 1985. Reliability of the emotions profile index, primate form, with *Papio hamadryas, Macaca fuscata*, and two *Saimiri* species. Primates 26:501–505.

Mason, W. A. 1965. The social development of monkeys and apes. *In* I. Devore, ed. Primate behavior: Field studies of monkeys and apes, pp. 514–543. Holt Rinehart and Winston, New York.

Mason, W. A. 1976. Primate social behavior: Pattern and process. *In* R. B. Masterton, C. B. G. Campbell, M. E. Bitterman, and N. Hotton, eds. Evolution of brain and behavior in vertebrates, pp. 425–455. Wiley, New York.

Mason, W. A. 1978. Ontogeny of social systems. *In* D. J. Chivers and J. Herbert, eds. Recent advances in primatology, pp. 5–14. Academic Press, New York.

McGrew, W. C. 1992. Chimpanzee material culture: Implications for human evolution. Cambridge University Press, Cambridge.

Melnick, D. J., and Pearl, M. C. 1987. Cercopithecines in multimale-groups: Genetic diversity and population structure. *In* B. B. Smuts, D. L. Cheney, R. Seyfarth, R. M. Wrangham, and T. T. Struhsaker, eds. Primate societies, pp. 121–134. University of Chicago, Chicago.

Mendoza, S. P., E. L. Lowe, and S. Levine. 1978. Social organization and social behavior in two subspecies of squirrel monkeys (*Saimiri sciureus*). Folia Primatologica 30: 126–144.

Mendoza, S. P., C. L. Coe, E. L. Lowe, and S. Levine. 1979. The physiological response to group formation in adult male squirrel monkeys. Psychoneuroendocrinology 1: 303–313.

Mendoza, S. P., M. B. Hennessy, and D. M. Lyons. 1989. Socially induced changes in adrenocortical activity: Immediate and long term responses to separation and pair formation in female squirrel monkeys. American Journal of Primatology 18:156.

Mendoza, S. P., D. M. Lyons, and W. S. Saltzman. 1991. Sociophysiology of squirrel monkeys. American Journal of Primatology 23:37–54.

Mendoza, S. P., and W. A. Mason. 1989. Primate relationships: Social dispositions and physiological responses. *In* P. K. Seth and S. Seth, eds. Perspectives in primate biology, vol. 2, pp. 129–143. Today & Tomorrow's Printers and Publishers, New Delhi.

Mendoza, S. P., and G. P. Moberg, G. P. 1985. Species differences in adrenocortical activity of New World primates: Response to dexamethasone suppression. American Journal of Primatology 8:215–224.

Mitchell, C. L. 1990. The ecological basis for female social dominance: A behavioral study of the squirrel monkey (*Saimiri sciureus*) in the wild (PhD dissertation). Princeton University, Princeton, NJ.

Mitchell, C. L. 1994. Migration alliances and coalitions among adult male South American squirrel monkeys (*Saimiri sciureus*). Behaviour 130:169–190.

Mitchell, C. L., S. Boinski, and C. P. van Schaik. 1991. Competitive regimes and female bonding in two species of squirrel monkeys (*Saimiri oerstedi* and *S. sciureus*). Behavioral Ecology and Sociobiology 28:55–60.

Mitchell, G. (ed). 1979. Behavioral sex differences in nonhuman primates. Van Nostrand Reinhold, New York.

Munck, A., P. Guyre, and N. Holbrook. 1984. Physiological functions of glucocorticoids during stress and their relation to pharmacological actions. Endocrine Review 5: 25–51.

Nagel, U., and Kummer, H. 1974. Variation in cercopithecoid aggressive behavior. *In* R. L. Holloway, ed. Primate aggression, territoriality, and xenophobia, pp. 159–184. Academic Press, New York.

Pereira, M. E., and J. Altmann. 1985. Development of social behavior in free-living nonhuman primates. *In* E. S. Watts, ed. Non-human primate models for human growth and development, pp. 217–309. Alan R. Liss, New York.

Plomin, R., M. C. Owen, and P. McGuffin. 1994. The genetic basis of complex human behaviors. Science 264:1733–1739.

Price, E. O. 1984. Behavioral aspects of animal domestication. Quarterly Review of Biology 59:1–32.

Rasmussen, K. L. R., and S. J. Suomi. 1989. Heart rate and endocrine responses to stress in adolescent male rhesus monkeys on Cayo Santiago. Puerto Rican Health Sciences Journal 8:65–71.

Reynolds, P. D., S. J. Pittler, and J. G. Scammell. 1997. Cloning and expression of the glucocorticoid receptor from the squirrel monkey (*Saimiri boliviensis boliviensis*), a glucocorticoid-resistent primate. Journal of Clinical Endocrinology and Metabolism 82:465–472.

Richard, A. F., S. J. Goldstein, and R. E. Dewar. 1989. Weed macaques: The evolutionary implications of macaque feeding ecology. International Journal of Primatology 10: 569–594.

Rowell, T. E. 1979. How would we know if social organization were not adaptive? *In* I. S. Bernstein and E. O. Smith, eds. Primate ecology and human origins, pp. 1–22. Garland STPM Press, New York.

Rylands, A. B., A. F. Coimbra-Filho, and R. A. Mittermeier. 1993. Systematics, geographic distribution, and some notes on the conservation staus of the Callitrichidae. *In* A. B. Rylands, ed. Marmosets and tamarins: Systematics, behavior, and ecology, pp. 11–77. Oxford University Press, Oxford.

Saltzman, W., S. P. Mendoza, and W. A. Mason. 1991. Sociophysiology of relationships in squirrel monkeys. I. Formation of female dyads. Physiology and Behavior 50: 271–280.

Sapolsky, R. M. 1990. Stress in the wild. Scientific American 262:11–123.

Sapolsky, R. M., and J. C. Ray. 1989. Styles of dominance and their endocrine correlates among wild olive baboons (*Papio anubis*). American Journal of Primatology 18:1–13.

Sapolsky, R. M., and L. J. Share. 1994. Rank-related differences in cardiovascular function among wild baboons: Role of sensitivity to glucocorticoids. American Journal of Primatology 32:261–275.

Scott, J. P., and J. L. Fuller. 1965. Dog behavior: The genetic basis. University of Chicago Press, Chicago.

Selye, H. 1937. Studies on adaptation. Endocrinology 21:169–188.

Selye, H. 1946. The general adaptation syndrome and the diseases of adaptation. Journal of Clinical Endocrinology 6:117–230.

Silk, J. 1987. Social behavior in evolutionary perspective. *In* B. B. Smuts, D. L. Cheney, R. Seyfarth, R. M. Wrangham, and T. T. Struhsaker, eds. Primate societies, pp. 318–329. University of Chicago Press, Chicago.

Silverman, A.-J., A. Hou-Yu, and D. D. Kelley. 1989. Modification of hypothalamic neurons by behavioral stress. *In* Y. Tache, J. E. Morley, and M. R. Brown, eds. Neuropetides and stress, pp. 23–38. Springer-Verlag, New York.

Smuts, B. B. 1985. Sex and friendship in baboons. Aldine, Hawthorne, NY.

Stanton, M. E., J. M. Patterson, and S. Levine. 1985. Social influences on conditioned cortisol secretion in the squirrel monkey. Psychoneuroendocrinology 10:125–134.

Stevenson-Hinde, J., and M. Zunz. 1978. Subjective assessment of individual rhesus monkeys. Primates 21:66–68.

Suomi, S. J. 1981. Genetic, maternal, and environmental influences on social development in rhesus monkeys. *In* A. B. Chiarelli and R. S. Corruccini, eds. Primate Behavior and Sociobiology, pp. 81–87. Springer-Verlag, New York.

Suomi, S. J. 1983. Social development in rhesus monkeys: Consideration of individual differences. *In* A. Oliverio and M. Zappella, eds. The behavior of human infants, pp. 71–92. Plenum Press, New York.

Suomi, S. J. 1987. Genetic and environmental contributions to individual differences in rhesus monkey biobehavioral development. *In* N. Krasnegor, E. Blass, M. Hofer, and W. Smotherman, eds. Perinatal development: A psychobiological perspective, pp. 397–420. Academic Press, New York.

Suomi, S. J. 1991. Uptight and laidback monkeys. *In* S. E. Brauth, W. S. Hall, and R. J. Dooling, eds. Plasticity of development, pp. 27–56. MIT Press, Cambridge, MA.

Suomi, S. J., J. M. Scanlan, K. L. R. Rasmussen, M. Davidson, S. Boinski, J. D. Higley, and B. Marriot. 1989. Pituitary-adrenal response to capture in Cayo Santiago-derived group M rhesus monkeys. Puerto Rican Health Science Journal 8:171–176.

Tattersall, I. 1993. Species and morphological differentiation in the Genus *Lemur*. *In* B. Kimbel and L. Martin, eds. Species, species concepts, and primate evolution, pp. 163–176. Plenum Press, New York.

Terborgh, J. 1983. Five new world primates. Princeton University Press, Princeton, NJ.

Thierry, B. 1985. Patterns of agonistic interactions in three species of macaque (*Macaca mulatta, M. fascicularis, M. tonkeana*). Aggressive Behaviour 11:223–233.

Thorington, R. W. 1985. The taxonomy and distribution of squirrel monkeys (*Saimiri*). *In* L. A. Rosenblum and C. L. Coe, eds. Handbook of squirrel monkey research, pp. 1–33. Plenum Press, New York.

van Schaik, C. P. 1989. The ecology of social relationships amongst female primates. *In* V. Standon and R. A. Foley, eds. Comparative socioecology: The behavioral ecology of humans and other animals, pp. 195–218. Blackwell, Oxford.

van Schaik, C. P., and M. A. van Noordwjik. 1983. Social stress and the sex ratio of neonates and infants among non-human primates. Netherlands Journal of Zoology 33: 249–265.

van Schaik, C. P., and M. A. van Noordwjik. 1985. Evolutionary effect of the absence of felids on the social organization of the macaques on the island of Simeulue (*Macaca fascicularis fusca*, Miller 1903). Folia Primatologia 44:138–147.

Visalberghi, E., and D. M. Fragaszy. 1990. Do monkeys ape? *In* S. T. Parker and K. R. Gibson, eds. "Language" and intelligence in monkeys and apes: Comparative and developmental perspectives, pp. 247–273. Cambridge University Press, Cambridge.

Vogt, J. L., C. L. Coe, and S. Levine. 1981. Behavioral and adrenocorticoid responsiveness to a live snake: Is flight necessarily stressful? Behavioral and Neural Biology 32: 391–405.

Wasser, S. K. 1983. Reproductive competition and cooperation among female yellow baboons. *In* S. K. Wasser, eds. Social behavior of female vertebrates, pp. 349–390. Academic Press, New York.

Whitten, P. L. 1983. Diet and dominance among female vervet monkeys (*Cercopithecus aethiops*). American Journal of Primatology 51:139–159.

Wiepkma, P. R., and W. G. P. Schouten. 1990. Mechanisms of coping in social situations. *In* R. Zayan and R. Dantzer, eds. Social stress in domestic animals, pp. 8–24. Kluwer Academic, Dordrecht.

Williams-Blangero, S., J. L. Vandeberg, J. Blangero, L. Konigsberg, and B. Dyke. 1990. Genetic differentiation between baboon subspecies: Relevance for biomedical research. American Journal of Primatology 20:67–81.

Wilson, D. S., A. B. Clark, K. Coleman, and T. Dearstyne. 1994. Shyness and boldness in humans and other animals. Trends in Ecology and Evolution 9:442–446.

Winter, P. 1969. Dialects in squirrel monkeys: vocalizations of the roman arch type. Folia Primatologica 10:216–229.

Wright, P. C. 1978. Home range, activity patterns, and agonistic encounters of a group of night monkeys (*Aotus trivirgatus*) in Peru. Folia Primatologia 29:43–55.

Yerkes, R. M., and Yerkes, A. W. 1929. The great apes: A study of anthropoid life. Yale University Press, New Haven, CT.

6

Ecology, Phenotype, and Character Evolution in Bluegill Sunfish

A Population Comparative Approach

TIMOTHY J. EHLINGER

Evolutionary ecology explores the intimate relationships between the mechanisms responsible for the production and maintenance of organismal form and the ecological function of the structures and behaviors that compose form (Arnold 1983). The analysis of diversity from this perspective is founded on the premise that variation in measured phenotypes can reflect the results of the process of natural selection (Williams 1966, 1992; Gould and Vrba 1982). However, because the fitness consequences of any particular phenotype are the result of complex interactions among an individual's genotype, morphology, behavior, and the environment within which it must function (Gould and Lewontin 1978, Endler 1986), a phenotype best suited for one set of environmental conditions may not perform best in another (e.g., Endler 1983, Rausher 1984, Ehlinger and Wilson 1988, Schluter 1993).

When making comparisons among populations, phenotypic variation due to underlying genetic differences that may reflect evolutionary responses to different environments must be distinguished from phenotypic variation that results from phenotypic plasticity and/or genotype–environment interactions (Stearns 1989). For example, regional environmental variation can result in different selective regimes *among populations* and produce "site-dependent" fitnesses for phenotypes (e.g., Reznick et al. 1990, Robinson and Wilson 1994; chapters in this volume). Likewise, varying social and trophic conditions on a local scale can result in "situation-dependent" performances and payoffs for different phenotypes *within populations* (Maynard-Smith 1974, Ehlinger 1990, Krebs and Kacelnik 1991). Both phenomena may influence patterns of geographic variation and must be considered when studying phenotypic differences among populations.

My aim in this chapter is to illustrate how population comparisons of bluegill sunfish (*Lepomis macrochirus*) can be used to study the evolution of behavioral and morphological variation. Critical features that shape bluegill trophic ecology (e.g., temperature, depth, substrate type, prey types, productivity, and predator abundance) vary among lakes in combination with forces that influence reproductive ecology (e.g., availability of spawning habitat and age or size structure of the population). Population comparisons provide unique opportunities for discerning the roles of sexual and trophic selection in bluegill phenotypic evolution. By comparing patterns of behavior, morphology, and performance among populations from lakes with diverse foraging and social environments, I attempt to discern the functional relationships between phenotypic variation and fitness. Next, by exploring differences between populations in the way that phenotypic variation correlates with fitness within populations, I illustrate the utility of geographic comparisons to gain insights into the integral role of behavior in evolutionary processes.

Ecophenotypic Hypotheses and Geographic Comparisons

The first step in the study of phenotypic evolution is to formulate mechanistic hypotheses for mapping morphology and behavior into performance and eventually fitness (Webb 1984, Wainwright 1991, Lauder et al. 1993). This requires the decomposition of phenotypic variation into components that are "functionally related" (sensu Motta and Kotrschal 1992) to variation in the performance of either trophic or reproductive tasks.

Comparisons among species (Harvey and Pagel 1991) can frequently provide insights into the function of intraspecific variation (McLennan 1991, Foster 1994). Although interspecific comparisons can provide valuable clues to the general patterns of adaptation among species (Lauder et al. 1993), there is seldom enough information on the genetic basis of form–function–performance attributes from a sufficient number of independent lineages to allow for the rejection of alternative hypotheses (Leroi et al. 1994).

Intraspecific population comparisons of freshwater fishes can avoid most of these problems because many species are abundant, they are subdivided into replicate populations, they are easily observed, and they are well studied ecologically (Foster and Bell 1994). In such systems, patterns of evolution across diverse abiotic and biotic environments can often be explored. Intraspecific ecophenotypic hypotheses may emphasize an ecological–morphological connection, as in the structure of fish gill rakers influencing the capture efficiency of different prey types (Robinson et al. 1993). Hypotheses may also stress an ecological–behavioral connection, as in predator searching speed influencing the detection of different prey types (Anderson 1981, Ehlinger 1989).

If the hypothesized phenotypic form–function–performance–fitness relationship has influenced the evolution of a character, then it should be possible to detect a correlative "fit" between variation in organismal traits, environmental conditions, performance, and success. The strength of such relationships can be examined between species (Motta 1988, Losos 1990) and between populations of the same species in different environments (Lavin and McPhail 1985, Smith 1987,

Schluter 1995). This ecophenotypic approach can be applied on numerous levels, including (1) laboratory experiments that manipulate biotic or abiotic conditions and examine the relationship between variation in phenotype and performance (e.g., Place and Powers 1979), (2) measurement of selection gradients that estimate the intensity of selection on particular traits (Moore 1990, Smith 1990, Warner and Schultz 1992), and (3) analysis of diversification among populations that are subject to different selective environments (Reznick et al. 1990, Schluter and McPhail 1992, Bell et al. 1993, Taylor and Bentzen 1993).

Comparisons of populations from differing selective regimes that are predicted to favor different individual trophic and reproductive phenotypes can also be useful in supporting the ecophenotypic–evolutionary hypothesis. By examining fish populations from lakes with different biotic and abiotic conditions, researchers can test whether variation in individual phenotypes correlates with the predicted magnitude and/or direction of selection for each lake. When applied to traits that potentially possess both trophic and reproductive function, it may be possible to find populations for which the hypotheses make different predictions for the same phenotypic trait. If so, close examination of these populations may allow us to assess the relative contributions of multiple selection pressures on the evolution of phenotype (Endler 1995).

Performance Attributes of Variation within Populations

The functional morphology (Wainwright and Lauder 1986, 1992), trophic ecology (Werner 1984, Ehlinger and Wilson 1988), and reproductive ecology (Gross 1982) of bluegill sunfish have been studied extensively. This wealth of background information permits formulation of specific hypotheses relating phenotypic variation to ecological function. Furthermore, because bluegill populations exist and thrive across a wide range of biotic and abiotic conditions, the evolution of phenotypic characters in response to varying environmental pressures can be examined.

With respect to foraging ecology, bluegill are facultative planktivores, possessing both generalist morphology (Webb 1984) and behavioral flexibility (Ehlinger 1989, 1990) that allow for movement among feeding niches depending on energetic return rates (Werner et al. 1981). Laboratory and field studies have shown that bluegill within single populations differ in morphological and behavioral traits that correlate directly with abilities to feed in either littoral-vegetated or limnetic-open water habitats of lakes (Ehlinger 1989, 1990). Stated simply, fish with deeper bodies and longer paired fins are better at the slow, precise maneuvering tactics that maximize feeding rates on the cryptic and hidden prey located in the structured, littoral niche. In contrast, individuals with more fusiform bodies and shorter paired fins are more efficient at cruising the distances required to forage on dispersed zooplankton in the structureless, limnetic niche. Individuals with these body forms feed most frequently in their respective habitats (Ehlinger and Wilson 1988).

With respect to reproductive ecology, nesting male bluegill aggressively establish territories and defend nesting sites within colonies in the shallow littoral habitat of lakes (Gross 1982). Shortly after colony establishment, schools of females move into the colony from offshore and commence spawning. Nesting males protect and care for the eggs for 7–10 days until larvae leave the nest. Males adopting this "parental" strategy typically mature at 7 years of age, although the age of maturity is highly plastic (Ehlinger et al. 1997). Males using an alternative "cuckolder" strategy mature during their second year (Gross 1982).

The outcome of male–male competition is influenced by functional relationships between morphology and behavior. Parental males in most *Lepomis* species flare their opercular ear tabs during aggressive encounters with other males while establishing and defending territories (Keenleyside 1971, Colgan and Gross 1977). Nesting male bluegill were collected from Wisconsin lakes and placed into 3-m diameter laboratory tanks (eight males per tank), where they established territories and began building nests. The number of aggressive encounters for each individual during replicate 5-min focal individual observation periods was tallied, and rank dominance was calculated as the difference between total number of aggressive acts initiated and the total number of aggressive acts received. Stable dominance hierarchies established readily in all tanks, and the relationships between dominance rank, age, size, and seven morphological features were analyzed. These experiments showed that ear-tab size was the only feature that correlated significantly with male rank in the dominance hierarchy (see example in fig. 6-1).

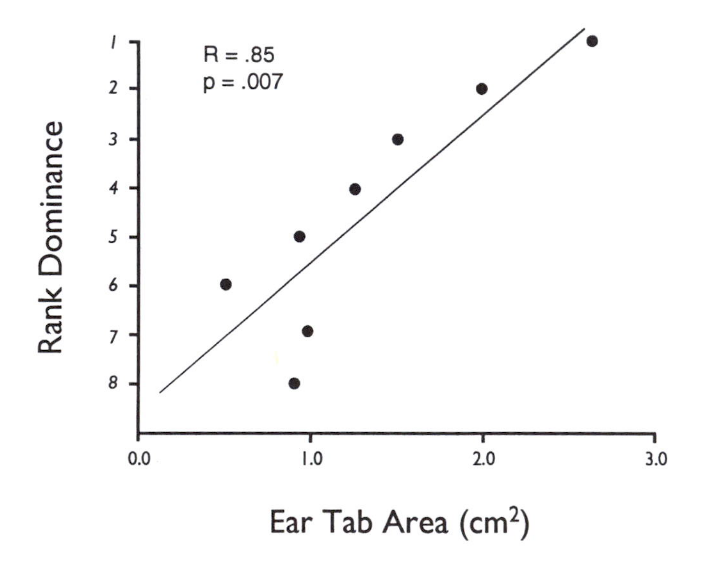

Figure 6-1 Example of the effect of parental male bluegill ear-tab area on position in laboratory dominance hierarchy. Data for males from Horn Lake, Wisconsin ($r = .85$, $p < .01$).

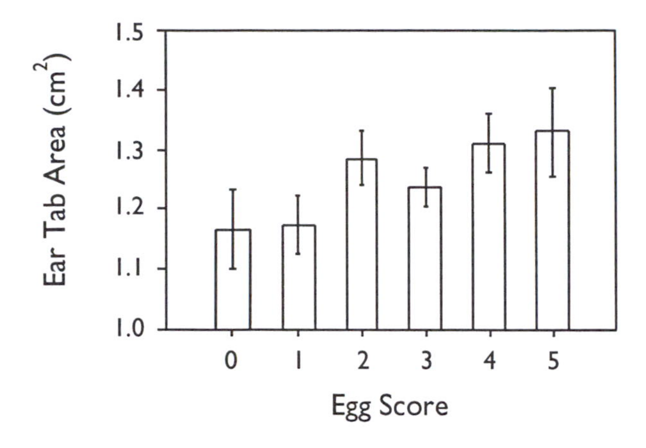

Figure 6-2 Relationship between ear-tab area and an index of number of eggs received for nesting parental male bluegill from Lake Opinicon, Ontario (from Cote et al., unpublished). Data are means ± 1 SE.

A field study of bluegill in Lake Opinicon, Ontario, conducted in collaboration with Isabelle Cote and Mart Gross (Cote et al. unpublished), examined whether selection on male morphology was operating primarily through female choice (intersexual selection) or through male competition for nest sites (intrasexual selection). Male reproductive performance was divided into two components: an intrasexual, male competition phase—the outcome being the acquisition of a central versus peripheral nest in the colony, and an intersexual, female choice phase—the outcome being the number of eggs spawned in a male's nest. The morphology and success of 190 bluegill males were measured, and selection gradient analysis (Lande and Arnold 1983) revealed that only ear-tab size was under current intrasexual selection. The results demonstrated that males with larger ear tabs were more successful in obtaining central nests, and, because females prefer to spawn in central nests due to lower egg predation and higher egg survival (Gross and MacMillan 1981, Cote and Gross 1993), males with larger ear tabs received more eggs (fig. 6-2).

On every level, it is critically important to distinguish between variation in behavioral performance correlated with morphological variation in "size" as opposed to "shape" of character traits (Atchley 1983, Bookstein et al. 1985). This is especially a concern when working with taxa that exhibit indeterminant growth, where environmental differences such as food availability can create size differences between populations independent of any adaptive explanation. Methods for decomposing phenotypes into components of size and shape are numerous (Rohlf and Bookstein 1990). Allometric effects must be measured and/or controlled for, both when testing for phenotypic correlations with individual performance as well as when comparing phenotypic differences among populations (Ehlinger 1991).

Population Comparisons: Teasing Apart Trophic and Reproductive Functions

Fitness must ultimately be gauged both by how well an organism grows and survives and by how well it reproduces. This creates a dilemma for researchers interested in the linkage between proximate measures of ecological success and evolutionary fitness. For example, a phenotype composed of traits optimal for nest site or mate acquisition, fertilization success, or parental care will be of little consequence unless individuals with that phenotype grow and survive to reproductive age. On the other hand, a phenotype with optimal design for foraging success, habitat choice, or predator avoidance has no chance to increase in frequency within a population unless members who possess it are able to pass on any heritable components of phenotype to future generations. These strong and sometimes divergent pressures may create trade-offs between the optimal phenotypes for trophic versus reproductive function; teasing apart their interaction is important for our understanding of how evolution proceeds under the direction of natural selection.

The studies described above demonstrate that morphological and behavioral differences among bluegill within populations correlate directly with variation in foraging and nesting performance. But to what extent does phenotypic variation reflect the concurrent demands of selection on sexual function versus trophic function? A study of bluegill reared in common pond environments showed that parental males develop deeper bodies and longer fins than females (Ehlinger 1991)—a pattern of morphological divergence that parallels the differences between littoral and limnetic trophic morphs (Ehlinger and Wilson 1988). Might morphological divergence between the sexes result from disruptive selection for habitat-specific foraging ability? This cannot be the case for all populations because trophic polymorphism has been observed in both sexes (Ehlinger and Wilson 1988). However, the correlated patterns of divergence in foraging and sexual morphology suggest a close functional linkage between traits important to reproductive and trophic success. Could sexual divergence result from selection acting on sexual function, with trophic divergence a coincidental, albeit adaptive, by-product?

I am currently investigating the patterns of morphological and behavioral variation among bluegill from lakes that provide different trophic and reproductive environments, chosen on the basis of the relative availability of littoral and limnetic foraging niches (Ehlinger 1997). Based on previous work correlating morphology with habitat-specific foraging ability, I predicted that bluegill from shallow, vegetated lakes would have body shapes associated with greater maneuverability (e.g., deeper bodies and longer paired fins) relative to populations from deep lakes with narrow littoral zones. If no aspects of reproductive ecology were found to differ significantly between lakes, I predicted that morphological features associated with sexual divergence would be conserved across lake types.

A comparison of bluegill populations from four lakes—two shallow, littoral lakes and two deep, limnetic lakes (table 6-1)—reveals that sexes respond differently to changes in lake trophic structure. To compare morphological differences

Table 6-1 Depth, bathymetric characteristics, and relative habitat availability in four Wisconsin study lakes.

Lake	Limnetic Habitat (%)	Littoral Habitat (%)	Surface Area (ha)	Mean Depth (m)	Colonies per Transect	Nests per Colony
Five	35.4	35.3	45.4	2.4	13.5 ± 4.7	32.0 ± 13.7
Ottawa	17.0	49.6	12.5	2.1	8.2 ± 2.1	28.4 ± 17.2
Green	61.2	18.2	31.6	5.2	3.8 ± 1.4	47.2 ± 18.7
Horn	66.0	16.4	5.3	4.9	3.5 ± 1.2	50.4 ± 17.4

Percent littoral and limnetic habitat were calculated as percent surface area with depth ≤2m and >4m, respectively. Number of active spawning colonies and number of nests per colony are means ± 1 SE occurring along 300 m transects in each study lake during the study period.

among populations, size adjustment was conducted separately for each sex (pooled among lakes) using a centroid size regression method and analysis of covariance (Ehlinger 1991). Female bluegill from shallow lakes with a high proportion of littoral habitat have deeper bodies and longer fins compared to females from deeper lakes with large limnetic habitats (fig. 6-3), supporting the prediction that

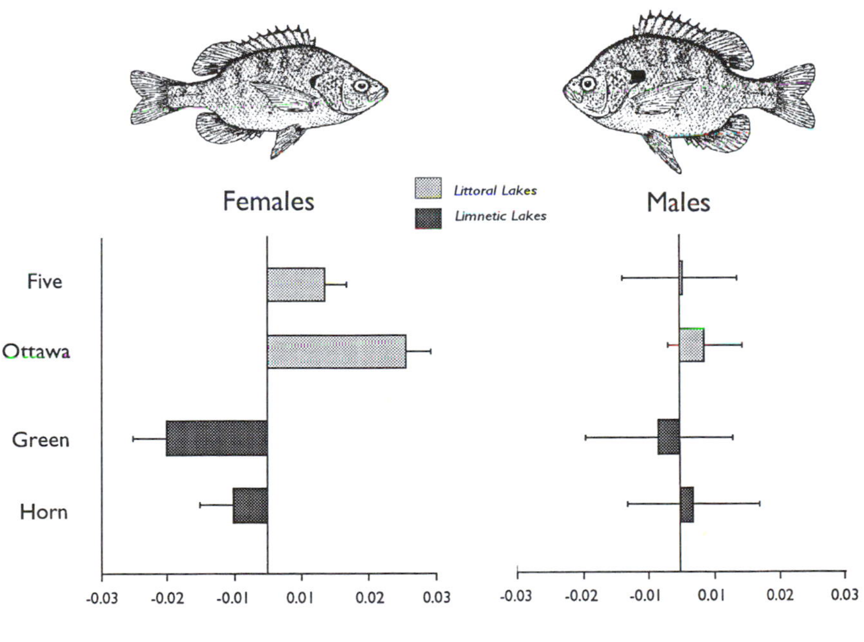

Figure 6-3 Differences in mean size-adjusted body depth among bluegill populations from littoral lakes (light-shaded bars) and limnetic lakes (dark-shaded bars) in Wisconsin for Spawning females and nesting parental males.

maneuvering and cruising phenotypes are subject to selection brought about by differences in lake trophic structure. Whereas female morphology appears to adapt to the trophic structure of lakes, males maintain deep bodied, long-finned maneuvering phenotypes in all lakes—perhaps indicating selective constraints due to the requirements of nesting and parental care.

In contrast to trophic characters, male secondary sexual characters vary greatly among lakes. Ear tabs of males from limnetic lakes are significantly larger than ear tabs of males from littoral lakes (fig. 6-4). This pattern in male morphology suggests that males are influenced by some difference in their reproductive environment that produces stronger intrasexual competition in limnetic lakes. Because bluegill spawn at similar depths among lakes (mean colony depth = 0.82 m, range = 0.3–1.9 m), lake morphometry may influence the intensity of male competition for nests if littoral lakes have more suitable space for nesting compared to limnetic lakes. To estimate nest-site availability, the location, depth, and number of nesting males per colony were recorded weekly along two 300-m transects along the perimeter of each lake during the 6-week spawning period (June–July 1993). This survey indicated that bluegill spawned at similar depths in all lakes but that in shallow littoral lakes they nested in more numerous, smaller colonies than in deeper, limnetic lakes (table 6-1). Because of this, the intensity of selection on secondary sexual characteristics related to the intensity of male competition for nest sites (e.g., ear-tab size) may covary with lake trophic structure.

Although the addition of data from more lakes is required to confirm these

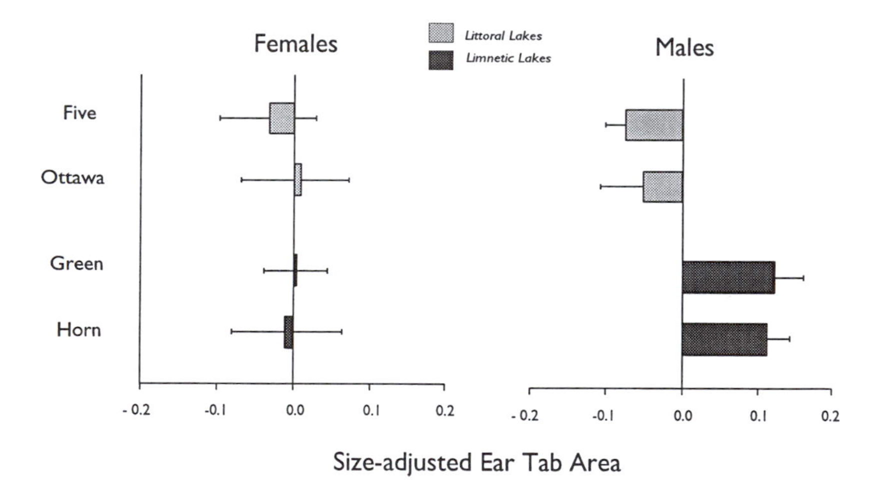

Figure 6-4 Differences in size-adjusted ear-tab area among bluegill populations from littoral lakes (light-shaded bars) and limnetic lakes (dark-shaded bars) in Wisconsin for spawning females and nesting parental males. Sizes were adjusted separately for each sex following procedures in Ehlinger (1991). Data are mean residuals (±1 SE) for 40 fish of each sex from each lake ($n = 160$ per sex) following regression of log-transformed body depth against log-transformed centroid body size.

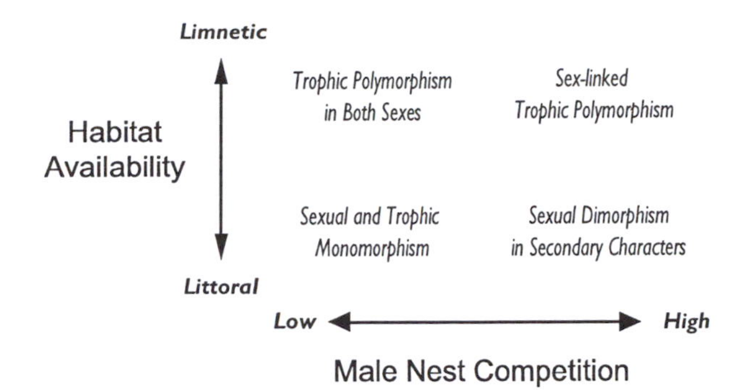

Figure 6-5 Predicted ecophenotypic responses to different levels of trophic and sexual selection in bluegill populations in littoral and limnetic lakes.

patterns, the results from these four populations suggest a tractable and intriguing scenario. The population comparisons suggest that changes in female morphology reflect differences in the availability of limnetic versus littoral feeding niches. On the other hand, male morphological variation among populations is consistent with adaptation to differences in the intensity of intrasexual selection between lakes, most likely related to the availability of and competition for breeding sites. These hypotheses can be combined in a model in which changes in bluegill sexual and trophic morphology are a function of changes in lake type (fig. 6-5). One axis represents differences in foraging niche availability, ranging from predominantly littoral lakes to lakes with well-defined and productive limnetic habitats. The other axis represents changes in the intensity of male sexual selection; ranging from lakes where competition for nests is low to lakes where only the most aggressive, competitive males acquire nests.

This model raises a number of testable predictions. For example, will males maintain a "littoral" feeding morphology except in lakes that have both a large limnetic niche and low levels of sexual selection (fig. 6-5)? Under relaxed sexual selection, can males that possess a "limnetic" feeding morphology still be successful in acquiring nesting sites? In this situation will trophic selection favor the development of a sex-independent trophic polymorphism? Will the sexes diverge trophically in limnetic lakes that have high male sexual competition (fig. 6-5), with parental males developing "littoral form" in the open-water lakes? Selection for sexual dimorphism in this case could be reinforced if maneuvering ability contributes to successful male parental care and aggressive territory defense. Conversely, will the sexes be monomorphic in littoral, low male competition lakes (fig. 6-5)?

At first look, one might expect variation in habitat availability among lakes to be strongly correlated with the abundance of suitable nesting sites, meaning that populations would only exhibit either sexual and trophic monomorphism or a sex-linked trophic polymorphism (fig. 6-5). However, differences in bluegill density caused by predation, disease, competition with other species, angling pressure,

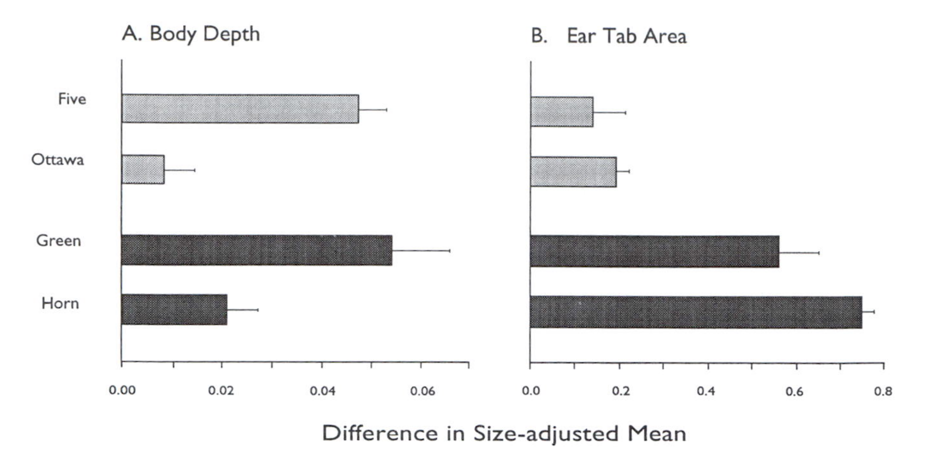

Figure 6-6 Magnitude of sexual dimorphism for bluegill populations in littoral lakes (light-shaded bars) and limnetic lakes (dark-shaded bars). Difference in adjusted means were determined by analysis of covariance of log-transformed data using log-transformed centroid size as a covariate separately for each lake.

etc., could shift the relative number of parental males per colony and thereby vary the intensity of male competition independently of lake trophic characteristics. These differences could generate predictions about geographic differences among lakes that would produce the other two outcomes in figure 6-5.

The observed pattern of sexual dimorphism within populations is strikingly different among my study populations. By adjusting for body size within each population separately (with both sexes pooled), the degree of sexual dimorphism within a population can be quantified and compared among populations (Ehlinger 1991). With respect to trophic morphology, males are deeper bodied than females in all lakes, but the magnitude of divergence varies (fig. 6-6A). Females are significantly more fusiform relative to males in the larger lakes (Five and Green) compared to the smaller lakes (Ottawa and Horn; fig. 6-6A; lake sizes in table 6-1). If this greater degree of trophic dimorphism is related to selection for limnetic feeding in females (fig. 6-5), then perhaps the limnetic niche must attain a minimum size (as opposed to a relative size) before a fusiform morphology becomes advantageous (Lavin and McPhail 1985).

Male ear tabs are larger than female ear tabs in all populations, but the magnitude of dimorphism is nearly fourfold greater in the limnetic lakes compared to the littoral lakes (fig. 6-6B). If the degree of dimorphism reflects the intensity of intrasexual selection, the most parsimonious explanation for the difference is that populations in limnetic lakes must compete more intensely to acquire access to females. Adult male densities do not differ significantly among these lakes, so it is likely that increased competition could be produced by a reduced availability of suitable nesting area in limnetic lakes, resulting in the crowding of males into fewer yet larger colonies (table 6-1).

Beyond Correlations: Fitness Consequences of Geographic Variation

Correlational comparisons of phenotypic divergence within and among lakes, like those described above, are a first step in understanding the evolution of geographic variation. However, hypotheses generated by these comparisons need to be supported by additional studies to examine how phenotypic variation contributes to performance, map variation in performance with its fitness consequences, and assess the heritable components of phenotype (Endler 1986).

Laboratory trials are useful for measuring the relationship between variation in ear-tab area and the outcome of social interactions among males from lakes with varying intensities of male sexual selection. These data can answer questions such as, will males from "more competitive" lakes dominate males from "less competitive" lakes? Predictions about correlations within lakes are less straightforward. For example, one could predict that males from lakes with high nest competition will exhibit a stronger relationship between morphology and dominance compared to fish from lakes where male competition is lower.

An important result from these experiments is the finding that phenotypic characters can contribute to performance differently in different populations. For example, there is a wide range among populations in the magnitude of the correlation between dominance rank and ear-tab area for males (table 6-2), and the strength of these correlations parallels the extent of sexual dimorphism in ear-tab size (fig. 6-6B). However, a different pattern emerges when dominance is broken down into the behavioral components of aggressive encounters initiated versus received. All four lakes show a significant positive correlation between male ear-tab area and initiated aggression. However, males with larger ear tabs are attacked significantly less frequently than males with smaller ear tabs in populations from the two most sexually selected lakes, as evidenced by a significant negative correlation between ear-tab size and aggression received (table 6-2). Thus, dominance status appears to be achieved by different means in different populations.

Table 6-2 Correlation coefficients and significance levels for the relationship between male ear-tab size and three measures of behavior in laboratory experiments.

Lake	Rank Dominance	Aggression Initiated	Aggression Received
Five	0.06^{ns}	0.34*	0.21^{ns}
Ottawa	0.32*	0.53*	0.28^{ns}
Green	0.63*	0.72*	–0.66*
Horn	0.72*	0.83*	–0.73*

Initiated aggression and received aggression were determined by the total number of chases, bites, and opercular flares directed toward or received from other males during 5-min observation periods. Dominance ranks were determined by the difference between initiated and received aggression. $*p < .05$; ns, not significant. Numbers are mean correlations for eight replicate experiments for each lake with eight fish per experiment (See fig. 6-1).

Table 6-3 Correlation coefficients for the relationship between measures of aggression and spawning success relative to male ear tab size.

Lake	Number of Adjacent Nests	Aggression Initiated	Aggression Received	n
Five	0.52*	0.48*	0.02^{ns}	18
Horn	–0.80*	0.23^{ns}	–0.77*	14

Aggression measures are weighted relative to the number of interacting males.
*$p < .05$; ns, not significant.

The next step is to assess whether the behavioral differences observed in the laboratory translate into tangible payoffs in the field. By following individually marked males across the spawning season, I can estimate three different measures of individual reproductive success: nest position, female visits, and eggs received. The number of adjacent nests within a 1-m radius measures whether a male is located centrally (i.e., has a high number of neighbors) versus peripherally (i.e., has a low number of neighbors). The average number of females visiting and spawning with a male during replicate 5-min sample periods indicates a male's attractiveness to females. The number of eggs deposited in nests is measured when spawning is completed by placing a 30-cm square with 5-cm grid over the nest and counting the number of cells containing eggs (resulting in an egg score from 0 to 36).

The value of using multiple populations for this type of field study is illustrated by the results collected to date from two populations. Nesting males in Lake Five (a shallow, littoral lake) show a significant positive correlation between ear-tab size and both nest position and aggression initiated but not aggression received (table 6-3). In contrast, nesting males in Horn Lake (a deeper, limnetic lake) show a strong negative correlation between ear-tab size and both nest position and aggression received, but not aggression initiated. In other words, nesting males with larger ear tabs in Lake Five obtain central nests by fighting more, whereas nesting males with larger ear tabs in Horn Lake are avoided by other males and end up with nests located peripheral to the main body of the colony. The payoff to getting a central nest in terms of female visits and the number of eggs received also shows a different pattern between lakes. Females in Lake Five show a preference for laying eggs in central nests, and males in the center initiate more aggression while acquiring and maintaining their nesting sites (table 6-4). Females in Horn Lake exhibit the reverse preference, choosing to spawn in peripheral nests. On average, males in Horn Lake partake in less total aggressive activity and receive more female visits and eggs than Lake Five males (table 6-5).

Conclusions: What Can Population Comparisons Tell Us?

The population comparisons described here show clearly that the functional linkages among morphology, behavior, and performance may change between popu-

Table 6-4 Correlation coefficients for the relationship between measures of aggression and spawning success relative to number of adjacent nests.

Lake	Number of Female Visits	Egg Score	Aggression Initiated	Aggression Received	n
Five	0.59*	0.21^{ns}	0.38*	0.07^{ns}	41
Horn	–0.48*	–0.36*	0.17^{ns}	0.21^{ns}	32

Aggression measures are weighted relative to the number of interacting males. *$p < .05$; ns, not significant.

lations that are subjected to different trophic and reproductive environments. We cannot assume that the relationship between phenotype and fitness will be the same among populations. However, the differences that we detect among populations can provide fertile ground for further research.

The comparisons of bluegill populations presented here raise a number of fascinating questions about the patterns of female choice and the evolution of male traits (Houde 1993). For example, in Horn Lake it appears that a trait used for fighting during intrasexual competition may have taken on the additional function of a signal or "badge," serving to reduce the incidence of aggression (Rohwer 1982, Kodric-Brown and Brown 1984). The evolution of such a signal in male competition could easily provide females with a good trait to select in choosing a mate (Kodric-Brown 1977, 1983). Why female bluegill might choose to mate with males with large ear tabs at the periphery of colonies in one population but not in another is unclear, as central nests are generally believed to provide safer conditions for developing eggs than do peripheral nests (Gross and MacMillan 1981, Cote and Gross 1993). If large ear tabs contribute to a reduction in the number of conspecific attacks directed against a male, males may be able to direct more time and energy into parental care activities (e.g., fanning and guarding from egg predators). However, nesting males frequently abandon females momentarily during spawning to chase neighboring males that encroach upon their territories. The intimidating effect of large ear tabs in combination with the peripheral location of nests may alternatively result in fewer interruptions, thus allowing males to retain individual females in their nests for longer periods of time once they arrive.

Clearly the power of geographic comparisons lies not only in the ability to provide replication to test predictions, but also in the ability to detect unexpected

Table 6-5 Population means and coefficients of variation (CV) for measures associated with nesting success of male bluegill from two lakes.

Lake	Aggression Initiated	CV	Aggression Received	CV	Female Visits	CV	Egg Score	CV
Five	8.1	62	1.1	191	0.3	203	16.3	45
Horn	3.2	93	0.20	250	0.8	164	28.2	29

patterns of behavior that may generate new hypotheses. As additional bluegill populations are included in this study, the ways in which sexual and trophic selection interact with different levels of male competition and niche composition should become clearer. Variation in bluegill population density may influence these interactions because of its indirect effects on food availability, growth rates, age of maturation, and breeding-site competition. It will be interesting to see whether the relationships between phenotypic variation, performance, and fitness are maintained under these diverse selection regimes.

It is tempting to speculate logically from processes that produce and maintain intraspecific geographic variation to infer potential mechanisms important in the evolution of reproductive isolation and speciation processes. For example, the finding that bluegill populations with larger sexually selected ear tabs exhibit lower total levels of aggression parallels patterns of interspecific differences in the dimorphism of status signals used in "honest advertisement" (Rohwer 1982, Plavcan and van Schaik 1992). However, if strong selection has been acting in a population for sufficient time, the level of variation might not be sufficient to detect a correlation between behavior and phenotypic variation (Falconer 1981, Trexler 1990). An important step in our research on the bluegill will be to examine the extent of year-to-year variation in the form and intensity of selection. The studies we have conducted indicate that sufficient variation exists in some lakes for selection to occur, but it is critical to consider the possibility that some populations may have been founded with little genetic variation and/or that variation has been reduced by selection.

In summary, geographic comparisons of phenotypic form and ecological function among populations can provide substantial questions and answers for evolutionary biologists. It is critical that studies comparing patterns of phenotypic character evolution *among* populations be interpreted in light of differences in the way traits correlate with performance and fitness *within* populations. The approach I have outlined here shows how population comparisons can be used to study how selection regimes direct phenotypic evolution, even when traits are subjected to multiple selection pressures. However, the bluegill system also illustrates the potential problem with assuming constant selection regimes among populations. Independently evolving populations exposed to different trophic and social environments do not necessarily share the same functional mapping of phenotype to performance to fitness. By combining comparative population studies with behavioral ecological studies of form and function, we can significantly increase our ability to understand fitness and detect the action of natural selection in the field.

Acknowledgments Many of the ideas presented here are the result of discussions with Isabelle Cote, for whose tutelage on the finer points of bluegill watching I am most grateful. Special thanks to Mart Gross and David Philipp for providing the opportunity to work with their Lake Opinicon research group. Laura Stremick, Dan Treloar, Theresa Hill, Liz House, Kasi Jackson, and Jenny Cutraro assisted in many aspects of laboratory and field work. An earlier version of this chapter was improved by the thoughtful comments of William Rowland, Jeff Walker, the editors of this book, and an anonymous reviewer. This research was funded by National Science Foundation grant IBN-9208295 to the author and support from the Wisconsin Department of Natural Resources Bureau of Fish Research and Sport Fish Restoration Fund under projects F83-R and F95-P.

References

Anderson, M. 1981. On optimal predator search. Theoretical Population Biology 19:58–86.

Arnold, S. J. 1983. Morphology, performance and fitness. American Zoologist 23:347–361.

Atchley, W. R. 1983. Some genetic aspects of morphometric variation. *In* J. Felsenstein, ed. Numerical taxonomy, pp. 346–363. Springer-Verlag, Berlin.

Bell, M. A., G. Orti, J. A. Walker, and P. Koenings. 1993. Evolution of pelvic reduction in threespine stickleback fish: A test of competing hypotheses. Evolution 47:906–914.

Bookstein, F., B. Chernoff, R. Elder, J. Humphries, G. Smith, and R. Strauss. 1985. Morphometrics in evolutionary biology, vol. 15. Special Publication. The Academy of Natural Sciences of Philadelphia, PA.

Colgan, P. W., and M. R. Gross. 1977. Dynamics of aggression in male pumpkinseed sunfish (*Lepomis gibbosus*) over the reproductive phase. Zeitschrift fur Tierpsychologie 43:139–151.

Cote, I. M., and M. R. Gross. 1993. Reduced disease in offspring: A benefit of coloniality in sunfish. Behavioral Ecology and Sociobiology 33:269–274.

Ehlinger, T. J. 1989. Learning and individual variation in bluegill foraging: Habitat-specific techniques. Animal Behaviour 38:643–658.

Ehlinger, T. J. 1990. Habitat choice and phenotype-limited feeding efficiency in bluegill: Individual differences and trophic polymorphism. Ecology 7:886–896.

Ehlinger, T. J. 1991. Allometry and analysis of morphometric variation in the bluegill, *Lepomis macrochirus*. Copeia 1991 (2):347–357.

Ehlinger, T. J. 1997. Male reproductive competition and sex-specific growth pattern in bluegill. North American Journal of Fisheries Management 17:508–515.

Ehlinger, T. J., M. R. Gross, and D. P. Philipp. 1997. Morphological and growth rate differences between bluegill males of alternative reproductive life histories. North American Journal of Fisheries Management 17:533–542.

Ehlinger, T. J., and D. S. Wilson. 1988. Complex foraging polymorphism in bluegill sunfish. Proceedings of the National Academy of Sciences USA 85:1878–1882.

Endler, J. A. 1983. Natural and sexual selection on color patterns in poeciliid fishes. Environmental Biology of Fishes 9: 173–190.

Endler, J. A. 1986. Natural selection in the wild. Princeton University Press, Princeton, NJ.

Endler, J. A. 1995. Multiple-trait co-evolution and environmental gradients in guppies. Trends in Ecology and Evolution 10:22–29.

Falconer, D. S. 1981. Introduction to quantitative genetics. Longman Scientific, Essex.

Foster, S. A. 1994. Inference of evolutionary pattern: Diversionary displays of three-spined sticklebacks. Behavioral Ecology 5:114–121.

Foster, S. A., and M. A. Bell. 1994. Evolutionary inference: The value of viewing evolution through stickleback-tinted lenses. *In* M. A. Bell and S. A. Foster, eds. The evolutionary biology of threespine stickleback, pp. 472–486. Oxford University Press, Oxford.

Gould, S. J., and R. C. Lewontin. 1978. The spandrels of San Marco and the Panglossian paradigm: A critique of the adaptationist programme. Proceedings of the Royal Society of London 205:581–598.

Gould, S. J., and E. S. Vrba. 1982. Exaptation—a missing term in the science of form. Paleobiology 1:4–15.

Gross, M. R. 1982. Sneakers, satellites and parentals—polymorphic mating strategies in North American sunfishes. Zeitschrift fur Tierpsychologie 60:1–26.

Gross, M. R., and A. M. MacMillan. 1981. Predation and the evolution of colonial nesting in bluegill sunfish (*Lepomis macrochirus*). Behavioral Ecology and Sociobiology 8: 163–174.

Harvey, P. H., and M. D. Pagel. 1991. The comparative method in evolutionary biology. Oxford University Press, Oxford.

Houde, A. E. 1993. Evolution by sexual selection: What can population comparisons tell us? American Naturalist 141:796–803.

Keenleyside, M. H. 1971. Aggressive behaviour of male longear sunfish (*Lepomis megalotis*). Zeitschrift fur Tierpsychologie 28:227–240.

Kodric-Brown, A. 1977. Reproductive success and the evolution of breeding territories in pupfish (Cyprinodon). Evolution 31:750–766.

Kodric-Brown, A. 1983. Determinants of male reproductive success in pupfish (*Cyprinodon pecosensis*). Animal Behaviour 31:128–137.

Kodric-Brown, A., and J. H. Brown. 1984. Truth in advertising: The kinds of traits favoured by sexual selection. American Naturalist 124:309–323.

Krebs, J. R., and A. Kacelnik. 1991. Decision-making. *In* J. R. Krebs and N. B. Davies, eds. Behavioural ecology: An evolutionary approach, pp. 105–136. Blackwell, Oxford.

Lande, R., and S. J. Arnold. 1983. The measurement of selection on correlated characters. Evolution 37:1210–1226.

Lauder, G. V., A. M. Leroi, and M. R. Rose. 1993. Adaptations and history. Trends in Ecology and Evolution 8:294–297.

Lavin, P. A., and J. D. McPhail. 1985. The evolution of freshwater diversity in the threespine stickleback (*Gasterosteus aculeatus*): Site-specific differentiation of trophic morphology. Canadian Journal of Zoology 63:2632–2638.

Leroi, A. M., M. R. Rose, and G. V. Lauder. 1994. What does the comparative method reveal about adaptation? American Naturalist 143:381–402.

Losos, J. B. 1990. The evolution of form and function: Morphology and locomotor performance in West Indian Anolis lizards. Evolution 44:1189–1203.

Maynard-Smith, J. 1974. The theory of games and the evolution of animal conflict. Journal of Theoretical Biology 47:209–221.

McLennan, D. A. 1991. Integrating phylogeny and experimental ethology: From pattern to process. Evolution 45:1773–1789.

Moore, A. J. 1990. The evolution of sexual dimorphism by sexual selection: The separate effects of intrasexual selection and intersexual selection. Evolution 44:315–331.

Motta, P. J. 1988. Functional morphology of the feeding apparatus of ten species of Pacific butterflyfishes (Perciformes: Chaetodontidae): An ecomorphological approach. Environmental Biology of Fishes 22:39–67.

Motta, P. J., and K. M. Kotrschal. 1992. Correlative, experimental, and comparative evolutionary approaches in ecomorphology. Netherlands Journal of Zoology 42:400–415.

Place, A. R., and D. A. Powers. 1979. Genetic variation and relative catalytic efficiencies: LDH-B allozymes of *Fundulus heteroclitus*. Proceedings of the National Acadademy of Science USA 76:2354.

Plavcan, J. M., and C. P. van Schaik. 1992. Intrasexual competition and canine dimorphism in anthropoid primates. American Journal of Physical Anthropology 87:461–477.

Rausher, M. D. 1984. Tradeoffs in performance on different hosts: Evidence from within- and between-site variation in the beetle *Deloyala guttata*. Evolution 42:582–595.

Reznick, D. N., H. Bryga, and J. A. Endler. 1990. Experimentally induced life-history evolution in a natural population. Nature 346:357–359.

Robinson, B. W., and D. S. Wilson. 1994. Character release and displacement in fishes: A neglected literature. American Naturalist 144:596–627.

Robinson, B. W., D. S. Wilson, A. S. Margosian, and P. T. Lotito. 1993. Ecological and morphological differentiation of pumpkinseed sunfish in lakes without bluegill sunfish. Evolutionary Ecology 7:451–464.

Rohlf, F. J., and F. L. Bookstein. 1990. Proceedings of the Michigan Morphometrics Workshop, vol. 2. Special Publication. University of Michigan Museum of Zoology, Ann Arbor.

Rohwer, S. 1982. The evolution of reliable and unreliable badges of fighting ability. American Zoologist 22:531–546.

Schluter, D. 1993. Adaptive radiation in sticklebacks: Size, shape and habitat use efficiency. Ecology 74:699–709.

Schluter, D. 1995. Adaptive radiation in sticklebacks: Trade-offs in feeding performance and growth. Ecology 76:82–90.

Schluter, D., and J. D. McPhail. 1992. Ecological character displacement and speciation in sticklebacks. American Naturalist 140:85–108.

Smith, T. B. 1987. Bill size polymorphism and intraspecific niche utilization in an African finch. Nature 329:717–719.

Smith, T. B. 1990. Natural selection on bill characters in the two bill morphs of the African finch *Pyrenestes ostrinus*. Evolution 44:832–842.

Stearns, S. C. 1989. The evolutionary significance of phenotypic plasticity. BioScience 39: 436–445.

Taylor, E. B., and P. Bentzen. 1993. Evidence for multiple origins and sympatric divergence of trophic ecotypes of smelt (*Osmerus*) in northeastern North America. Evolution 47:813–832.

Trexler, J. C. 1990. Genetic architecture of behavior in fishes and the response to selection. Annales Zoologici Fennici 27:149–164.

Wainwright, P. C. 1991. Ecomorphology: Experimental functional anatomy for ecological problems. American Zoologist 31:680–693.

Wainwright, P. C., and G. V. Lauder. 1986. Feeding biology of sunfishes: Patterns of variation in the feeding mechanism. Zoological Journal of the Linnean Society 88: 217–228.

Wainwright, P. C., and G. V. Lauder. 1992. The evolution of feeding biology in sunfishes (Centrarchidae). *In* R. L. Mayden, ed. Systematics, historical ecology, and North American freshwater fishes, pp. 472–491. Stanford University Press, Stanford, CA.

Warner, R. R., and E. T. Schultz. 1992. Sexual selection and male characteristics in the bluehead wrasse, *Thalassoma bifasciatum*: Mating site acquisition, mating site defense and female choice. Evolution 46:1421–1442.

Webb, P. A. 1984. Body form, locomotion, and foraging in aquatic vertebrates. American Zoologist 24:107–120.

Werner, E. E. 1984. The mechanisms of species interactions and community organization in fish. *In* D. R. Strong Jr., D. Simberloff, L. G. Abele, and A. B. Thistle, eds. Ecological communities: Conceptual issues and evidence, pp. 360–382. Princeton University Press, Princeton, NJ.

Werner, E. E., G. G. Mittelbach, and D. J. Hall. 1981. The role of foraging profitability and experience in habitat use by the bluegill. Ecology 62:116–125.

Williams, G. C. 1966. Adaptation and natural selection: A critique of some current evolutionary thought. Princeton University Press, Princeton, NJ.

Williams, G. C. 1992. Natural selection: Domains, levels, and challenges. Oxford University Press, New York.

7

The Causes and Consequences of Geographic Variation in Antipredator Behavior

Perspectives from Fish Populations

ANNE E. MAGURRAN

Predators are extremely effective agents of selection. After all, if an individual member of a prey species does not survive long enough to reproduce, it will have lost its chance (kin selection considerations apart) to bequeath its genes to future generations. It is not surprising, therefore, that many cases of population difference have been attributed to geographic variation in risk. These population differences can take a variety of forms and may, for example, involve modifications to morphology or to life-history traits. The correlation between armor and predation in the three-spined stickleback, *Gasterosteus aculeatus*, is one case that has been well documented (see Reimchen 1994 for a review and discussion), while another is the association between reproductive allotment and risk (Reznick and Endler 1982) and male color pattern and risk (Endler 1980) in the Trinidadian guppy, *Poecilia reticulata*. However, such adaptations can be futile if they are not accompanied by effective antipredator behavior. For instance, a cryptic color pattern confers no advantage if its holder chooses the "wrong" background or behaves in a conspicuous manner. Behavior is also flexible in a way that life histories or morphology may not be, and it allows moment-to-moment changes in response as risk increases or decreases. Because it is such an important weapon in the evolutionary arms race, antipredator behavior provides important insights into the causes and consequences of natural selection.

Some of the best examples of geographically variable antipredator responses occur in populations of freshwater fish (see, e.g., Bell and Foster 1994). The predation regime of these populations is relatively easy to classify—at least in terms of the presence and absence of predatory species—and the distribution of key predators can explain much of the documented variation in antipredator behavior (see p. 140). Covariance in predation regime and antipredator responses is compelling evidence for natural selection. Moreover, because predation regimes can change (or be manipulated) over relatively short periods of time, there is an opportunity to record heritable changes in antipredator responses—in other words, to watch evolution in action. The fact that predator–prey communities may be replicated in a number of habitats means that independent measures of the consequences of predation are possible, and because some predators consume a variety of species, an interspecific perspective on the evolutionary significance of predation is also feasible. Fish are excellent subjects for behavioral investigations both in the field and in the laboratory (Barlow 1993). Natural populations can be used to study behavioral decision-making as well as to test ideas about evolution in the wild. Developmental questions, such as the role of early experience in shaping population differences in behavior, are effectively addressed using fish.

The aim of this chapter is to use examples from freshwater fish to demonstrate how variation in predation risk can lead to the evolution of behavioral and, ultimately, biological diversity (May 1994). I begin by describing adaptive variation in antipredator behavior in populations of freshwater fish. The causes of nonadaptive variation in antipredator behavior are also explored. I then turn to the genetic and developmental bases of variation and ask how quickly population differences evolve and conclude by considering the links between antipredator responses, sexual selection, and speciation.

Adaptive Variation in Antipredator Behavior

Early ethologists viewed behavioral differences among individuals of the same species as uninteresting or inconvenient (see Magurran 1993 for a discussion) and took steps to eliminate this unwelcome variation from their investigations. Gradually, however, it became clear that a rich seam of behavioral variation was waiting to be unearthed. The history of the exploration of the guppy in Trinidad illustrates this well. Guppies occur naturally in northeast South America and the nearby Caribbean Islands including Trinidad (fig. 7-1). The species was first de-

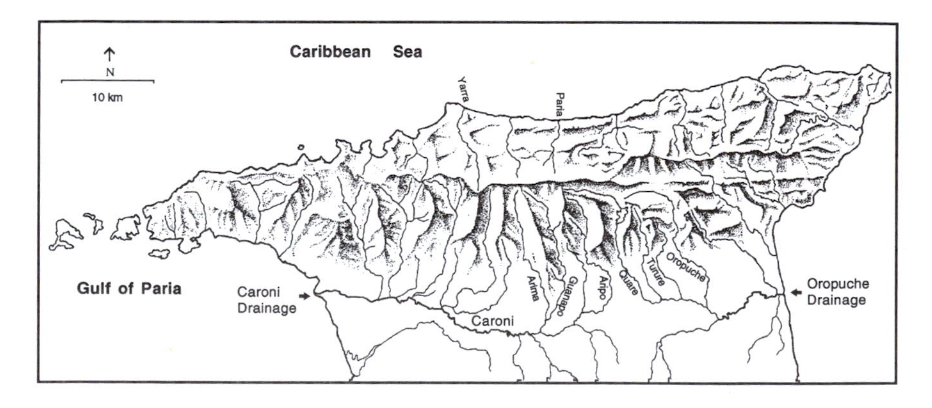

Figure 7-1 Map of northern Trinidad showing the location of the major guppy populations mentioned in this chapter.

scribed in 1859 (Winer and Boos 1991). The bright and highly variable coloration of males, as well as the ease of care in captivity, made the guppy a popular aquarium fish. By the 1920s, considerable steps had already been made toward understanding the genetic basis of male color patterns (Winge 1922, 1927). In 1936 Haskins and his colleagues turned their attention to variation in the coloration of natural populations of the guppy and made significant (and still pertinent) contributions to our understanding of sexual selection (Haskins and Haskins 1950, Haskins et al. 1961; see also Liley 1966). Haskins (following Fisher 1930) argued that male color patterns represent a balance between natural selection and sexual selection and that predators, such as the pike cichlid, *Crenicichla alta*, play a pivotal role in the process.

This pioneering work prompted further studies on the extent of variation among natural populations of the guppy. Seghers (1973, 1974) was the first to demonstrate (for any species) that schooling tendency (and other antipredator tactics) varies adaptively with predation (fig. 7-2), while Endler (1978, 1980, 1991) significantly advanced the investigation of color polymorphism. Since then the number of studies of population variation in the Trinidadian guppy has increased rapidly (Seghers 1992, Magurran et al. 1995). It is now evident that selection by predators in this system has far-reaching consequences. We know, for instance, that there are links between predation pressure and life history (Reznick and Endler 1982), foraging decisions (Fraser and Gilliam 1987), male mating tactics (Farr 1975, Luyten and Liley 1985, 1991), female choice (Houde 1987, Houde and Endler 1990, Endler and Houde 1995), and sexual conflict (Magurran and Seghers 1994c). Indeed, it is hard to find behavioral, morphological or life-history traits that vary independently of predation regime. Molecular traits and allozymes provide some of the few examples of the latter.

Experimental manipulations (both in the field and in the laboratory) confirm that population variation arises as a consequence of selection exerted by predators (Endler 1980, Reznick et al. 1990, Magurran et al. 1992). Population variation in antipredator behavior has also been extensively explored in two other species, the European minnow, *Phoxinus phoxinus*, and the three-spined stickleback, *Gasterosteus aculeatus*. In each case, populations that coexist with specialized piscivores, such as the pike, *Esox lucius* (in the case of the minnows and sticklebacks), and the pike cichlid (in the case of the guppies) behave in a similar fashion (Magurran et al. 1993, Huntingford et al. 1994). The parallel responses in these three species testify to the significance of predation as an evolutionary force.

Schooling and Predator Inspection

Fish from high-risk localities form larger schools (fig. 7-2, Seghers 1974, Magurran and Seghers 1994b), even in the absence of direct threat, and are thus more likely to detect a predator should one appear, as vigilance increases with school size (Magurran et al. 1985). If a predator is sighted, the fish begin to inspect it by swimming slowly toward the predator before turning and retreating (George 1960, Magurran 1986). The purpose of inspection is to confirm identification of the predator and assess its motivation (Magurran and Girling 1986, Licht 1989,

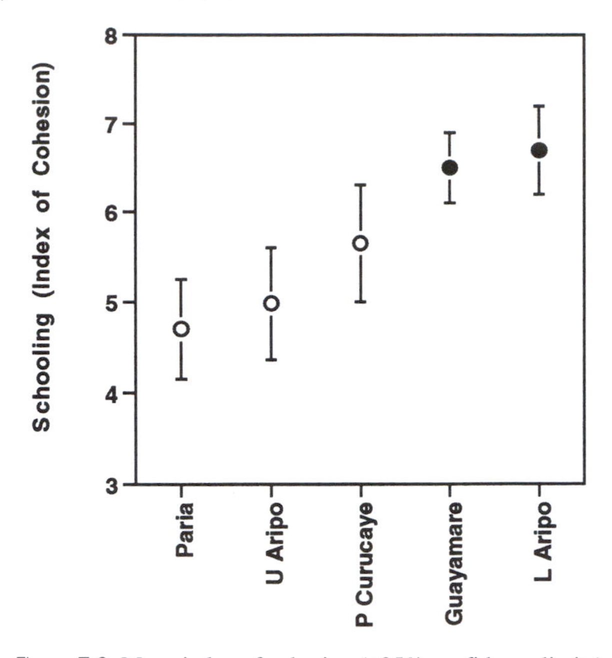

Figure 7-2 Mean index of cohesion (+95% confidence limits) for five laboratory stocks of guppies. Groups of 10 fish, 5 of each sex, were placed in an aquarium (48 × 110 cm with water 3 cm deep). The tank floor was covered by a grid of 10 squares each measuring 528 cm^2. Fish were allowed to explore the tank. Their schooling behavior was recorded after 5 h. The index of cohesion was calculated by scoring the maximum density of fish (in a 1-min period) for any of the 10 squares. Scores were averaged over the 30 observations. The index has a theoretical minimum of one and a maximum of 10. Five populations were compared. There were 10 replicates for each of these 5 stocks. Filled circles denote those (high-risk) sites where the pike cichlid, *Crenicichla*, also occurs. Low-risk (from fish predators) populations are indicated by open circles. Redrawn from Seghers (1974).

Magurran 1990a, Pitcher 1992). Fish that normally coexist with predators not only dilute risk by inspecting in larger groups but also minimize danger by maintaining a greater distance between themselves and their adversary (fig. 7-3, Magurran 1986, Magurran and Seghers 1994b). Cooperative behavior during inspection seems more likely in guppies (Dugatkin and Alfieri 1992) and sticklebacks (Huntingford et al. 1994) from high-risk populations than in those experiencing little risk. Inspection, particularly by larger groups, may inhibit attack, though it is clear that this is not the sole function of the behavior (Magurran 1990a, Dugatkin and Godin 1992, Godin and Davis 1995; see also Milinski and Bolthauser 1995). If fish from a high-risk population are attacked, they integrate escape tac-

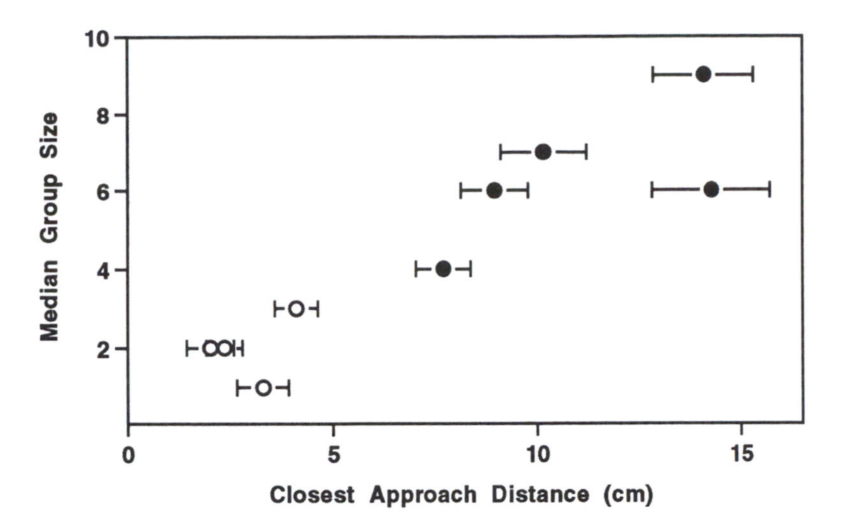

Figure 7-3 Inspection behavior by wild guppies from 9 populations in Trinidad. The inspection response was tested by placing a realistically painted predator model in the rivers. Guppies were abundant at all sites and started inspecting the predator model as soon as it appeared. Inspections could be distinguished from normal swimming behavior because the inspecting fish remained visually fixated on the predator during their approach and often engaged in "avoidance drift" (Seghers 1973), during which they turned laterally to the model and swam in an arclike trajectory along its body. Members of an inspecting group were almost always within three body lengths of their nearest neighbor. Group size was recorded at the point at which the fish were closest to the predator. Fifty separate inspections were recorded at each site. No fish was observed more than once. The graph shows the relationship between the mean (±95% limits) closest approach distance (of the leading fish) to the predator model and the median group size while inspecting. Populations from high-risk (*Crenicichla*) sites are denoted by filled circles, those from low-risk (*Rivulus*) sites by open circles. Drawn from data in Magurran and Seghers (1994b).

tics more effectively than do their counterparts from low-risk environments (Giles and Huntingford 1984, Magurran and Pitcher 1987, Huntingford et al. 1994).

Tactics that revolve around schooling behavior are not appropriate in all circumstances. Certain populations of guppy in Trinidad coexist with freshwater prawns in the genus *Macrobrachium*. These prawns are predators on guppies (Chace and Hobbs 1969, Endler 1983), and because they hunt using tactile and olfactory cues, rather than primarily visual ones, schooling behavior offers few antipredator advantages. Schooling could even be disadvantageous because schools may provide stronger olfactory cues and thus attract predators to their prey. In addition, these freshwater prawns are probably less susceptible than vertebrate predators to the confusion effect (Landeau and Terborgh 1986, Magurran and Seghers 1990a). Guppies from the Paria population in Trinidad (fig. 7-1), where prawns of the species *M. crenulatum* are abundant, have a reduced schooling tendency (Seghers 1974, Magurran and Seghers 1994b). However, these fish are much more wary than their comrades from sites where pike cichlids are common when inspecting a prawn, and they suffer fewer attacks as a consequence (Magurran and Seghers 1990a).

Other piscivores have evolved means of counteracting schooling tactics. For example, the conspicuous coloration of penguins in the genus *Spheniscus* seems to depolarize schools of pelagic prey, thus facilitating the capture of separated individuals (Wilson et al. 1987). Schooling may also be an ineffective defense against certain avian, mammalian, and fish predators with cooperative hunting tactics (Norris and Dohl 1980, Pitcher and Parrish 1993) as well as against some solitary predators such as herons and kingfishers.

Different populations of a single species are exposed to a variety of predators with a diversity of hunting modes. Reimchen (1994) notes that at least 67 species of predators, ranging from jellyfish to seals, include the three-spined stickleback in their diet. The overall severity of predation, as well as the combination of predators present at a particular site, dictate the adaptive behavior for each population. So far, little emphasis has been placed on studying the escape responses of fish that co-occur with predators that have different pursuit strategies (Lima 1992; but see Giles and Huntingford 1984), although such investigations could increase our understanding of behavioral variation.

Individual Antipredator Behavior

Group-based tactics such as schooling constitute effective defensive behavior but are not the only way of evading predators. Solitary fish adopt a variety of tactics to increase their chances of survival. For instance, three-spined sticklebacks have a range of antipredator responses to draw on as the risk escalates (see Huntingford et al. 1994 for a review). In the initial stages of an encounter, the prey fish raises its spines and inspects the predator. Should the attack continue, a number of evasive maneuvers such as a slow retreat or a rapid jump may be used. If all else fails the stickleback locks its spines in a vertical position to increase the handling difficulties for the predator (Hoogland et al. 1957, Reimchen 1994). Other species have a similar range of defenses. For example, when under attack guppies will "surface jump": they leap out of the water and reenter it at a place where the predator is not expecting them (Seghers 1973).

These individual tactics, like the group defenses described earlier, vary adaptively among populations. Surface jumps and other individual antipredator behaviors are more prevalent in guppies from populations that are sympatric with predators (Seghers 1973). Similarly, there is population variation in individual responses by sticklebacks to predators (Huntingford 1976, Giles and Huntingford 1984, Huntingford and Giles 1987, Foster and Ploch 1990). Foster and Ploch (1990) investigated the behavior of territorial males in two British Columbia lakes. They found a high degree of risk sensitivity to a preserved prickly sculpin, *Cottus asper*, by Garden Bay Lake sticklebacks. Sticklebacks co-occur with sculpin in this lake. In Crystal Lake, where sculpin are not found, the sticklebacks responded to the model by approaching it and nipping it.

The Indirect Costs of Effective Antipredator Behavior

Investment in defenses can be costly; time devoted to predator vigilance or avoidance is time lost from feeding, courtship, and other important activities (Lima and Dill 1990, Magnhagen 1991). In addition, the adaptations required for successful predator evasion may curtail an individual's options, even in circumstances where there is no immediate threat of predation. Thus, fish from populations where antipredator responses are elevated may bear costs not experienced by other conspecifics. This evolutionary trade-off between defense and other activities plays an important role in shaping geographic variation in behavior.

Schooling is an example of such a compromise. It is now clear that schooling functions most effectively as a defense against predation when the individual school members are of a similar size and appearance and behave in a coordinated fashion (Ohguchi 1981, Magurran and Pitcher 1987, Theodarkis 1989). However, selection for uniformity of behavior and morphology implies selection against distinctiveness and thus limits scope for resource defense.

Guppies from populations where schooling tendency has been heightened in response to predation display low levels of aggression at a food patch. Conversely, guppies from populations where schooling tendency is low vigorously defend feeding sites (Magurran and Seghers 1991, Magurran et al. 1995). These population differences in aggression are evident in the wild as well as in the laboratory. Levels of intraspecific aggression in three-spined sticklebacks are also inversely correlated with the predation risk to which the population is exposed. Individuals from sites where predators are abundant show least aggression to sticklebacks intruding into their breeding territory (Huntingford 1982).

Giles and Huntingford (1984) found that female sticklebacks have much stronger fright responses than males. Female guppies are similarly inclined to devote greater effort to avoiding predation than their male conspecifics (Magurran and Seghers 1994c). There are a number of explanations for this asymmetry in risk taking, the most important of which relates to the relative probability of reproductive success (Magurran and Nowak 1991). As long as males are available for mating, female reproductive success is guaranteed by survival. However, because of female choice (Houde 1988, Houde and Endler 1990) and male–male competition (Kodric-Brown 1992), mating opportunities are distributed unequally among males. Males may thus father few or no offspring irrespective of longevity.

The potential for marked individual variation in reproductive success may explain why male guppies constantly pursue females in order to court them (Luyten and Liley 1985). Male guppies have two means of securing a mating, and most males use the two tactics to a greater or lesser extent (Magurran and Seghers 1990c, Reynolds et al. 1993). They may either perform a sigmoid display in the hope of persuading a receptive female to copulate with them or opt for a sneaky mating and thrust their gonopodium toward the female's genital pore (Liley 1966). Because females are receptive only as virgins or for short periods after the birth of a brood (Liley 1966), most male advances are unwelcome. Males take advantage of females who are otherwise preoccupied to launch a sneaky mating attempt. For instance, when females approach a potential predator to inspect it, males follow behind attempting sneaky matings as they swim (Magurran and Nowak 1991). Female guppies in high-risk sites in Trinidad (i.e., those where the pike cichlid is present) receive approximately one sneaky mating per minute, which is twice the rate seen in low-risk populations (Magurran and Seghers 1994c). The female recipients of sneaky matings suffer a range of costs. Not only does this activity prejudice their choice of partner, but it also restricts a female's ability to forage efficiently (Magurran and Seghers 1994a). These examples illustrate how selection by predators can have far-reaching consequences and why populations experiencing contrasting predation regimes differ in a wide range of traits.

Is All Variation Adaptive?

In her chapter in this volume, Susan Reichert asks whether population differences in the behavior of the funnel-web spider, *Agelenopis aperta*, are adaptive and concludes that this is not invariably the case. She argues that the apparent maladaptation of one population has arisen as a consequence of gene flow and not through phylogenetic inertia. Fish populations also provide examples that do not readily fit the adaptive model. In Trinidad, two major river systems drain the southern slopes of the northern range. One of these, the Caroni, flows westward into the Gulf of Paria; the other, the Oropuche, flows eastward into the Atlantic Ocean (see fig. 7-1). Most investigations of fish behavior have been in the streams and rivers that form the Caroni system and, in all cases, these populations show a good relationship between predation regime and schooling tendency. Only recently have populations from the Oropuche system been included in behavioral surveys. Unexpectedly, guppies from the lower, high-risk reaches of the Quare and Oropuche Rivers (both belonging to the Oropuche system; fig. 7-1) have a much lower schooling tendency than their compatriots from the Caroni drainage (Magurran et al. 1992, Seghers and Magurran 1995). Because fish from *Crenicichla* sites in both drainages are equally wary when inspecting a model pike cichlid, it seems unlikely that this behavioral difference is indicative of impoverished antipredator behavior in Oropuche system guppies (Magurran et al. 1992, Magurran et al. 1995, Seghers and Magurran 1995).

There are a number of possible, and not mutually exclusive, explanations for the interdrainage differences in behavior. First, the *Crenicichla* in the Oropuche drainage may adopt different hunting tactics. This seems unlikely, although a recent laboratory investigation of *Crenicichla* feeding behavior revealed differences between pike cichlids from the Aripo (Caroni drainage) and Oropuche Rivers (Mattingly and Butler 1994). Second, Oropuche fish may be trading off schooling tendency against some other behavior to arrive at an equally successful but novel solution to the game of life. Finally, suboptimal behavior could be retained in certain populations as a genetic legacy from the founding fish (Magurran et al. 1995). The behavioral differences in the two drainages are paralleled by marked genetic differences (Carvalho et al. 1991, Shaw et al. 1991, Fajen and Breden 1992). A similar behavioral anomaly occurs in two north coast guppy populations that are also genetically distinct (Magurran et al. 1995). This circumstantial correlation between genetic and behavioral divergence leaves open the possibility that there are historical influences on the evolution of guppy populations. There is now evidence that founder events are implicated in the differentiation of Trinidadian guppy populations (Carvalho et al. 1996). Of course, founder effects could also be involved in the different but equal solution proposed above.

The Genetic and Developmental Bases of Variation

Because population differences in antipredator behavior are so dramatic, it is easy to assume that they invariably derive from genetic differences. A recent series of studies has confirmed that geographic variation in antipredator behavior does have a genetic component. However, these studies have also revealed that there is an important interplay between genes and the environment and that subtle developmental events can have far-reaching behavioral consequences.

The Development of Population Differences

Intuitively it is obvious that young fish must be able to respond appropriately to a predator the first time they encounter one. Because this requirement apparently leaves little scope for learning, it is often assumed that most, if not all, antipredator behavior is inherited. Experiments confirm that fish raised without exposure to predators exhibit many of the characteristic antipredator skills of their own species or population. A study on European minnows provides an apt example. As noted earlier, wild-caught minnows from a high-risk population (i.e., one that co-occurs with pike) have a higher schooling tendency and inspect predators more often than minnows from a low-risk (pike-free) population (Magurran 1986, Magurran and Pitcher 1987). These population differences in behavior are evident in inexperienced (laboratory-reared) fish during their first exposure to a predator model at 10 weeks of age (Magurran 1990b). The fact that guppies (Seghers 1974, Breden et al. 1987) and sticklebacks (Huntingford and Wright 1993) likewise retain their population-specific antipredator behavior when raised in the laboratory over one or more generations implies that these responses are genetically determined. Indeed, population differences in the schooling behavior of Trinidadian guppies are already apparent in newborn fish (Magurran and Seghers 1990b).

Yet other studies show that early experience can play an important role in fine-tuning the antipredator response. Hatchery-reared salmonids are notoriously vulnerable to predation after release (Isakson 1980, Piggins 1980), and it is evident that appropriate early experience can increase their chances of survival (Patten 1977, Suboski and Templeton 1989). Juvenile *Crenicichla* frequent the edges of Trinidadian streams and pursue guppies. The smallest *Crenicichla*, which are most likely to attack juvenile guppies, often miss their target (D. F. Fraser, personal communication). Such failures allow the guppies to sharpen their antipredator skills. The pike cichlids presumably benefit from the experience too.

Minnows will also modify their antipredator behavior after early experience. Brief exposure to a realistic model of a pike at 10 weeks leads to an increase in the inspection rate of 2-year-old fish, even though they have not had any encounters with predators in the intervening period (Magurran 1990b).

Experience need not be in the form of predator attack. Goodey and Liley (1986) found that cannibalistic attacks improved the escape responses of guppies, while Benzie (1965) and later Huntingford and her colleagues (Tulley and Huntingford 1987, Huntingford and Wright 1993) demonstrated that parental care by male three-spined sticklebacks contributes to the development of antipredator behavior in their offspring (fig. 7-4). (Fry that attempt to escape from the nest are chased and retrieved by the parental males.) These studies indicate that the qualitative components of the antipredator response such as schooling and inspection

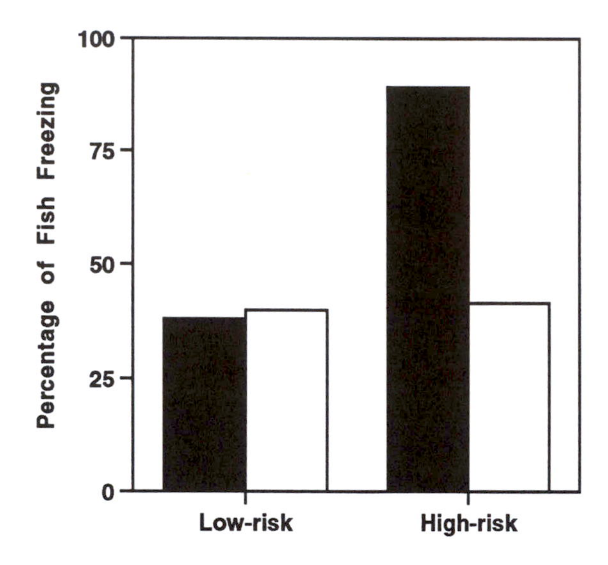

Figure 7-4 The effect of rearing on the first response of laboratory-reared sticklebacks to a model piscivorous fish. The sticklebacks came from populations occurring in two distinct habitats (low risk and high risk) and were reared either normally (solid bars) or as orphans (open bars). The sticklebacks were 8 weeks old when they were exposed to the model predator. Normally reared, high-risk, sticklebacks were most likely to freeze when they first saw the predator. However, in the absence of paternal care, the responses of fish from the high-risk population resembled those from the low-risk population (raised under either regime). Apart from freezing, sticklebacks either jumped or did nothing. Redrawn from Huntingford et al. (1994).

are determined genetically but that the quantitative levels of these behaviors may be subject to modification in the light of experience. They also illustrate the dangers of concluding too quickly that antipredator behavior is inflexible. We do not know, for example, whether the schooling behavior of newborn guppies (Magurran and Seghers 1990b) is in any way influenced by prenatal events.

Phenotypic Plasticity and Predispositions to Learn

Although antipredator behavior may be plastic, at least within limits, it is not immediately obvious whether the effects of experience should be additive, with all fish of a particular species improving their antipredator skills to a more or less equal degree given equivalent opportunities to learn, or interactive, with certain populations predisposed to benefit more from the same event (Fuiman and Magurran 1994). Two studies, one on minnows the other on sticklebacks, help resolve the issue. Recall the minnow populations discussed already in this chapter. One of these came from a high-risk site, the other from a low-risk one. As already indicated, the high-risk fish have superior antipredator tactics irrespective of whether they are wild caught or raised in the laboratory. In addition, both populations, from quite different predation regimes, show some behavioral adjustment after early experience of predation threat. However, the change in antipredator behavior as a result of this experience is significantly greater in the population that occurs sympatrically with predators (fig. 7-5, Magurran 1990b).

As mentioned earlier, three-spined sticklebacks retain their population-specific antipredator response whether they mature in the wild or are raised in the laboratory with normal parental care. When sticklebacks from two Scottish mainland populations (one high risk, the other low risk) were reared as orphans, the behavioral differences vanished; both sets of fish showed the responses normally seen in the low-risk population (fig. 7-4, Huntingford and Wright 1993, Huntingford et al. 1994). Detailed analyses of father–offspring interactions revealed that even on the day of hatching, the fry from the high-risk population reacted more strongly to their father's approach. Over the period of parental care these responses strengthened so that by the time the fry were ready to leave the nest they possessed many of the skills seen in wild fish of this population. In contrast, the fry from the low-risk population responded only weakly to their father's retrieval attempts (Huntingford and Wright 1993). Both examples suggest that fish from a high-risk population are predisposed to refine their antipredator behavior to a greater extent after experience.

How is this behavioral flexibility achieved? Experiments by Huntingford and Wright (1989, Wright and Huntingford 1992) have revealed that sticklebacks from high-risk populations learn more quickly to avoid dangerous locations. In their experiments Huntingford and Wright offered individual sticklebacks a choice of two (equally profitable) feeding compartments. Each individual soon came to prefer one of the compartments and fed there almost exclusively. The experimenters then started to threaten each stickleback (with a simulated predator attack) whenever it visited its preferred compartment. They discovered that individuals from high-risk sites needed a shorter time and fewer attacks before they

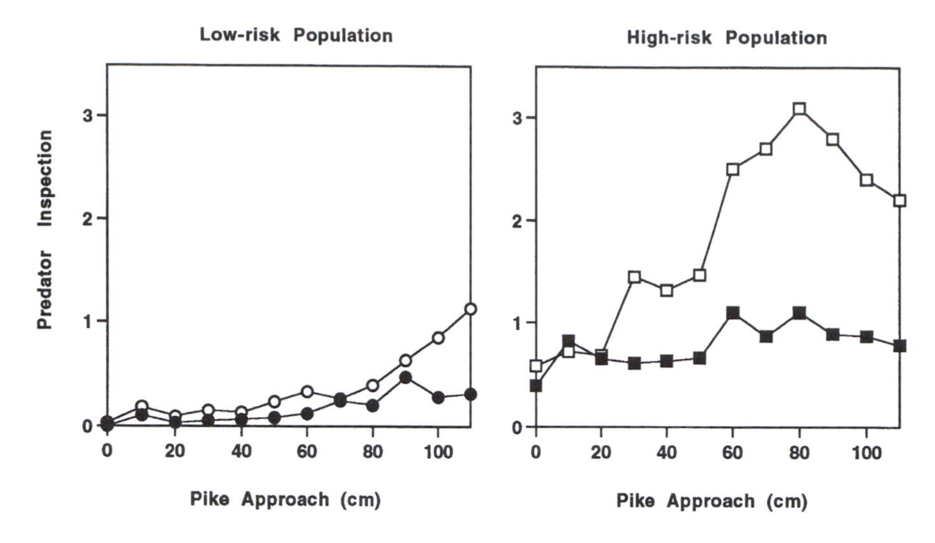

Figure 7-5 The inspection behavior of adult (2-year-old) European minnows bred and raised under controlled conditions in the laboratory. The figure compares two populations: Gwynedd minnows (left) came from a low-risk (pike-free) site in Wales, the Dorset minnows (right) were taken from a high-risk (pike present) site in England. The solid points on each graph denote the mean inspection behavior of the (naive) fish who had their first exposure to a realistic model pike as adults. These contrast with the open symbols, which indicate the mean inspection response of (experienced) minnows who were given a brief exposure to a threatening model predator at 10 weeks of age. The effect of experience is clearly greater in the minnows from the high-risk population. The x-axis traces the approach of the pike from the point at which it began to stalk (0 cm) until it reached the minnows' feeding site (110 cm). The measure of predator inspection is the mean frequency of approaches (inspections) per individual minnow at each phase of the pike's stalk. Redrawn from Magurran (1990b).

switched their attention to the previously spurned but now safe feeding site (fig. 7-6). The same learning differences were found in wild-caught sticklebacks (Huntingford and Wright 1989) and in predator-naive ones (Wright and Huntingford 1992).

These results suggest that the greater phenotypic plasticity of fish in high-risk environments arises through improved learning skills, at least in the context of predator recognition. There is also evidence that fish from high-risk populations are slower to lose their antipredator responses. Huntingford and Coulter (1989) examined the habituation of inspection behavior in sticklebacks from two populations. Sticklebacks from the risky locality maintained their inspection rate over successive encounters with a larger but benign fish to a greater extent than did their conspecifics from a safer environment.

Fish use olfactory cues as well as visual cues when assessing and evading predators (Csányi and Dóka 1993). For example, the fright response of bluntnose minnows, *Pimephales promelas*, exposed to pike odor is higher when the fish occur sympatrically with the predator (fig. 7-7, Mathis et al. 1993). The reaction

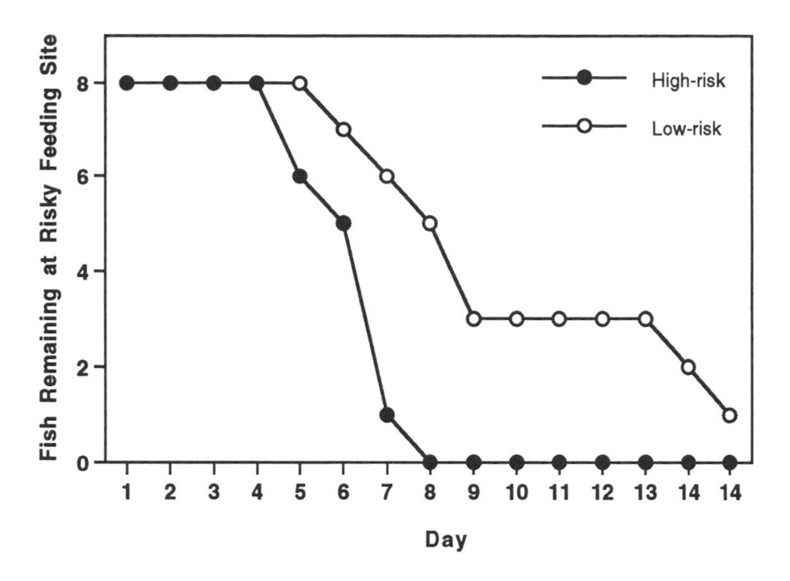

Figure 7-6 The number of fish remaining at a risky feeding site on successive days of avoidance training. The sticklebacks were predator naive (i.e., they were raised in the laboratory) but originated from two populations, a high-risk one (denoted by closed circles) and a low-risk one (open circles). Fish from the high-risk source took a median of 6 days to switch to the safer patch. This contrasted with the median of 8 days observed in the sticklebacks from the low-risk population. Redrawn from Huntingford and Wright (1992).

to predator odor is a conditioned response that arises when the odor is experienced in conjunction with some other frightening event (Magurran 1989, Mathis and Smith 1993). A constraint on learning also exists. Fish do not initially react to the odors of either piscivorous or nonpiscivorous fish. However, once conditioned they respond preferentially to the odor of natural predators (Magurran 1989). This constraint, evidently rooted in evolutionary history, illustrates the subtleties of learning that can exist within a population. Given the population differences in the response of the wild-caught minnows to the pike odor and the population specific learning skills uncovered in the experiments with sticklebacks and (European) minnows, it would be surprising if there were not adaptive variation in the acquisition of olfactory predator recognition. This assertion, however, remains to be tested.

Unlike predator odor, the response to Schreckstoff (alarm substance; Smith (1992) appears to be innate (Göz 1941). European minnows from high-risk populations respond most vigorously to alarm substance (Magurran and Pitcher 1987, Levesley and Magurran 1988). Nevertheless, it remains possible that learning can influence the reaction to Schreckstoff and that such learning capacity will vary across populations.

Why should learning skills vary? After all, it might seem optimal for all popu-

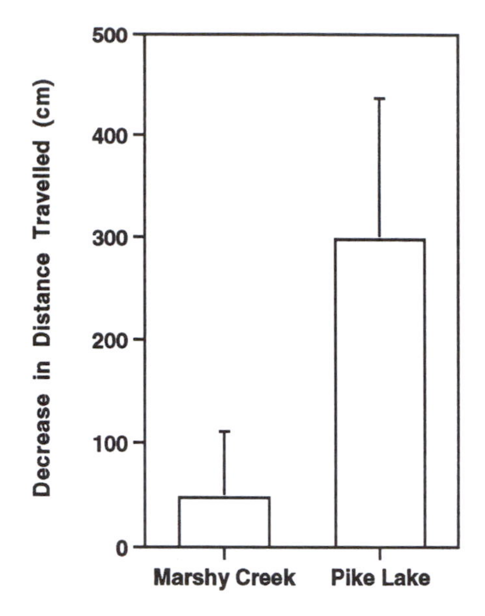

Figure 7-7 Population differences in the responses of fathead minnows to chemical stimuli from pike. Marshy Creek is a site without endemic piscivorous fish. Fathead minnows in Pike Lake coexist with large numbers of *Esox lucius*. In this experiment wild-caught minnows from the two sites were exposed to water drawn from a tank which a pike had occupied. The graph shows the mean (+SE) decrease in activity following exposure to pike odor. This decrease was significant only in minnows from Pike Lake. Redrawn from Mathis et al. (1993).

lations to show equal plasticity. Fish could then develop the appropriate antipredator response for the environment in which they are placed. However, as we saw earlier, efficient antipredator behavior has both direct and indirect costs. It may be the case that the inherited predisposition to modify behavior after experience is also costly. If individuals in a low-risk environment were to shift their antipredator response dramatically after a single threatening encounter, they could be disadvantaged in future foraging or courtship situations.

How Quickly Can Population Differences Evolve?

Guppy populations in Trinidad represent an important case study in evolution. In addition to providing correlations between predation regime and the expression of particular traits, they also offer scope for experimental manipulation. One such manipulation recently and fortuitously came to light. As noted earlier, guppies from the Oropuche and Caroni systems show low genetic identity. However, an allozyme survey of one river in the Oropuche drainage, the Turure River (fig. 7-1), yielded puzzling results. Rather than resembling guppies in other streams in the Oropuche, fish from the Turure seemed to cluster genetically with the populations in the Caroni (Shaw et al. 1991). The mystery was solved when it emerged through correspondence that, in 1957, Haskins had introduced 200 guppies from a high-risk Caroni site into a previously guppy- (and predator-) free section of the Turure River. This portion of the river contained just one other fish species, *Rivulus hartii* (a minor predator, unlikely to target adult guppies; Liley and Seghers 1975). The transplant had never been published and might otherwise have remained undetected. Further genetic analyses confirmed that the descendants of the founders had colonized the upper section of the Turure River and even moved downstream beyond the barrier waterfall to replace the indigenous population (Shaw et al. 1992). The Lower Turure is a high-risk site for guppies.

The Turure introduction offered a valuable opportunity to assess behavioral evolution after the relaxation of predation risk. Thirty-four years (more than 100 generations) had elapsed since the fish were moved from a *Crenicichla* habitat to the upper reaches of the river. Guppies were collected from the upper and lower Turure River and from two control sites in the Aripo River in the Caroni drainage and returned to the laboratory. The offspring of these fish were raised under standard conditions with no experience of predation. When the fish were tested as adults, it became clear that the upper Turure individuals (like the low-risk upper Aripo ones) had a reduced schooling tendency and moved close to a predator during inspection, whereas the high-risk lower Turure and lower Aripo guppies devoted more time to schooling and showed greater caution when inspecting (fig. 7-8). Incidentally, the lower Turure fish, which were derived from Caroni stock, show much greater levels of schooling than their unrelated contemporaries in other high-risk populations of the Oropuche system.

The Turure system thus provides evidence for the evolution of behavior in the wild. However, it is not the only transplant experiment that has been undertaken in Trinidad. In 1976 Endler collected 200 guppies from the (high risk) lower Aripo (fig. 7-1) and placed them in a low-risk tributary farther upstream (Endler 1980, Reznick et al. 1990). An insurmountable barrier water fall had prevented other fish species (with the exception of *Rivulus*) from reaching it. The transplanted guppies flourished and, as with the Turure introduction, their descendants soon colonized the tributary. Endler (1980) monitored the population for changes in male color pattern. An increase in male conspicuousness due to a greater number and density of color spots was apparent within 2 years. Equally rapid modification of life history, including phenotypic shifts in reproductive allotment and offspring size, was observed (Reznick and Endler 1982). The heritability of these changes was confirmed within 4 years from the initial introduction when guppies from the site were bred and raised under standard laboratory conditions (Reznick and Bryga 1987). Subsequent investigations showed that the population continued to diverge from founding over an 11-year period (Reznick et al. 1990).

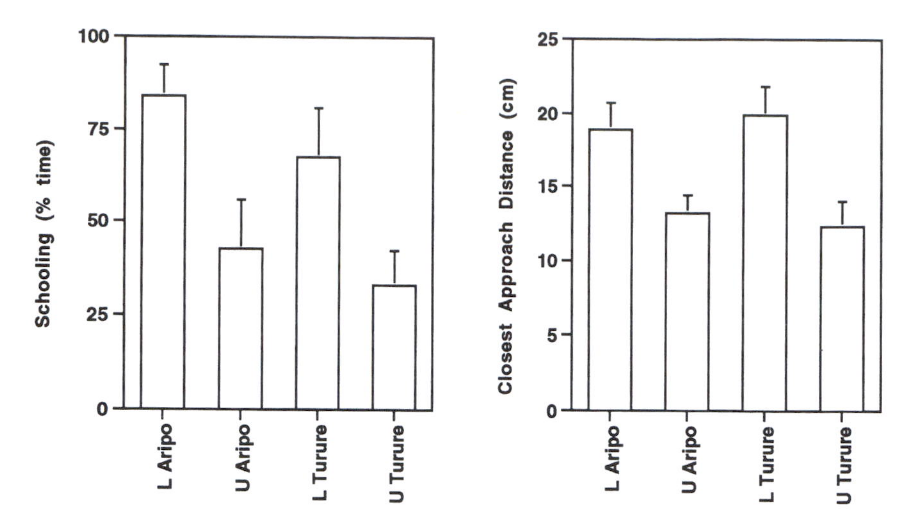

Figure 7-8 The mean percentage time (+95% confidence limits) spent schooling (left) and the closest approach to a realistic predator (right) by female guppies in a laboratory test. All fish had been bred and raised in controlled conditions in the laboratory. The Turure guppies were descendants of 200 individuals transplanted from a high-risk Caroni site in 1957. The graph contrasts the behavior of females from the Upper Turure (a low-risk site) and Lower Turure (high-risk) with high-risk (Lower Aripo) and low-risk (Upper Aripo) sites in the Caroni drainage and shows that the behavior of guppies in both sections of the Turure is appropriate for their present predation regime. Redrawn from Magurran et al. (1992).

Field observations of guppies in the transplanted population in 1992 suggested that behavior had also evolved in the 16 years following the introduction (Magurran et al. 1995). Female guppies spent only a small portion of their time schooling, and fish of both sexes displayed little caution when inspecting a predator model (fig. 7-9a,b). However, these phenotypic shifts could not be replicated when fish from the population were collected so that their offspring would be born and raised in the laboratory. Rather than displaying behavior typical of guppies from low-risk habitats, these laboratory-reared individuals resembled the ancestral Lower Aripo population in their schooling tendency and response to predators (fig. 7-9c,d, Magurran et al. 1995).

There are a number of possible reasons why schooling behavior evolves more slowly than do other traits. One possibility is that genetic variation in the founding population was more limited for behavior than for color patterns or life histories. Yet analysis of allozyme frequencies suggests that little genetic variation was lost during the founding of the new population (Magurran et al. 1995, Carvalho et al. 1996). Furthermore, half of the transplanted fish were large, pregnant females, multiply mated by an average of 2.3 males (J. A. Endler personal communication), with the result that the effective population size was considerably larger than 200 (the number of fish in the introduction).

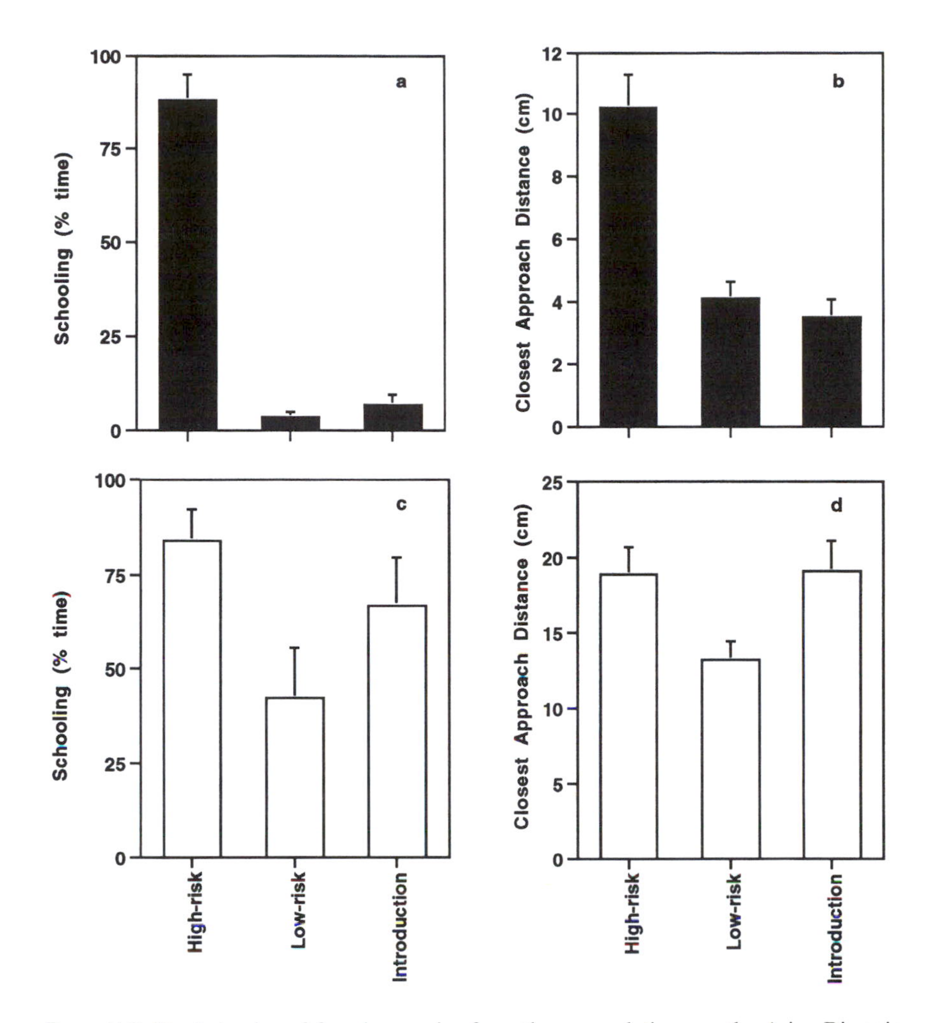

Figure 7-9 The behavior of female guppies from three populations on the Aripo River in Trinidad. The sites are the high-risk Lower Aripo (*Crenicichla* present), the low-risk Upper Aripo (*Rivulus* present), and the introduction site (*Rivulus* present) where Endler transplanted 200 Lower Aripo guppies in 1976. Data collected in the wild (in 1992) suggest that female behavior including (a) mean schooling tendency (+95% confidence limits and measured as the percentage of time schooling with one or more individual) and (b) the mean closest approach distance during inspection (+95% confidence limits) resemble a low-risk population in the same river. However, when fish from these populations were bred under standard conditions in the laboratory, the schooling response (c) and inspection behavior (d) of females in the introduction site matched that of the ancestral population. Redrawn from Magurran et al. (1995).

A second explanation concerns the lower heritability of behavioral as opposed to morphological traits. The heritability of color patterns ranges from 0.7 to 1.0, while the heritability of behavior is probably in the order of 0.3–0.5 or less (J. A. Endler personal communication). With time the antipredator behavior of guppies in Endler's Aripo introduction site will probably moderate, as has already happened in descendants of the fish that Haskins introduced into the upper Turure.

Additionally, the conservation of ancestral behavior in the transplanted fish may be a consequence of the relaxed predation regime under which they now exist. The direct and indirect costs of antipredator behavior in low-risk situations are substantially less than the costs of responding inappropriately to a predator attack. It thus seems likely that antipredator behavior is gained much more quickly than it is lost after a change in predation regime. Indeed, it may be that the costs of antipredator behavior are so low that the repertoire diminishes only as a consequence of nonadaptive decay. This occurs when accumulating mutations gradually reduce the incidence of antipredator genes in the population and inevitably takes a long time. Richard Coss's chapter in this volume illustrates the extent to which predator recognition can be retained in populations where predation risk is reduced (see also Curio 1993).

The presence of antipredator behavior in the laboratory-reared guppies from the Aripo introduction site is also evidence for phenotypic plasticity. Phenotypic plasticity can be defined as "the ability of a single genotype to produce more than one alternative form of morphology, physiological state, and/or behavior in response to environmental conditions" (West-Eberhard 1989, p. 249) and is a particularly valuable asset in a variable environment (Via 1993). Whether plasticity evolves as a target or as a by-product of selection is a question that remains to be resolved (Via 1987, 1993; Thompson 1991).

From Population Differentiation to Speciation

Population differentiation in behavior (and other traits) has been linked to variation in predation risk in a number of fish species. Field manipulations demonstrate that this relationship is not merely one of correlation and that fish populations do evolve as a consequence of shifts in predator pressure. However, behavior not only reflects the course of natural selection, it may also facilitate speciation. Lande (1981) has shown how the Fisherian "runaway" process of sexual selection can lead to rapid speciation (see Butlin and Ritchie 1994 for a review). This occurs when female choice influences male mating success and when there is a genetic correlation between female preference and male characters. If subsets of a species are differentiated, for whatever reason, then there is scope for sexual selection to operate in this way. Population differentiation, arising as a consequence of predation pressure, can provide the raw material for this process. To what extent does the predator-driven population differentiation documented in this chapter lead to speciation?

There is evidence of rapid speciation in a number of fish groups. Perhaps the most remarkable example is provided by the cichlid species flocks in the African great lakes of Victoria, Malawi, and Tanganyika (Meyer et al. 1990, Meyer 1993). Each of these lakes contains large numbers of endemic cichlids and there is an indication that some of the species have arisen in the recent past. For example, Lake Nabugabo has been isolated from Lake Victoria for only about 4000 years, and yet it contains five unique cichlids (Greenwood 1965). There is even the suggestion that some Lake Malawi cichlids speciated within the last few hundred years (Owen et al. 1990). The diversification of cichlids in large lakes is often attributed to sexual selection (see, e.g., Dominey 1984), and the fact that these species vary primarily in male coloration rather than in other morphological traits (Sturmbauer and Meyer 1992) provides strong support for this idea.

Sexual selection may amplify differences to such an extent that new species are produced. The underlying variation, upon which sexual selection works, arises separately—for example, by random genetic processes or through natural selection. Consequently, geographic variation in antipredator responses could, in certain circumstances, represent the first steps toward the formation of new species.

The Trinidadian guppy system seems to provide all the necessary ingredients. There is, as this chapter has shown, marked variation in many traits. Much of this variation can be attributed to variation in predation risk. A change in risk leads to rapid heritable change in morphology and behavior. Males vary in color pattern, females exert choice (Endler and Houde 1995), and there is a genetic correlation between the two (Houde 1994). It is therefore perplexing to discover that, despite this potential, there is little evidence for the reproductive isolation of guppy populations, even among those that have been separated for hundreds of thousands of years (Shaw et al. 1991, Magurran et al. 1995). Endler (1995) argues that the strong gene flow and weak isolation (Endler and Houde 1995) that characterize guppy populations may be sufficient to prevent speciation. The fact that the selection regimes operate over a small geographical scale relative to the scale of gene flow is likely to be an important contributory factor.

The stickleback genus *Gasterosteus* provides another intriguing case. Despite the well-documented ability of populations of the three-spined stickleback to diverge, again as a consequence of predation risk (Bell and Foster 1994), this ancient genus is represented by a mere two species (McPhail 1994). These cases underline our poor understanding of the links between population differentiation and speciation. Yet, as pointed out in the Introduction to this book, population differentiation is a topic that has only recently begun to fascinate biologists. It is clear that much remains to be done and that there is still considerable scope for using geographic variation in behavior to elucidate evolutionary processes.

Acknowledgments This chapter has drawn extensively on the work and insights of Ben Seghers and Felicity Huntingford, and I particularly wish to acknowledge their contribution to the field. John Endler, Susan Foster, Doug Fraser, Siân Griffiths, and an anonymous referee provided many helpful comments on earlier versions of the chapter. I am grateful to the Royal Society (London) for its generous support.

References

Barlow, G. W. 1993. Fish behavioural ecology: pros, cons and opportunities. *In* F. A. Huntingford and P. Torricelli, eds. Behavioural ecology of fishes, pp. 7–27. Harwood, Chur, Switzerland.

Bell, M. A., and S. A. Foster. 1994. The evolutionary biology of the threespine stickleback. Oxford University Press, Oxford.

Benzie, V. L. 1965. Some aspects of the anti-predator responses of two species of sticklebacks (PhD dissertation). Oxford University, Oxford.

Breden, F., M. Scott, and E. Michel. 1987. Genetic differentiation for anti-predator behaviour in the Trinidad guppy, *Poecilia reticulata*. Animal Behaviour 35:618–620.

Butlin, R. K., and M. G. Ritchie. 1994. Behaviour and speciation. *In* P. J. B. Slater and T. R. Halliday, eds. Behaviour and evolution, pp. 43–79. Cambridge University Press, Cambridge.

Carvalho, G. R., P. W. Shaw, A. E. Magurran, and B. H. Seghers. 1991. Marked genetic divergence revealed by allozymes among populations of the guppy, *Poecilia reticulata* (Poeciliidae) in Trinidad. Biological Journal of the Linnean Society 42:389–405.

Carvalho, G. R., P. W. Shaw, B. H. Seghers, and A. E. Magurran. 1996. Artificial introductions, evolutionary rates and population differentiation in Trinidadian guppies (*Poecilia reticulata*: Poeciliidae). Biological Journal of the Linnean Society. 57:219–234.

Chace, F. A., and H. H. Hobbs 1969. The freshwater and terrestrial decapod crustaceans of the West Indies with special reference to Dominica. Bulletin of the U.S. National Museum 292:1–258.

Csányi, V., and A. Dóka, A. 1993. Learning interactions between prey and predator fish. *In* F. A. Huntingford and P. Torricelli, eds. Behavioural ecology of fishes, pp. 63–78. Harwood, Chur, Switzerland.

Curio, E. 1993. Proximate and developmental aspects of antipredator behaviour. Advances in the Study of Behavior 22:135–238.

Dominey, W. J. 1984. Effects of sexual selection and life history on speciation: Species flocks in African cichlids and Hawaiian *Drosophila*. *In* A. A. Echelle and I. Kornfield, eds. Evolution of fish species flocks, pp. 231–249. University of Maine Press, Orono.

Dugatkin, L. A., and M. Alfieri. 1992. Interpopulational differences in the use of the tit-for-tat strategy during predator inspection in the guppy, *Poecilia reticulata*. Evolutionary Ecology 6:519–526.

Dugatkin, L. A., and J.-G. J. Godin. 1992. Prey approaching predators: A cost-benefit perspective. Annales Zoologica Fennici 29:233–252.

Endler, J. A. 1978. A predator's view of animal colour patterns. Evolutionary Biology 11: 319–364.

Endler, J. A. 1980. Natural selection on color patterns in *Poecilia reticulata*. Evolution 34:76–91.

Endler, J. A. 1983. Natural and sexual selection on color patterns in poeciliid fishes. Environmental Biology of Fishes 9:173–190.

Endler, J. A. 1991. Variation in the appearance of guppy color patterns to guppies and their predators under different visual conditions. Vision Research 31:587–608.

Endler, J. A. 1995. Multiple-trait coevolution and environmental gradients in guppies. Trends in Ecology and Evolution 10:22–29.

Endler, J. A., and A. E. Houde. 1995. Geographic variation in female preferences for male traits on *Poecilia reticulata*. Evolution 49:456–468.

Fajen, A., and F. Breden. 1992. Mitochondrial DNA sequence variation among natural populations of the Trinidad guppy, *Poecilia reticulata*. Evolution 46:1457–1465.

Farr, J. A. 1975. The role of predation in the evolution of social behavior of natural populations of the guppy, *Poecilia reticulata* (Pisces: Poeciliidae). Evolution 29:151–158.

Fisher, R. A. 1930. The evolution of dominance in certain polymorphic species. American Naturalist 64:385–406.

Foster, S., and S. Ploch. 1990. Determinants of variation in antipredator behavior of territorial male threespine stickleback in the wild. Ethology 84:281–294.

Fraser, D. F., and J. F. Gilliam. 1987. Feeding under predation hazard: Response of the guppy and Hart's rivulus from sites with contrasting predation hazard. Behavioral Ecology and Sociobiology 21:203–209.

Fuiman, L. A., and A. E. Magurran. 1994. Development of predator defences in fishes. Review of Fish Biology and Fisheries 4:145–183.

George, C. J. W. 1960. Behavioral interaction of the pickerel (*Esox niger* and *Esox americanus*) and the mosquitofish (*Gambusia patruelis*) (PhD dissertation). Harvard University, Cambridge, MA.

Giles, N., and F. A. Huntingford. 1984. Predation risk and interpopulation variation in antipredator behaviour in the three-spined stickleback, *Gasterosteus aculeatus* L. Animal Behaviour 32:264–275.

Godin, J.-G. J., and S. A. Davis. 1995. Who dares, benefits: Predator approach behaviour in the guppy (*Poecilia reticulata*) deters predator approach. Proceedings of the Royal Society of London B 259:193–200.

Goodey, W., and N. R. Liley. 1986. The influence of early experience on escape behavior in the guppy (*Poecilia reticulata*). Canadian Journal of Zoology 64:885–888.

Göz, H. 1941. Über den art- und individualgeruch bei Fischen. Zeitschrift für vergleichende Physiologie 29:1–45.

Greenwood, P. H. 1965. The cichlid fishes of Lake Nabugabo, Uganda. Bulletin of the British Museum of Natural History (Zoology) 12:315–357.

Haskins, C. P., and E. F. Haskins. 1950. Factors affecting sexual selection as an isolating mechanism in the poeciliid fish, *Lebistes reticulatus*. Proceedings of the National Academy of Science USA 36:464–476.

Haskins, C. P., E. F. Haskins, J. J. A. McLaughlin, and R. E. Hewitt. 1961. Polymorphism and population structure in *Lebistes reticulatus*, an ecological study. *In* W. F. Blair, ed. Vertebrate speciation, pp. 320–395. University of Texas Press, Austin.

Hoogland, R. D., D. Morris, and N. Tinbergen. 1957. The spines of sticklebacks (*Gasterosteus* and *Pygosteus*) as a means of defense against predators (*Perca* and *Esox*). Behaviour 10:205–236.

Houde, A. E. 1987. Mate choice based upon naturally occurring color pattern variation in a guppy population. Evolution 41:1–10.

Houde, A. E. 1988. Genetic difference in female choice between two guppy populations. Animal Behaviour 36:510–516.

Houde, A. E. 1994. Effect of artificial selection on male colour patterns on mating preferences of female guppies. Proceedings of the Royal Society of London B 256:125–130.

Houde, A. E., and J. A. Endler. 1990. Correlated evolution of female mating preferences and male color pattern in the guppy, *Poecilia reticulata*. Science 248:1405–1408.

Huntingford, F. A. 1976. The relationship between anti-predator behaviour and aggression among conspecifics in three-spined stickleback, *Gasterosteus aculeatus*. Animal Behaviour 24:694–697.

Huntingford, F. A. 1982. Do inter- and intra-specific aggression vary in relation to predation pressure in sticklebacks? Animal Behavior 30:909–916.

Huntingford, F. A., and R. M. Coulter. 1989. Habituation of predator inspection in the three-spined stickleback, *Gasterosteus aculeatus* L. Journal of Fish Biology 35:153–154.

Huntingford, F. A., and N. Giles. 1987. Individual variation in anti-predator responses in the three-spined stickleback (*Gasterosteus aculeatus* L.). Ethology 74: 205–210.

Huntingford, F. A., J. Lazarus, B. D. Barrie, and S. Webb. 1994. A dynamic analysis of cooperative predator inspection in sticklebacks. Animal Behavior 47:413–423.

Huntingford, F. A., and P. J. Wright. 1989. How sticklebacks learn to avoid dangerous feeding patches. Behavioral Processes 19:181–189.

Huntingford, F. A., and P. J. Wright. 1993. The development of adaptive variation in predator avoidance in freshwater fishes. *In* F. A. Huntingford and P. Torricelli, eds. Behavioral ecology of fishes, pp. 45–61. Harwood, Chur, Switzerland.

Huntingford, F. A., P. J. Wright, and J. F. Tierney. 1994. Adaptive variation in antipredator behavior in threespine stickleback. *In* M. A. Bell and S. A. Foster, eds. The evolutionary biology of the threespine stickleback, pp. 277–296. Oxford University Press, Oxford.

Isakson, A. 1980. Salmon ranching in Iceland. *In* J. E. Thorpe, ed. Salmon ranching, pp. 131–156. Academic Press, London.

Kodric-Brown, A. 1992. Male dominance can enhance mating success in guppies. Animal Behavior 44:165–167.

Lande, R. 1981. Models of speciation by sexual selection on polygenic traits. Proceedings of the National Academy of Science USA 78:3721–3725.

Landeau, L., and Terborgh, J. 1986. Oddity and the 'confusion effect' in predation. Animal Behaviour 34:1372–1380.

Levesley, P. B., and A. E. Magurran. 1988. Population differences in the reaction of minnows to alarm substance. Journal of Fish Biology 32:699–706.

Licht, T. 1989. Discriminating between hungry and satiated predators: The response of guppies (*Poecilia reticulata*) from high and low predation sites. Ethology 82:238–242.

Liley, N. R. 1966. Ethological isolating mechanisms in four sympatric species of Poeciliid fishes. Behaviour (supplement) 13:1–197.

Liley, N. R., and B. H. Seghers. 1975. Factors affecting the morphology and behaviour of guppies in Trinidad. *In* G. P. Baerends, C. Beer, and A. Manning, eds. Function and evolution in behaviour, pp. 92–118. Clarendon Press, Oxford.

Lima, S. L. 1992. Life in a multi-predator environment: Some considerations for anti-predatory vigilance. Annales Zoologica Fennici 29:217–226.

Lima, S. L., and L. M. Dill. 1990. Behavioural decisions made under risk of predation: A review and prospectus. Canadian Journal of Zoology 68:619–640.

Luyten, P. H., and N. R. Liley. 1985. Geographic variation in the sexual behaviour of the guppy, *Poecilia reticulata* (Peters). Behaviour 95:164–179.

Luyten, P. H., and N. R. Liley. 1991. Sexual selection and competitive mating success of male guppies (*Poecilia reticulata*) from four Trinidad populations. Behavioral Ecology and Sociobiology 28:329–336.

Magnhagen, C. 1991. Predation risk as a cost of reproduction. Trends in Ecology and Evolution 6:183–186.

Magurran, A. E. 1986. Predator inspection behaviour in minnow shoals: Differences between population and individuals. Behavioral Ecology and Sociobiology 19:267–273.

Magurran, A. E. 1989. Acquired recognition of predator odor in the European minnow (*Phoxinus phoxinus*). Ethology 82:216–223.

Magurran, A. E. 1990a. The adaptive significance of schooling as an anti-predator defense in fish. Annales Zoologica Fennici 27:51–66.

Magurran, A. E. 1990b. The inheritance and development of minnow anti-predator behaviour. Animal Behaviour 39:834–842.

Magurran, A. E. 1993. Individual differences and alternative behaviours. *In* T. J. Pitcher, ed. Behaviour of teleost fishes, pp. 441–477. Chapman and Hall, London.

Magurran, A. E., and S. L. Girling. 1986. Predator recognition and response habituation in shoaling minnows. Animal Behaviour 34:510–518.

Magurran, A. E., and M. N. Nowak. 1991. Another battle of the sexes: The consequences of sexual asymmetry in mating costs and predation risk in the guppy, *Poecilia reticulata*. Proceedings of the Royal Society of London B 246:31–38.

Magurran, A. E., W. J. Oulton, and T. J. Pitcher. 1985. Vigilant behaviour and shoal size in minnows. Zeitschrift fur Tierpsychologie 67:167–178.

Magurran, A. E., and T. J. Pitcher. 1987. Provenance, shoal size and the sociobiology of predator evasion behaviour in minnow shoals. Proceedings of the Royal Society of London B 229:439–465.

Magurran, A. E., and B. H. Seghers. 1990a. Population differences in predator recognition and attack cone avoidance in the guppy, *Poecilia reticulata*. Animal Behaviour 40: 443–452.

Magurran, A. E., and B. H. Seghers. 1990b. Population differences in the schooling behaviour of newborn guppies, *Poecilia reticulata*. Ethology 84:334–342.

Magurran, A. E., and B. H. Seghers. 1990c. Risk sensitive courtship in the guppy (*Poecilia reticulata*). Behaviour 112:194–201.

Magurran, A. E., and B. H. Seghers. 1991. Variation in schooling and aggression amongst guppy (*Poecilia reticulata*) populations in Trinidad. Behaviour 118:214–234.

Magurran, A.E., and B. H. Seghers. 1994a. A cost of sexual harassment in the guppy, *Poecilia reticulata*. Proceedings of the Royal Society of London B 258:89–92.

Magurran, A. E., and B. H. Seghers. 1994b. Predator inspection behaviour covaries with schooling tendency amongst wild guppy, *Poecilia reticulata*, populations in Trinidad. Behaviour 128:121–134

Magurran, A. E., and B. H. Seghers. 1994c. Sexual conflict as a consequence of ecology: Evidence from guppy, *Poecilia reticulata*, populations in Trinidad. Proceedings of the Royal Society of London B 255:31–36.

Magurran, A. E., B. H. Seghers, G. R. Carvalho, and P. W. Shaw. 1992. Behavioral consequences of an artificial transplant of guppies, *Poecilia reticulata* in N. Trinidad: Evidence for the evolution of anti-predator behaviour in the wild. Proceedings of the Royal Society of London B 248:117–122.

Magurran, A. E., B. H. Seghers, G. R. Carvalho, and P. W. Shaw. 1993. Evolution of adaptive variation in antipredator behaviour. *In* F. A. Huntingford and P. Torricelli, eds. Behavioural ecology of fishes, pp. 29–44. Harwood Academic, Chur, Switzerland.

Magurran, A. E., B. H. Seghers, P. W. Shaw, and G. R. Carvalho. 1995. Behavioral diversity and evolution of guppy, *Poecilia reticulata*, populations in Trinidad. Advances in the Study of Behavior 24:155–202.

Mathis, A., D. P. Chivers, and R. J. F. Smith. 1993. Population differences in responses of fathead minnows (*Pimephales promelas*) to visual and chemical signals from predators. Ethology, 93:31–40.

Mathis, A., and R. J. F. Smith. 1993. Fathead minnows, *Pimephales promelas*, learn to recognize northern pike, *Esox lucius*, as predators on the basis of chemical stimuli from minnows in the pike's diet. Animal Behaviour 46:645–656.

Mattingly, H. T., and M. J. Butler. 1994. Laboratory predation on the Trinidadian guppy: Implications for the size-selective predation hypothesis and guppy life history evolution. Oikos 69:54–64.

May, R. M. 1994. Biological diversity: Differences between land and sea. Philosophical Transactions of the Royal Society of London B 343:105–111.

McPhail, J. D. 1994. Speciation and the evolution of reproductive isolation in the sticklebacks (*Gasterosteus*) of south-western British Columbia. *In* M. A. Bell and S. A. Foster, eds. The evolutionary biology of the threespine stickleback, pp. 399–437. Oxford University Press, Oxford.

Meyer, A. 1993. Phylogenetic relationships and evolutionary processes in East African cichlid fishes. Trends in Ecology and Evolution 8:279–284.

Meyer, A., T. D. Kocher, P. Basasibwaki, and A. C. Wilson. 1990. Monophyletic origin of Lake Victoria cichlid fishes suggested by mitochondrial DNA sequences. Nature 347:550–553.

Milinski, M., and P. Bolthauser. 1995. Boldness and predator deterrence: A critique of Godin & Davis. Proceedings of the Royal Society of London B 262:103–105.

Norris, K. S., and T. P. Dohl. 1980. The structure and function of cetacean schools. *In* L. M. Herman, ed. Cetacean behavior: Mechanisms and processes, pp. 211–168. Wiley, New York.

Ohguchi, O. 1981. Prey density and the selection against oddity by three-spined sticklebacks. Zeitschrift fur Tierpsychologie (supplement) 23:1–78.

Owen, R. B., R. Crossley, T. C. Johnson, D. Tweddle, I. Kornfield, S. Davison, D. H. Eccles, and D. E. Engstrom. 1990. Major low lake levels in lake Malawi and their implications for speciation rates in cichlid fishes. Proceedings of the Royal Society of London B 240:519–553.

Patten, B. G. 1977. Body size and learned avoidance as factors affecting predation on coho salmon fry, *Oncorhynchus kisutch*, by torrent sculpin, *Cottus rhotheus*. Fisheries Bulletin 75:457–459.

Piggins, D. J. 1980. Salmon ranching in Ireland. *In* J. E. Thorpe, ed. Salmon ranching, pp. 187–198. Academic Press, London.

Pitcher, T. J. 1992. Who dares, wins: The function and evolution of predator inspection behaviour in shoaling fish. Netherlands Journal of Zoology 42:371–391.

Pitcher, T. J., and J. K. Parrish. 1993. Functions of shoaling behaviour in teleosts. *In* T. J. Pitcher, ed. Behaviour of teleost fishes, pp. 363–439. Chapman and Hall, London.

Reimchen, T. E. 1994. Predators and morphological evolution in threespine stickleback. *In* M. A. Bell and S. A. Foster, eds. The evolutionary biology of the threespine stickleback, pp. 240–276. Oxford University Press, Oxford.

Reynolds, J. D., M. R. Gross, and M. J. Coombs. 1993. Environmental conditions and male morphology determine alternative mating behaviour in Trinidadian guppies. Animal Behaviour 45:145–152.

Reznick, D. N., and H. Bryga. 1987. Life-history evolution in guppies (*Poecilia reticulata*): 1. Phenotypic and genetic changes in an introduction experiment. Evolution 41: 1370–1385.

Reznick, D. N., H. Bryga, and J. A. Endler. 1990. Experimentally induced life-history evolution in a natural population. Nature 346:357–359.

Reznick, D. N., and J. A. Endler. 1982. The impact of predation on life history evolution in Trinidadian guppies (*Poecilia reticulata*). Evolution 36:160–177.

Seghers, B. H. 1973. An analysis of geographic variation in the antipredator adaptations of the guppy, *Poecilia reticulata* (PhD dissertation). University of British Columbia, Vancouver.

Seghers, B. H. 1974. Schooling behavior in the guppy (*Poecilia reticulata*): An evolutionary response to predation. Evolution 28:486–489.

Seghers, B. H. 1992. The rivers of northern Trinidad: Conservation of fish communities for research. *In* P. J. Boon, P. Calow, and G. E. Petts, eds. River conservation and management, pp. 81–90. Wiley, London.

Seghers, B. H., and A. E. Magurran. 1995. Population differences in the schooling behaviour of the Trinidad guppy, *Poecilia reticulata*: Adaptation or constraint? Canadian Journal of Zoology 73:1100–1105.

Shaw, P. W., G. R. Carvalho, A. E. Magurran, and B. H. Seghers. 1991. Population differentiation in Trinidadian guppies (*Poecilia reticulata*): Patterns and problems. Journal of Fish Biology 39 (supplement A):203–209.

Shaw, P. W., G. R. Carvalho, B. H. Seghers, and A. E. Magurran. 1992. Genetic consequences of an artificial transplant of guppies, *Poecilia reticulata* in N. Trinidad. Proceedings of the Royal Society of London B 248:111–116.

Smith, R. J. F. 1992. Alarm signals in fish. Reviews in Fish Biology and Fisheries 2: 33–63.

Sturmbauer, C., and A. Meyer. 1992. Genetic divergence, speciation and morphological stasis in a lineage of African cichlid fishes. Nature 358:578–581.

Suboski, M. D., and J. J. Templeton. 1989. Life skills training for hatchery fish: Social learning and survival. Fisheries Research 7:343–352.

Theodarkis, C. W. 1989. Size segregation and the effects of oddity on predation risk in minnow schools. Animal Behaviour 38:496–502.

Thompson, J. D. 1991. Phenotypic plasticity as a component of evolutionary change. Trends in Ecology and Evolution 6:246–249.

Tulley, J. J., and F. A. Huntingford. 1987. Parental care and the development of adaptive variation in anti-predator responses in sticklebacks. Animal Behaviour 35:1570–1572.

Via, S. 1987. Genetic constraints on the evolution of phenotypic plasticity. *In* V. Loeschcke, ed. Genetic constraints on adaptive evolution, pp. 47–71. Springer-Verlag, Berlin.

Via, S. 1993. Adaptive phenotypic plasticity: Target or by-product of selection in a variable environment? American Naturalist 142:352–365.

West-Eberhard, M. J. 1989. Phenotypic plasticity and the origins of diversity. Annual Review of Ecology and Systematics 20:249–278.

Wilson, R. P., P. G. Ryan, A. James, and M. P.-T. Wilson. 1987. Conspicuous coloration may enhance prey capture in some piscivores. Animal Behaviour 35:1558–1560.

Winer, L., and H. E. Boos. 1991. Agouti to zandoli: fauna in the dictionary of Trinbagonian. Living World: Journal of the Trinidad and Tobago Field Naturalists' Club 1991–1992: 25–28.

Winge, O. 1922. One-sided masculine and sex-linked inheritance in *Lebistes reticulatus*. Journal of Genetics 12:145–162.

Winge, O. 1927. The location of eighteen genes in *Lebistes reticulatus*. Journal of Genetics 18:1–42.

Wright, P. J., and F. A. Huntingford. 1992. Inherited population differences in avoidance conditioning in three-spined sticklebacks (*Gasterosteus aculeatus*). Behaviour 122: 164–173.

8

Geographic Variation and the Microevolution of Avian Migratory Behavior

PETER BERTHOLD

The study of migratory behavior is, at a fundamental level, the study of geographic variation in behavior. This is necessarily the case when residency grounds at either end of the migratory route differ in spatial extent, because different directional movement patterns will be required of individuals from different parts of the two ranges. As many migratory animals are widespread, substantial differences in migratory routes exist among populations of single species (examples in Dadswell et al. 1987, Baker 1991, Dingle 1991, Groot and Margolis 1991, Berthold 1993). Because the individuals that migrate often have not done so before and often do not have older migration-experienced individuals to follow (e.g., Baker 1991, Berthold 1996), the study of navigational mechanisms and their genetic underpinnings is essential to understanding migratory behavior.

Naturalists have long been captivated by the problem of control mechanisms in migratory behavior. As early as 1702, von Pernau suggested that birds were "driven at the proper time by a hidden drive." In modern terms, this amounted to suggesting that migratory behavior was triggered by innate or genetically programmed stimuli, rather than by environmental factors alone. This speculation was supported by the discovery of endogenous annual cycles more than 260 years later (see Gwinner 1986 for a review). Our understanding of the control mechanisms of migratory behavior has expanded rapidly during the last 90 years. Research in the field has elucidated, for example, genetic and endocrine control mechanisms and their interface with environmental cues such as photoperiod, relationships between environmental conditions, such as weather and food availability and the timing of migration, and unique physiological and morphological correlates of migratory behavior (for reviews, see Berthold and Terrill 1991, Berthold 1996).

Birds, fishes, and insects have proven especially valuable subjects for the study of migratory behavior, and all show substantial population differentiation in migratory patterns within species (Dadswell et al. 1987, Dingle 1991, Groot and Margolis 1991, Berthold 1993). In this chapter I focus on migratory behavior in birds, especially that in the blackcap, *Sylvia atricapilla*, an Old World warbler that has been the subject of extensive research in my laboratory since the early 1970s. My reason for restricting the taxonomic focus of this chapter is that I intend not to provide a review of migratory behavior, but instead to illustrate the ways in which careful study of geographically disparate populations can provide insights into both the causation and evolution of migratory behavior.

Not only is migration a uniquely geographical phenomenon, it should also result in substantial genetic structuring of populations across widespread species. This phenomenon has recently been documented in the dunlin (*Calidris alpina*), a widely distributed, migratory shorebird (Wenink et al. 1993). Such studies, in combination with extensive tagging studies and, more recently, satellite tracking programs, are providing unequivocal evidence that many avian species comprise populations that traditionally use geographically disparate, well-defined breeding grounds and wintering areas (e.g., Moreau 1972, Zink 1973–1985). Because selection pressures differ among sites, adaptive differentiation of migratory behavior is expected and has been demonstrated (e.g., Salomonsen 1955, Herrera 1978, Mead 1983).

All of these assets for research on the evolution of geographic variation in migratory behavior characterize the blackcap. In addition, more is known of the genetic bases of geographic variation in several aspects of migratory behavior in the blackcap than in any other avian species (Berthold 1996). This is a critical component of evolutionary studies because demonstration of evolutionary change requires evidence of underlying genetic differentiation (e.g., Fisher 1930, Endler 1986). I first describe, in general terms, geographic variation in the migratory behavior of the blackcap. I then present evidence for genetic differentiation of several elements of migratory behavior and provide evidence that populations of obligate partial migrants harbor sufficient genetic variation to respond extremely rapidly to divergent selection for several aspects of migratory behavior. Finally, I describe an example of rapid microevolutionary change in the migratory behavior of a wild blackcap population. I conclude with a discussion of the merits of geographic comparison for elucidating evolutionary pattern when there are both within- and between-population variation in traits such as those behavioral characters that together make up migratory behavior.

The Blackcap

Sylvia is a genus of Old World warblers comprising about 17 species that range in migratory behavior from exclusive, long-distance migrants (traveling from Siberia to southeast Africa) to resident populations on Atlantic islands. Among these species is the blackcap, *Sylvia atricapilla*, which exhibits sufficient population differentiation that almost all the migration strategies observed in Eurasian passerines migrating within Eurasia or to Africa are present in at least one population (Berthold et al. 1990b). The species has proven amenable to laboratory experimentation and breeding, facilitating research into patterns of migratory behavior and their causes.

Some blackcap populations are characterized by exclusive, long-distance migration. For instance, birds from north European and Asian breeding areas migrate to southeast Africa, and to a lesser extent from central Europe to western Africa (fig. 8-1). Most central European populations are middle-distance migrants wintering in the Mediterranean area or, recently, on the British Isles. In southern

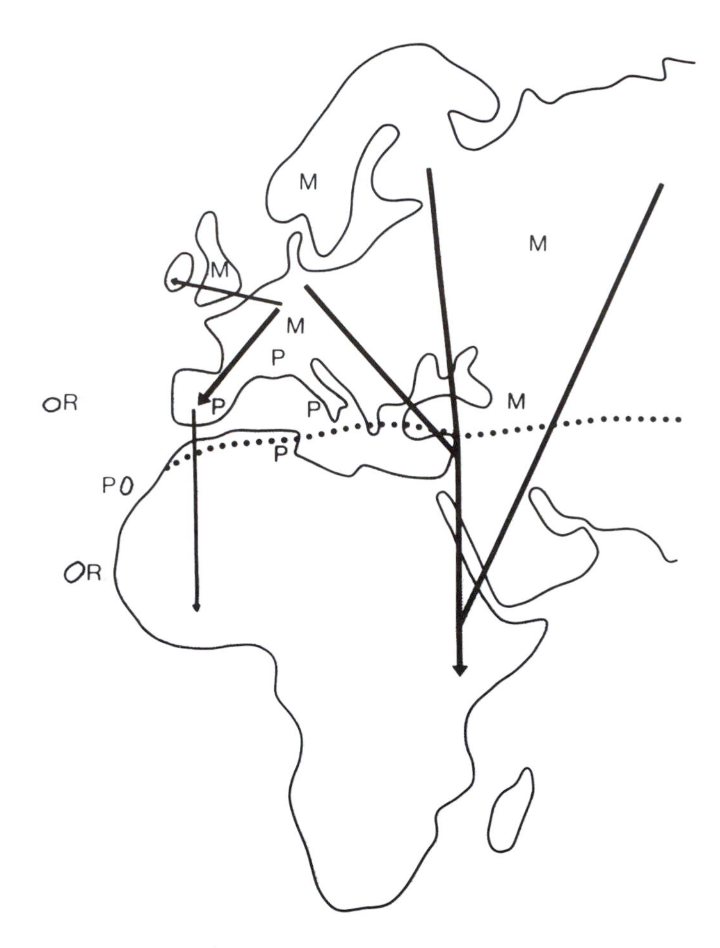

Figure 8-1 The blackcap (*Sylvia atricapilla*) breeds in Europe and west Africa as far north as about 69° latitude and in Africa as far south as about 15° (on the Cape Verde Islands). Populations of this species are exclusively migratory (M), partially migratory (P), or resident (R). Thick arrows indicate main migration routes (to African and Mediterranean winter quarters), thin arrows show by-routes (to west African and northern wintering areas), the dotted line shows the southern border of the continental breeding grounds (after Berthold 1988a).

Europe, northern Africa, and on the Canary Islands, blackcaps are obligate partial migrants (i.e., part of the population migrates every year) and short-distance migrants. On the Cape Verde Islands (Berthold et al. 1990a), on Madeira (P. Berthold unpublished data), and certainly also on the Azores, resident populations have evolved. Blackcaps have two migration divides. One is between European populations migrating in an easterly or westerly direction to southern Europe and Africa, and another one is in central western Europe, splitting birds into those which migrate to the south and others that fly to the northwest to wintering grounds on the British Isles.

Southern populations are small and have short, round wings. Individuals from northeastern populations are larger and have more pointed wings that are longer (up to 5%). These characters are apparently under quantitative control, in that hybrids produced by crossing two populations, one from either end of the morphological spectrum, expressed intermediate phenotypes (Berthold and Querner 1982).

Migratory populations show considerable body mass increase (up to about 50% lean weight) during migratory periods due to fat deposition. The resident population on the Cape Verdes is characterized by two annual breeding periods and, accordingly, by a two-peaked gonadal cycle. Fat deposition in this population occurs during the autumnal breeding period, which often falls in a rather dry period. These patterns were again shown to reflect underlying genetic differences in the populations (Berthold et al. 1990b, Berthold and Querner 1993). Finally, the timing of juvenile molt is genetically controlled and is adjusted to the time of migration (Berthold and Querner 1982).

Thus, the blackcap is highly structured geographically. As has been suggested for the dunlin (Wenink et al. 1993), the diversity of population types may, in part, trace its origin to Pleistocene population fragmentation. As described below, this diversity extends to behavioral phenotypes associated with migration.

Geographic Variation in the Components of Migratory Behavior

Components of migratory behavior that have been studied in detail include (1) the urge to migrate, as indicated by migratory activity (i.e., migratory restlessness or *Zugunruhe* in captive individuals kept in registration cages; Berthold 1990b), (2) migratory direction, as indicated by directional preferences in orientation cages, and (3) spatiotemporal programs, as indicated by shifts in orientation preferences during the typical period of migration. In each case, general background on the behavior is presented, and then population differences in expression of the behavior in blackcaps is described along with information on the genetic underpinnings of the differences.

Control of the Urge to Migrate

In many late-migrating bird species, especially facultative migrants which migrate in some seasons but not in all, departure in autumn or winter is most likely to occur as a direct response to deteriorating environmental factors (see Berthold 1993 for a review). This is obviously not true for early migrants, many of which depart at the peak of the summer season. These latter migrants were those which attracted the attention of Von Pernau in 1702, and it is in these birds that pronounced endogenous rhythms have been documented (Gwinner 1986).

These innate rhythms, termed "circannual rhythms," were first demonstrated in the genera *Phylloscopus* and *Sylvia* (Gwinner 1986, Berthold 1988a). They have subsequently been shown to control many annual processes including the initiation and temporal courses of migratory disposition and migratory activity. They are self-sustained, of life-long efficacy, and free-running under constant experimental conditions (Berthold 1978a, Gwinner and Dittami 1990). Because these circannual rhythms are self-started, they are probably entirely innate, as is the initiation of migration they mediate. In two migrants, blackcaps and to some extent European robins (*Erithacus rubecula*), cross-breeding and selective breeding experiments have demonstrated the expected genetic control of the essential migratory characteristics (Berthold 1988a).

Crosses between populations of blackcaps with extremely different habits have proven particularly useful in elucidating the genetic control of migratory activity. When hand-reared blackcaps from an exclusively migratory central European population (southern Germany) were crossed with hand-reared birds from a nonmigratory population from the Cape Verde Islands, 40% of the F_1 hybrids exhibited migratory activity. Thus, the urge to migrate is heritable.

Comparisons of population-specific onsets of migratory activity in captive individuals with dates of initiation of autumn migration in free-living conspecifics strongly support the view that the inherited urge to move is the fundamental releaser for departure in many migrants (Berthold 1990b). In 19 central European passerines, including 9 *Sylvia* populations, the dates of onset of autumn migration in wild birds were strongly correlated with the dates on which migratory activity was first observed in captive individuals from the same populations ($r = .967$). Again, population comparisons provide evidence of innate differences in migratory activity.

Other evidence for the inheritance of the urge to migrate has come from selective breeding studies using obligate partial migrants. When migrant and nonmigrant individuals from a partially migratory Mediterranean blackcap population were selectively bred, both selected lines responded strongly (fig. 8-2, Berthold et al. 1990a). Similar results were obtained from laboratory and field studies of European robins (Biebach 1983), European blackbirds (*Turdus merula*, Schwabl 1983), and song sparrows (*Melospiza melodia*, Nice 1937, Berthold 1984a).

In combination, these results clearly demonstrate that at least in some migrating passerine genera, including *Sylvia, Erithacus*, and *Turdus*, geographic variation in migratory urge and in the date of departure for autumn migration depend on genetically programmed, endogenous rhythms. These genetically based differences in migratory tendency, amount of migratory activity, and timing are found not only between populations, but also within populations, suggesting that these traits could respond rapidly to changes in selective regime, thereby producing the observed population differences (Berthold and Pulido 1994).

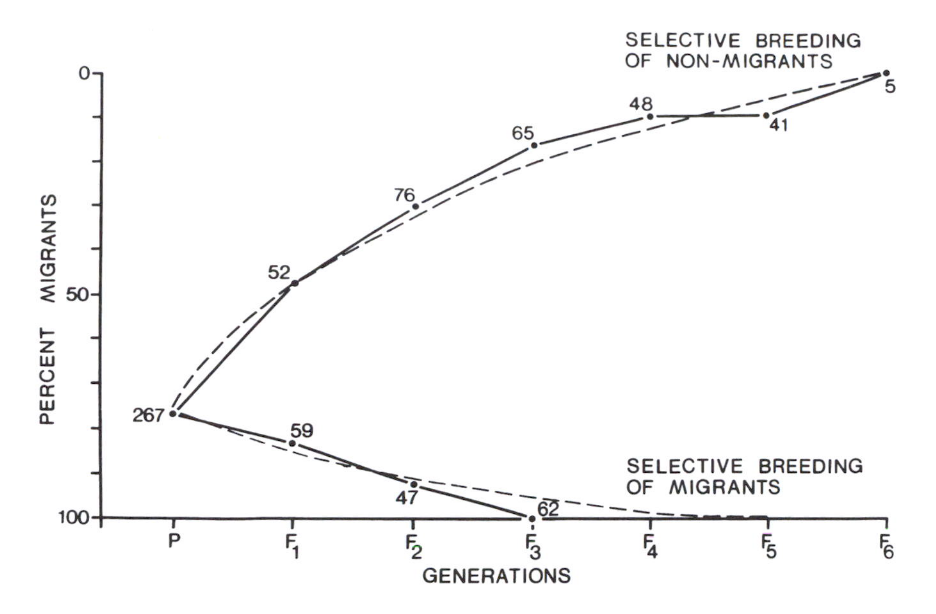

Figure 8-2 Results of a two-way selective breeding experiment with partially migratory blackcaps from southern France. Nonmigrants were bred up to the F_6 generation, migrants up to the F_3 generation. Numbers indicate individuals tested in each generation. The broken lines represent mathematical functions that best fit the selection response (after Berthold et al. 1990a).

Control of Migratory Direction

Genetic programming of migratory behavior is not restricted to the timing of the onset of migratory activity. Evidence of preprogramming of migratory flight direction is also accumulating rapidly. Recoveries of banded individuals indicate generally consistent species- and population-specific migration routes, often spanning decades (Zink 1973–1985). Migration-inexperienced individuals from such populations regularly display directional preferences consistent with those shown by ringing recoveries when tested in orientation cages, suggesting that the preference is innate (see Berthold 1991b, 1996 for reviews). Individuals from populations that shift flight direction during migration display orientation shifts at appropriate dates in captivity. Garden warblers (*Sylvia borin*) and Blackcaps from European populations first migrate to the Mediterranean in southwesterly and southeasterly directions, respectively, but then shift direction of migration to the south, toward African winter quarters. Both perform similar directional shifts under experimental conditions (Gwinner and Wiltschko 1978, Helbig 1991).

Displacement experiments, in which birds are moved from their normal ranges, also support the existence of innate directional programming. Transplanted individuals from a number of species, including European starlings (*Sturnus vulgaris*) and white storks (*Ciconia ciconia*) traveled in directions nearly parallel to those they would normally have flown (see Berthold 1991b, 1993, for reviews).

Blackcaps also use population-specific migration routes. Again, innate differences in directional preferences for migration were revealed in migration-inexperienced individuals from different populations, and the phenotypes of population hybrids provided evidence of underlying genetic differences. For example, when fully migratory blackcaps from southern Germany were crossed with nonmigratory conspecifics from the Cape Verdes, hybrids displayed the directional preference of the migratory German population (northwest–southeast axis, fig. 8-3,

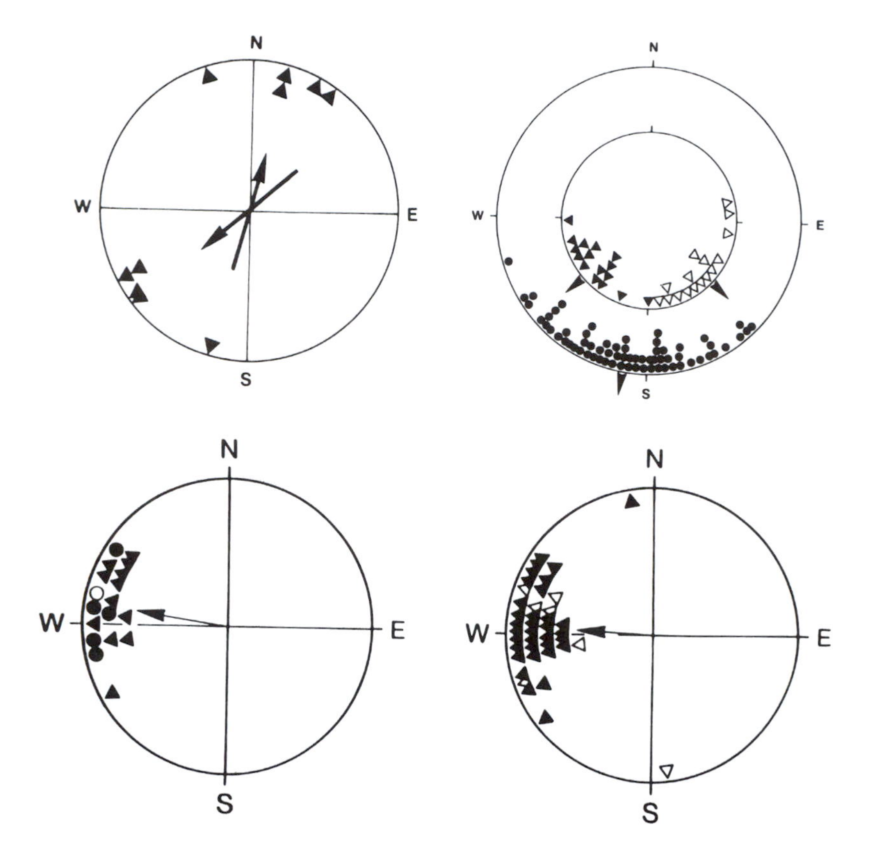

Figure 8-3 (Top left) Directional preference of F_1 hybrids of blackcaps from migrants from central Europe and nonmigrants from the Cape Verdes. The northern ends of the spring axes of individual birds are given in the upper semicircle, the southern ends of the autumn axes in the lower, with arrows in the center representing the mean vectors (after Berthold et al. 1990c). (Top right) Individual means for directional choices during the autumn migratory period and overall means (arrows) of blackcaps from southern Germany (solid triangles) and eastern Austria (open triangles). Inner circle shows data from parents, outer circle (filled circles) shows those from F_1 hybrids (after Helbig 1991). (Below) Individual means of directional choices during the autumn migratory period (symbols) and overall means (arrows) of blackcaps caught in winter in Britain and tested the following autumn in southern Germany (left) and of their F_1 offspring (right). Filled symbols: vector significant (after Berthold et al. 1992).

Berthold et al. 1990a). In another experiment, blackcaps from both sides of the central European migration divide were crossed (i.e., southwesterly migrants from southern Germany and southeasterly migrants from eastern Austria), and F_1 hybrids displayed an intermediate orientation (fig. 8-3, Helbig 1991). Finally, birds of continental European origin, which have recently evolved a novel migratory direction to the west–northwest were trapped in England (Berthold et al. 1992). Their offspring, tested in orientation cages, showed a corresponding west–northwest directional preference in autumn (fig. 8-3). Clearly, the directional orientation behavior of blackcaps is population-specifically preprogrammed and innate.

Population-specific Spatiotemporal Programs

Early comparative studies of the migratory activity of caged *Phylloscopus* and *Sylvia* warblers indicated that long-distance migrants displayed greater amounts of activity than did middle-distance migrants (Gwinner 1968, Berthold et al. 1972). Later comprehensive studies involving 13 species and populations of *Sylvia* that ranged from long- to short-distance migrants have consistently provided evidence of a positive relationship between the amount of migratory activity and the length of the migration route (Berthold 1984b).

This relationship is also evident among populations of the blackcap (Berthold 1978b), and, again, the pattern of differences persists under experimental conditions, indicating endogenous control mechanisms. Not unexpectedly, a cross-breeding experiment between middle-distance migrants from Europe (southern Germany) and short-distance migrants from Africa (Canary Islands) resulted in intermediate patterns and amounts of migratory activity in the offspring (Berthold and Querner 1981). Thus, the patterns of migratory activity in the blackcap, and probably in many other bird species, are population-specific, inherited characteristics.

Detailed comparisons of patterns of migratory restlessness in caged birds with distances covered and types of migration performed by wild conspecifics have shown that migratory restlessness closely mirrors the migration pattern of conspecifics in the wild with respect to duration of the migratory period and the speed of migration, although intensity varies temporally (normally depending on environmental circumstances). Long-distance migrants usually exhibit pronounced migratory restlessness and short-distance migrants small amounts, while medium-distance migrants are intermediate in behavior. Second, two different methods of analysis have revealed that the amount of migratory restlessness would, if exhibited as migratory activity in the wild, carry migrants from their breeding grounds to their winter quarters. From banding recoveries of *Phylloscopus* warblers, Gwinner (1968) calculated distance performances for different stages of the autumn migration, correlated these results with corresponding values for migratory restlessness, and derived theoretical migration distances by extrapolating from the total amount of restlessness during the whole autumn migration period. The distances covered would lead the birds to the center of the species-specific winter quarters. In the garden warbler, migratory restlessness was recorded and analyzed using video recordings under infrared light conditions at night. Restlessness is almost exclusively expressed by whirring (the generation of high-frequency wingbeats of low amplitude while perching). If the total time of whirring activity is multiplied by the average flight speed of the species, the resulting distance would again lead the birds to the center of their species-specific winter quarters (Berthold 1991b).

These results, together with those on innate migratory directions, strongly support the view that migrants like blackcaps are equipped with inherited spatiotemporal programs for migration. Such programs appear to provide an essential basis for determining the distance between breeding grounds and unknown winter quarters in inexperienced first-time migrants. These programs seem to lead inexperienced migrants "automatically" to their population-specific wintering grounds on a genetically determined time-and-directon vector or a bearing-and-distance basis (vector-navigation hypothesis, Berthold 1991b).

Partially Migratory Populations: A Source of Genetic Variation for Microevolutionary Change?

Partially migratory populations are of two types. They are described as obligate when part of a population migrates every year. In contrast, in some years, no members of facultatively migratory populations migrate. The distinction is critical because the underlying control mechanisms may well differ between these two kinds of populations (Terrill and Able 1988, Berthold 1993).

The control of partial migration has been discussed on a purely theoretical basis for many years. In the course of these discussions, two main contradictory and controversial hypotheses have been proposed: the "genetic" hypothesis and the "behavioral-constitutional" or "dominance rank" hypothesis. The genetic hypothesis, favored by Nice (1937) and Lack (1943), holds that genetic control mechanisms are responsible for the coexistence of migratory and nonmigratory individuals within populations. The second hypothesis, proposed by Kalela (1954) and Gauthreaux (1978), following Miller (1931), suggests that the decision to migrate is condition dependent, such that strong individuals tend to win conflicts in the autumn, and stay on the breeding grounds as residents, while losers are forced to depart.

We have demonstrated in one obligate, partially migratory blackcap population from southern France that the urge to migrate is heritable and thus not simply the outcome of contests between individuals with genetically similar migratory tendencies. When migrant and nonmigrant individuals from the population were selectively bred, the responses to selection were strong, and measured heritabilities were reasonably high (fig. 8-2, Berthold 1988b). Corresponding results have been obtained in studies of other obligate, partially migratory species. Biebach (1983) documented genetic influences on migratory tendencies in the European robin using selective breeding, and Schwabl (1983) found that migratory parents produced more migratory-active young than did resident pairs of European blackbirds when young were hand reared. Genetic factors were also involved in the control of obligate partial migration in a wild population of song sparrows studied by Nice (1937; for reevaluation see Berthold 1984a).

The conclusion that genetic factors play an important role in obligate, partial migration is also indicated by the fact that habits of migratoriness and sedentariness are highly stable within individuals. Changes from sedentariness to migratoriness are extremely rare, whereas shifts in the opposite direction are more common. These shifts, however, are often age-dependent under controlled laboratory conditions and may therefore be endogenously controlled by the maturation processes as well as by environmental factors (Berthold 1990a, Schwabl and Silverin 1990).

The inheritance of the urge to migrate is best explained as a multilocus system with a threshold for expression in both the blackcap and the European robin (Berthold 1988b, Berthold et al. 1990c). Such a control mechanism suggests that partially migratory populations are composed of individuals with programs for migratoriness and sedentariness differing in strength. Individuals with strong programs either migrate or fail to do so consistently according to their genotype. In weakly programmed individuals, external factors may also influence decision making (see last section of this chapter).

Regardless of the mode of genetic determination of migratory tendency, the weight of available evidence suggests that there are high levels of genetic variation in populations of obligate partial migratory populations. Thus, they should have the potential to respond rapidly to selection on appropriate elements of migratory behavior. This was demonstrated clearly in a two-way selective breeding experiment using obligate partially migratory blackcaps from southern France (fig. 8-2). When migratory parents were selectively bred, the resulting lines became completely migratory in just three generations. Those in which nonmigratory individuals were bred became almost exclusively nonmigratory in four to six generations. Hence, selection responses are rapid, indicating a high evolutionary potential. Similar results have been obtained in a selection experiment using a population of European robins that also displayed obligate partial migration (Biebach 1983).

These results demonstrate that at least some populations of obligate partial migrants harbor the variation necessary to permit rapid microevolutionary change in migratory behavior in response to appropriate changes in selection regimes. This potential may have permitted the survival and evolution of the many populations of the blackcap and of other species that dispersed into novel winter or summer habitats and survived as viable populations with novel spatial and temporal patterns of migratory activity. Recent observations of microevolutionary change in the migratory behavior of blackcaps suggests that even exclusively migratory populations may harbor sufficient variation to foster rapid change.

Recent Microevolutionary Change in Migratory Behavior

Over the last 30 years, increasingly large proportions of exclusively migratory blackcaps in central European populations have migrated toward the west–northwest in autumn, establishing a novel migratory direction and a new wintering area in the British Isles. Until the 1950s, the blackcap rarely wintered in the British

Isles. Today, wintering birds number in the thousands. Banding recoveries have demonstrated that these winter residents are migrants from central European populations, the members of which had previously migrated exclusively in a southwesterly direction to the western Mediterranean region (Berthold et al. 1992). The novel migratory habit has presumably been favored by a number of advantages of the new winter quarters. These include improved winter food (primarily at bird feeders), climate amelioration, shorter migration distance, less intraspecific competition in the wintering area, earlier return to breeding areas due to an earlier termination of the photorefractory period and an advanced spring photoperiod, physiological preadaptation to potentially harsh conditions on the breeding grounds in early spring, occupancy of best breeding territories, and possible assortative mating among early arrivals from the British Isles.

We have trapped blackcaps that wintered in England and determined their directional preferences and those of their offspring bred in our aviaries. Both displayed the novel west–northwest orientation in autumn (fig. 8-3). The concordance between the adults and their offspring shows that the novel migratory direction is heritable (Berthold et al. 1992, Berthold 1995). The rate of evolutionary change could have been enhanced by assortative mating on shared breeding grounds, as individuals wintering in the British Isles reach their central European breeding areas about 2–3 weeks earlier, on average, than do their conspecifics returning from Mediterranean wintering grounds (Berthold 1994), but this possibility remains to be examined. Even if some degree of assortative mating does occur, mating between pairs with different wintering sites must be expected. We found, crossing individuals trapped in southern Germany with those trapped in England, that F_1 offspring oriented significantly toward 253° or west–southwest. This directional preference was intermediate to, and significantly different from, those of previously tested birds from south Germany and those wintering in Britain. Our results indicated that about one-third of the young of the mixed pairs would migrate to the British Isles to winter (Helbig et al. 1994).

These results demonstrate that novel directional preferences for migration can increase rapidly in frequency in blackcap populations when favored by natural selection. This is the best documented case of rapid microevolutionary change in the migratory behavior of an animal species and, to our knowledge, the first case in any wild vertebrate in which a drastic and recent evolutionary change of behavior has been documented and its genetic basis established (Berthold et al. 1992). Recent, rapid loss of migratory tendencies is indicated in the great crested grebe (*Podiceps cristatus*) in the Netherlands. In recent years, the proportion of locally wintering Dutch grebes has increased, apparently because human activities have increased lake area, enhancing habitat suitability for overwintering (Adriaensen et al. 1993). Whether this change will reflect genetic change in the population as hypothesized remains to be tested. Given the remarkable number of cases in which the behavior of migrants appears to have an underlying genetic basis, the hypothesis is certainly plausible. This view is supported by recent heritability estimates for migratory activity measured as regressions of offspring on midparent values for parents captured in the southern Germany study population.

Heritability estimates (h^2) ranged from 0.37 to 0.46, demonstrating substantial evolutionary potential for migratory behavior within populations. Under moderate selection intensities, the southern German blackcap population could evolve from a middle-distance into a short-distance migrant in 10–20 generations given these heritability values (Berthold and Pulido 1994).

Prospects

The study of *Sylvia* warblers and above all, of blackcaps, has shown that pronounced geographic variation in migratory behavior is generally based on genetic variation. Differences between migratory and resident populations, varying degrees of migratoriness in obligate partially migratory populations, and differences in the direction and spatiotemporal patterns of migration were all shown to be under at least partial genetic control. Research on several other migrants has provided similar insights, although population differentiation has typically been studied in less detail. Given that many migrants leave breeding grounds during the summer before the environmental changes occur that trigger departure in later migrants, and because young of the year often migrate before migration-experienced adults, widely distributed heritable migration programs are to be expected.

In the blackcap population from southern Germany in which the heritability of migratory activity has been estimated (Berthold and Pulido 1994), the values fall on the high end of typical measures for both behavior and life history (Mousseau and Roff 1987), including traits associated with migration (Dingle 1991). This seems reasonable, insofar as migratory behavior straddles the two classes of traits and should therefore fall within the ranges of both. The fact that the estimates were on the high end of the distribution (approximately 80% of life-history estimates and 65% of behavior estimates were lower; Mousseau and Roff 1986) may reflect the dichotomy in autumn migration patterns observed in the population from southern Germany. Lower values might be expected for populations of obligate migrants with a single, population-wide migratory route.

Regardless of these considerations, there clearly exists sufficient genetic variation within populations of blackcaps and other migrants to facilitate the evolution of novel migratory tendencies and pathways. That this is so is indicated by the remarkable shift of some central European migrants to a novel wintering area in the British Isles that is one-third closer to the breeding grounds than the traditional Mediterranean wintering area. This shift has taken place in the last 30 years and was followed in detail because both scientists and amateurs collect far more complete distributional data than are available for other taxa. In combination with banding studies, these data afford a source of invaluable information for detecting the early stages of microevolutionary change in migratory behavior. Although we have only one well-documented case of such a shift, others may be expected, especially as anthropogenically induced changes in habitat and climate continue (Berthold 1994, Berthold and Pulido 1994). The more detailed the information available before these events, the better we should be able to predict and interpret the courses of future microevolutionary changes in migratory behavior (Berthold and Pulido 1994).

One of the most remarkable elements of the change in migratory route observed in central European blackcaps during the last 30 years is the speed with which the novelty has spread. Typically, evolutionary biologists are restricted to estimating rates of change on the basis of outside dates such as the last glacial recession yielding the first available habitat for occupation (e.g., Bell and Foster 1994, Coss this volume). Such estimates are likely to indicate slower rates of change than actually occurred because the evolutionary transition may have have spanned only a small segment of the total time available (Gingerich 1983).

In some instances, however, recent introductions of known date, followed by changes in character states of the introduced species or of a species with which it interacts have permitted better estimation of the rates of microevolutionary change (see, e.g., Carroll and Boyd 1992, Coss this volume). Estimates of rates of change based on the timing of introductions suffer the same limitations as those described above, but the narrower window of time is likely to provide a more accurate measure of rate of change. Only when the initiation of the microevolutionary event is detected and the process followed, as has been possible with the blackcaps, are the estimates of rates of change really accurate. Both introductions and our work indicate the rates of microevolutionary change can be rapid indeed.

Studies of geographic variation in migratory behavior are providing us with a wealth of information both on microevolutionary processes and on the specifics of the evolution of migratory behavior. The power associated with the ability to cross differentiated populations to explore patterns of inheritance, along with the ease of experimental study of passerines, makes them excellent subjects for such research. Ideally, in the future we should be able to use population comparisons not only to evaluate the patterns of evolutionary change in migratory behavior and the role of genetic variation in facilitating that change, but also to understand the physiological and developmental links between genotype and behavior, both of which are poorly understood at present (Schwabl and Silverin 1990, Wingfield et al. 1990). Another important aspect will be to elucidate the genetic–environmental interactions involved in the phenotypic expression of avian migratory behavior. An interplay between genotype and environment was suggested by field observations on a population of partially migratory European robins in Belgium (Adriaensen and Dhondt 1990), a species in which Biebach (1983) has demonstrated a genetic basis for the control of migratory behavior. Almost all females proved to be migrants, as were 70% of the males in the woodland, whereas most males in parks and gardens were resident. Similarly, in a population of blackcaps from southern France in which there were more migratory females than males, residents were more common in evergreen woods, and migrants in deciduous forests. In this instance, however, sex-linked inheritance of migratory tendency was indicated (Berthold 1986, 1993). It will be an exciting task to disentangle the varying importance of genetic and environmental input into the expression of individual migratory behavior.

References

Adriaensen, F., and A. A. Dhondt. 1990. Population dynamics and partial migration of the European robin (*Erithacus rubecula*) in different habitats. Journal of Animal Ecology 59:1077–1090.

Adriaensen, F., P. Ulenaers, and A. A. Dhondt. 1993. Ringing recoveries and the increase in numbers of European great crested grebes (*Podiceps cristatus*). Ardea 81:59–70.

Baker, R. R. 1991. Fantastic journeys. The marvels of animal migration. Weldon Owen, Syndey.

Bell, M. A., and S. A. Foster. 1994. Introduction to the evolutionary biology of the threespine stickleback. *In* M. A. Bell and S. A. Foster, eds. The evolutionary biology of the threespine stickleback, pp. 1–27. Oxford University Press, Oxford.

Berthold, P. 1978a. Circannuale Rhythmik: Freilaufende selbsterregte Periodik mit lebenslanger Wirksamkeit bei Vögeln. Naturwissenschaften 65:546.

Berthold, P. 1978b. Endogenous control as a possible basis for varying migratory habits in different bird populations. Experientia 34:1451.

Berthold, P. 1984a. The control of partial migration in birds: A review. Ring 10:253–265.

Berthold, P. 1984b. The endogenous control of bird migration: A survey of experimental evidence. Bird Study 31:19–27.

Berthold, P. 1986. Wintering in a partially migratory Mediterranean blackcap (*Sylvia atricapilla*) population: Strategy, control, and unanswered questions. *In* A. Farina, ed. Proceedings of the First Conference on Birds Wintering in the Mediterranean Region. Supplement to Ricerche di Biologia della Selvaggina 10:33–45.

Berthold, P. 1988a. The control of migration in European warblers. Proceedings of the International Ornithological Congress XIX:215–249.

Berthold, P. 1988b. Evolutionary aspects of migratory behavior in European warblers. Journal of Evolutionary Biology 1:195–209.

Berthold, P. 1990a. Genetics of migration. *In* E. Gwinner, ed. Bird migration: Physiology and ecophysiology, pp. 259–280. Springer-Verlag, Berlin.

Berthold, P. 1990b. Wegzugbeginn und Einsetzen der Zugunruhe bei 19 Vogel-populationen—eine vergleichende Untersuchung. Journal of Ornithology 131 (supplement): 217–222.

Berthold, P. 1991a. Patterns of avian migration in light of current global 'greenhouse' effects: A central European perspective. Acta XX Congress International Ornithology 780–788.

Berthold, P. 1991b. Spatiotemporal programmes and genetics of orientation. *In* P. Berthold, ed. Orientation in birds, pp. 86–105. Birkhäuser-Verlag, Berlin.

Berthold, P. 1993. Bird migration. Oxford University Press, Oxford.

Berthold, P. 1994. Verhaltensänderungen von Zugvögeln. Tumult. Schriften zur Verkehrswissenschaft 38–49.

Berthold, P. 1995. Microevolution of migratory behaviour illustrated by the blackcap *Sylvia atricapilla*. 1993 Witherby Lecture. Bird Study 42:89–100.

Berthold, P. 1996. Control of bird migration. Chapman and Hall, London.

Berthold, P., E. Gwinner, H. Klein, and P. Westrich. 1972. Beziehungen zwischen Zugunruhe und Zugablauf bei Garten- und Mönchsgrasmücke (*Sylvia borin und S. atricapilla*). Zeitschrift fur Tierpsychologie 30:26–35.

Berthold, P., A. J. Helbig, G. Mohr, and U. Querner. 1992. Rapid microevolution of migratory behaviour in a wild bird species. Nature 360:668–669.

Berthold, P., G. Mohr, and U. Querner. 1990a. Steuerung und potentielle Evolutionsgeschwindigkeit des obligaten Teilzieherverhaltens: Ergebnisse eines Zweiweg-Selektionsexperiments mit der Mönchsgrasmücke (*Sylvia atricapilla*). Journal of Ornithology 131:33–45.

Berthold, P., and F. Pulido. 1994. Heritability of migratory activity in a natural bird population. Proceedings of the Royal Society of London B 257:311–315.

Berthold, P., and U. Querner. 1981. Genetic basis of migratory behavior in European warblers. Science 212:77–79.

Berthold, P., and U. Querner. 1982. Genetic basis of moult, wing length, and body weight in a migratory bird species, *Sylvia atricapilla*. Experientia 38:801–802.

Berthold, P., and U. Querner. 1993. Genetic and photoperiodic control of an avian reproductive cycle. Experientia 49:342–344.

Berthold, P., and S. B. Terrill. 1991. Recent advances in studies of bird migration. Annual Review of Ecology and Systematics 22:357–378.

Berthold, P., U. Querner, and R. Schlenker. 1990b. Die Mönchsgrasmücke, *Sylvia atricapilla*. Die Neue Brehm-Bücherei. Ziemsen Verlag, Wittenberg Lutherstadt.

Berthold, P., W. Wiltschko, H. Miltenberger, and U. Querner. 1990c. Genetic transmission of migratory behavior into a nonmigratory bird population. Experientia 46:107–108.

Biebach, H. 1983. Genetic determination of partial migration in the European robin, *Erithacus rubecula*. Auk 100:601–606.

Carroll, S. P., and C. Boyd. 1992. Host race radiation in the soapberry bug: Natural history with the history. Evolution 46:1052–1069.

Dadswell, M. J., R. J. Klauda, C. M. Moffitt, R. L. Saunders, R. A. Rulifson, and J. E. Cooper. 1987. Common strategies of anadromous and catadromous fishes. American Fisheries Society Symposium 1. American Fisheries Society, Bethesda, MD.

Dingle, H. 1991. Evolutionary genetics of animal migration. American Zoologist 31:253–264.

Endler, J. A. 1986. Natural selection in the wild. Princeton University Press, Princeton, NJ.

Fisher, R. A. 1930. The genetical theory of natural selection. Oxford University Press, Oxford.

Gauthreaux, S. A. 1978. The ecological significance of behavioural dominance. *In* P. P. G. Bateson and P. H. Klopfer, eds. Perspectives in ethology, pp. 17–54. New York.

Gingerich, P. D. 1983. Rates of evolution: Effects of time and temporal scaling. Science 222:159–161.

Groot, C., and L. Margolis. 1991. Pacific salmon life histories. University of British Columbia Press, Vancouver.

Gwinner, E. 1968. Cirannuale Periodik als Grundlage des jahreszeitlichen Funktionswandels bei Zugvögeln—Untersuchungen am Fitis (*Phylloscopus trochilus*) und am Waldlaubsänger (*P. sibilatrix*). Journal of Ornithology 109:70–95.

Gwinner, E. 1986. Circannual rhythms. Springer-Verlag, Berlin.

Gwinner, E., and J. Dittami. 1990. Endogenous reproductive rhythms in a tropical bird. Science 249:906–908.

Gwinner, E., and W. Wiltschko. 1978. Endogenously controlled changes in migratory direction of the garden warbler, *Sylvia borin*. Journal of Comparative Physiology 125: 267–273.

Helbig, A. J. 1991. Inheritance of migratory direction in a bird species: A cross-breeding experiment with SE- and SW-migrating blackcaps (*Sylvia atricapilla*). Behavioral Ecology and Sociobiology 28:9–12.

Helbig, A. J., P. Berthold, G. Mohr, and U. Querner. 1994. Inheritance of a novel migratory direction in central European blackcaps. Naturwissenschaften 81:184–186.

Herrera, C. M. 1978. On the breeding distribution pattern of European migrant birds: MacArthur's theme reexamined. Auk 95:496–509.

Kalela, O. 1954. Populationsökologische Gesichtspunkte zur Entstehung des Vogelzuges. Annales Zoologici Societatis Zoologicae Botanicae Fennicae '*Vanamo*' 16:1–30.

Lack, D. 1943. The problem of partial migration. British Birds 37:122–130, 143–150.

Mead, C. J. 1983. Bird migration. Newness, Feltham.

Miller, A. H. 1931. Systematic revision and natural history of the American shrikes. University of California Publications in Zoology 38:11–242.

Moreau, R. E. 1972. The palaeartic-African bird migration systems. Academic Press, London.

Mousseau, T. A., and D. A. Roff. 1987. Natural selection and the heritability of fitness components. Heredity 59:181–197.

Nice, M. M. 1937. Studies in the life history of the song sparrow. 1. Transactions of the Linnean Society 4:1–247.

O'Connor, R. J. 1990. Some ecological aspects of migrants and residents. *In* E. Gwinner, ed. Bird migration, pp. 175–182. Springer-Verlag, Berlin.

Salomonsen, F. 1955. The evolutionary significance of bird migration. Danske Biologiske Meddelelset 22:1–62.

Schwabl, H. 1983. Ausprägung und Bedeutung des Teilzugverhaltens einer südwestdeutschen Population der Amsel *Turdus merula*. Journal of Ornithology 124:101–116.

Schwabl, H., and B. Silverin. 1990. Control of partial migration and autumnal behavior. *In* E. Gwinner, ed. Bird migration: Physiology and ecophysiology, pp. 144–155. Springer-Verlag, Berlin.

Terrill, S. B., and K. P. Able. 1988. Bird migration terminology. Auk 150:205–206.

van Noordwijk, A. J. 1984. Quantitative genetics in natural populations of birds illustrated with examples from the great tit, *Parus major*. *In* K. Wöhrmann and V. Loeschcke, eds. Population biology and evolution, pp. 67–79. Springer-Verlag, Berlin.

von Pernau, F. A. 1702. Unterricht. Was mit dem lieblichen Geschöpff, denen Vögeln, auch ausser dem Fang, nur durch die Ergründung deren Eigenschafften und Zahmmachung oder anderer Abrichtung man sich vor Lust und Zeitvertreib machen könne. Nürnberg.

Wenink, P. W., A. J. Baker, and M. G. J. Tilanus. 1993. Hypervariable-control-region sequences reveal global population structuring in a long-distance migrant shorebird, the Dunlin (*Calidris alpina*). Proceedings of the National Academy of Science USA 90:94–98.

Wingfield, J. C., H. Schwabl, and P. W. Mattocks, Jr. 1990. Endocrine mechanisms of migration. *In* E. Gwinner, ed. Bird migration: Physiology and ecophysiology, pp. 232–256. Springer-Verlag, Berlin.

Zink, G. 1973–1985. Der Zug europäischer Singvögel. Vogelzug-Verlag, Moeggingen.

9

Effects of Relaxed Natural Selection on the Evolution of Behavior

RICHARD G. COSS

Theoretical discussion of the role of natural selection in shaping behavioral variation in different habitats has been an integral part of the study of animal behavior since the late 19th century. Herbert Spencer (1888) was among the first to argue that migrating populations that fail to adjust to environmental circumstances "are the first to disappear." A common rationale for comparing populations or related species is the desire to identify behavioral differences that correspond with habitat properties providing different patterns of selection (Tuomi 1981, Riechert 1993, this volume). Behavioral similarities are often ignored or are treated as less interesting because the thrust of the research program emphasizes behavioral differences as an empirical test of the theory of natural selection. Nevertheless, these similarities can be as revealing of evolutionary process as are differences when they reflect behavioral convergence or slow disintegration of behavior under relaxed selection (Coss and Goldthwaite 1995).

When populations invade novel habitats, they not only experience new selective regimes; they can also experience relaxed selection on specific behavioral phenotypes. This is particularly common when the new habitat is missing a class of predators that was abundant in the ancestral habitat (e.g., Curio 1975, Pressley 1981). Under relaxed selection, characters may disintegrate, presumably because mutations that result in loss of the phenotype are not at a selective disadvantage. Disintegration is not always observed, however. Instead, behavioral characters are sometimes retained for long periods of time after selection has been relaxed (Coss 1991b, Kaneshiro 1989).

Inferring relaxed selection requires that the history of the contrasted populations be relatively well known. Both ancestral selective regimes and behavioral characters must be known if character polarity is to be established. Character polarity must be established to distinguish disintegration from parallel evolution of novel behavior patterns. This often is a problem in population contrasts because differentiation is usually too recent to have resulted in the evolution of enough derived characters for the use of standard cladistic methods of phylogenetic reconstruction, although recent advances in statistical and molecular techniques are promising (Foster 1994, Foster and Cameron 1996). Instead, inference of character polarity has typically relied on geological evidence and comparison with closely related species.

This chapter examines patterns of evolutionary disintegration and persistence of behavior under relaxed selection as revealed by population contrasts. I begin with a brief review of the literature that focuses on vertebrates and then provide an example with detailed discussion of variation in the antisnake behavior of ground squirrel populations, illustrating the ways in which cases of disintegration or persistence can be inferred. Although I provide examples of the ways in which information on character polarity can aid in understanding certain means by which natural selection produces adaptive phenotypes, the primary focus of the closing discussion is to explore the less often considered role of genetic and epigenetic processes in maintaining traits as mutations and other genetic changes accumulate (maintenance of characters under conditions of relaxed selection).

A Review of the Literature

Here I describe cases in which populations are thought to have invaded habitats devoid of predators with which the ancestors of the population were historically associated. A problem commonly encountered is determining whether selection is indeed relaxed. The absence of one selective agent does not rule out parallel directional selection on the trait. An excellent example is found in the stickleback from Trout Lake, British Columbia. Although sculpin, present in ancestral environments, were missing in this lake, the antipredator response typically elicited by this predator was retained (Pressley 1981). This could have been either because the behavior was retained in response to the presence of predatory trout in the lake or could instead reflect evolutionary persistence of a historical adaptation. In this instance, unequivocal inference of relaxed selection would have been inappropriate. I note these uncertainties when encountered.

Mimetic Egg Rejection: Disintegration under Relaxed Selection

Although not a typical predator–prey interaction, discrimination by birds of their own eggs from those of interspecific brood parasites provides some of the best examples of rapid disintegration in response to relaxed selection. In sub-Saharan Africa, the native village weaver, *Ploceus cucullatus*, readily distinguishes eggs of the didric cuckoo, *Chrysococcyx caprius*, from its own. Village weavers from Hispaniola, introduced during the 18th century slave trade, experienced no nest parasitism from the time of introduction until the arrival of the shiny cowbird (*Molothrus bonariensis minimus*) from South America in the 1970s. Village weavers from Hispaniola are less likely to reject eggs of parasitic species introduced to their nests than are those from Senegal, Africa, suggesting disintegration of the ability to recognize foreign eggs (cf. Victoria 1972, Cruz and Wiley 1989). Similarly, pied wagtails (*Motacilla alba yarrelii*) from Britain, where they are exposed to nest parasitism by the cuckoo (*Cuculus canorus*), are more likely to reject foreign eggs than are white wagtails (*M. a. alba*) from cuckoo-free Iceland, where massive deforestation by Vikings in the ninth century apparently decimated cuckoo habitat (Davies and Brooke 1989a).

The rapid loss of ability to discriminate foreign eggs could result from disintegration of an innate schema for egg markings (Davies and Brooke 1989b) or from loss of ability to learn and/or reject particular egg markings (see Rothstein 1974, 1978; Moksnes and Røskaft 1989). Indirect support for whole-clutch learning as the process for detecting foreign eggs is provided by the positive correlation between interclutch variation in egg appearance, but not intraclutch variation, and known egg rejection rates in a wide variety of European passerines (Øien et al. 1995).

Predator Recognition

In a classical study, Curio (1964, 1966, 1969) presented evidence that Galápagos finches retain alarm responses to snakes, owls, and hawks even on islands from which these predators are missing. Because these finches were derived from mainland stocks, predator–prey relationships with snake and avian predators are undoubtedly ancient. Dissipation of antipredator responses certainly could have occurred in the 3–4 million years since the Galápagos Islands first appeared, although the time of arrival of the finches is unknown (Hall 1983, Hickman and Lipps 1985). As one might have predicted, on snake-free Pinta Island, snakes elicit a fear response from *Geospiza fuliginosa minor* that is attenuated relative to that observed in *G. f. fuliginosa* on islands such as Santa Cruz where snakes are found.

In contrast, the Galápagos finch (*G. difficilis septentrionilis*) that inhabits predator-free Wenman Island is only slightly less fearful when confronted with hawk and owl models than is *G. fuliginosa*, a species that encounters both predators naturally. Although it is sympatric with short-eared owls (*Asio flammeus galapagoensis*), *G. difficilis acutirostris* on Genovesa Island exhibits the weakest alarm response to hawk and owl models. These results are difficult to interpret clearly but are worth mentioning because Curio then presented the birds with a headless hawk and owl, and with hawk and owl heads alone with and without eyes. Levels of alarm calling by members of all three populations were only slightly lower when presented with a stuffed owl without eyes, or with an owl head, than they were when presented with an entire owl. Only Wenman finches responded with reduced calling rates to an eyeless head, suggesting the uncoupling of the meaningful eye schema from other owl features with relaxed selection.

Morphometric and genetic analyses have been applied to Darwin's finches that are reasonably congruent in identifying the progression of species divergence (cf. Polans 1983, Schluter 1984). Application of Nei's D ($1\ D = 5$ Myr) to the genetic distance of *G. difficilis* and *G. fuliginosa* as a calibrator to time provides rough estimates of 45,000 years (Yang and Patton 1981) and 90,000 years (Polans 1983). Coupled with the uncertainty of island colonization, these estimates of species divergence cannot be used to generate a time scale for relaxed selection from hawks and owls on Wenman finches presumably associated with subspeciation of *G. difficilis* more recently in time.

In a similar vein, Curio (1961, 1975) demonstrated that the Spanish subspecies of the pied flycatcher (*Ficedula hypoleuca iberiae*) had largely lost the behavioral response to shrikes that characterizes *F. hypoleuca hypoleuca* in Germany. One female of this subspecies, which does not experience shrike predation, responded to presentation of a model redback shrike with a fully developed alarm calling response, whereas the others responded with little fear. All members of both subspecies exhibited similar alarm calling responses to models of pygmy and tawny owls, with which each is sympatric. Despite a small sample size and lack of good information on character polarity, shrike recognition is seemingly intact as a relic of previous selection in a small proportion of Spanish pied flycatchers. Curio (1975, 1993) has suggested that the distinguishing feature is the black mask of the shrike.

Population variation in antipredator responses of the threespine stickleback (*Gasterosteus aculeatus*) in northwestern North America also provides evidence of loss of recognition of predators to which ancestral marine populations were exposed. Prickly sculpin (*Cottus asper*) prey both on young stickleback in the nests and on adults (Foster and Ploch 1990). In Crystal Lake, British Columbia, where sculpin are missing, territorial male stickleback have lost the ancestral, cautious approach to predator inspection. In Trout Lake, where sculpins are also missing, the cautious behavior has been retained (Pressley 1981). These differences could reflect lower lake elevation and hence more recent separation of Trout Lake fish from marine populations or, possibly, the fact that Trout Lake fish have been exposed continuously to predation by trout, whereas trout were missing from Crystal Lake until recently. If similar perceptual cues are used to recognize both predators (see below), predation by trout could have maintained the ability to recognize sculpin as a threat even where sculpin were missing. Because these freshwater populations occupy habitat covered by ice during the last glacial maximum, loss of predator recognition in the Crystal Lake population must have occurred in the last 12,000 years (see Hughes 1985, Broecker and Denton 1989).

The deer mouse (*Peromyscus maniculatus*) is found in postglacial habitat in the same region. Weasels, major predators on the mainland, are absent from Moresby Island, which could have been invaded by deer mice no longer than 12,000 years ago. Nociception was used to determine whether deer mice from Moresby Island and the mainland respond differently to the scents of rabbits (nonpredators) and weasels (Kavaliers 1990). The mice have been shown to defer foot licking for longer periods when placed on a hot plate in the presence of a scent indicative of a dangerous predator than when in the presence of a nonthreatening scent. Latency to foot licking was highest for mainland mice exposed to the scent of weasel, but low for Moresby mice exposed to weasel scent and for both kinds of mice exposed to rabbit scent. Fifteen-minute exposures to weasel odor produced an opioid-mediated analgesic response in both populations of deer mice, a finding that could indicate latent recognition of weasel odor in Moresby mice unveiled by the longer exposure time.

Many species of *Peromyscus* are also targets of snake predation, a relationship that probably extends from the first appearance of *Peromyscus* during the late Miocene (Korth 1994). *Peromyscus maniculatus gambeli* encounters the Great Basin gopher snake, whereas *P. m. austerus* lives well outside its range (Nussbaum et al. 1983). Only *P. m. gambeli* distinguishes this dangerous snake from the nonthreatening racer snake (*Coluber constrictor*); its behavior resulted in longer survival than was observed for *P. m. austerus* (Hirsch and Bolles 1980). In this instance, *P. m. austerus* does appear to have lost most of its snake-adapted behavior. The time frame of relaxed selection is unclear, but it presumably was affected by paleoclimatic changes during the Pleistocene, perhaps as recently as the last Ice Age.

In summary, reports in the literature suggest that the responses to relaxed selection from major predators are somewhat unpredictable. In some cases, predator recognition and appropriate antipredator behavior patterns have been retained even over apparently long periods of relaxed selection. In other instances they have been lost. In an effort to better understand the pattern of variation and to provide a more precise time frame that could give us greater insight into the causes of these two divergent outcomes of relaxed selection, below I provide a detailed example of 12 populations of California ground squirrels with different histories of relaxed selection from snakes.

Population Variation in California Ground Squirrel Antisnake Behavior

North American ground squirrels live in burrows in which they rear young and take shelter from predators and adverse environmental conditions. These burrows are also used for refuge and thermoregulation by the rattlesnakes and gopher snakes, with which they co-occur throughout much of their range. Both types of snakes prey heavily on young ground squirrels, and in some cases upon adults as well (see Fitch 1948, 1949). The high intensity of predation imposed by snakes has fostered the evolution of a suite of behavioral and physiological responses that promote survival of the ground squirrels. Originally considered species-typical characters, these defenses have recently been shown to vary across populations of California ground squirrels in a pattern that appears to reflect local predation regimes. However, different characters have diverged at different rates, and certain behavioral characters have disintegrated slowly or not at all in response to reduced intensity of predation (Coss and Goldthwaite 1995, Coss and Biardi 1997).

In this section I focus on population differentiation of antipredator traits in the California ground squirrel, *Spermophilus beecheyi*, which has been particularly well studied. I illustrate the ways in which population contrasts can be used to describe evolutionary restraints on character disintegration under conditions of relaxed selection. Before presenting the differences among populations, I describe the suite of characters thought to represent the ancestral state and provide a brief explanation of the reasons they are thought to be ancestral. I then describe the pattern of antipredator character states across populations and examine the relationships in terms of predation regime and evolutionary persistence. I conclude with a brief discussion of the genetic and neurological constraints that can prevent disintegration under conditions of relaxed selection, drawing on data from California ground squirrels and from research on other taxa.

The Predator–Prey Relationship

Through much of their range, California ground squirrels co-occur with both Pacific gopher snakes (*Pituophis melanoleucus catenifer*) and northern Pacific rattlesnakes (*Crotalus viridis oreganus*). Both snakes prey on ground squirrels extensively, but the venom of the northern Pacific rattlesnake makes it a greater threat to the squirrels than is the Pacific gopher snake that kills by constriction. To denote this differential risk, Fitch (1948, 1949) estimated that California ground squirrels in the western Sierra Nevada foothills made up by weight 69% of the diet of the northern Pacific rattlesnake and 44% of the diet of the Pacific gopher snake. Juvenile and adult California ground squirrels have serum proteins that confer strong resistance to the venom of northern Pacific rattlesnakes (Poran and Coss 1990). However, there is substantial population variation in the degree of venom resistance, as indicated by both *in vivo* and *in vitro* bioassays (Poran et al. 1987, Towers and Coss 1990, Coss et al. 1993). Resistance is greater at sites where rattlesnakes are abundant, suggesting rapid adaptive evolution of this character.

Newborn ground squirrels are entirely dependent on adults for defense from snakes in burrows. Pups that have just emerged from natal burrows are at risk, first because they investigate snakes like adults and second because they can be taken easily due to their small body size, which delimits evasive leaping and venom neutralization. Adult California ground squirrels have a set of defensive behavior patterns that protect them and their young. Both above and below ground, adults employ behaviors that engender species-specific displays from the snakes, permitting species recognition and subsequent use of appropriate defensive behavior. Prolonged staring, conspicuous tail flagging, substrate throwing, and biting are used to deter snakes and often provoke defensive displays such as hissing or rattling that permit recognition of rattlesnakes (Owings and Coss 1977, Hennessy and Owings 1978, Coss and Owings 1985, 1989, Rowe and Owings 1990, Coss 1991a). Close-range inspection above and below ground also permits discrimination of a rattlesnake or gopher snake via olfaction (Hennessy and Owings 1978, Towers and Coss 1990). Once identified, ground squirrels change tactics for investigating and harassing snakes in keeping with their level of dangerousness (Towers and Coss 1990).

Laboratory-reared California ground squirrels from populations sympatric with both species of snakes have an innate ability to recognize snakes that is expressed several days before the pups emerge from natal burrows (approximately 45 days of age). Gopher snake odor engenders marked fear and alarm calling around the time the pups' eyes open (Coss 1991a). Both a gopher snake, especially its head region, and a static strip with snakelike speckling are startling at 40 or 41 days of age, coincident with the first day a pup uses vision to avoid obstacles (Coss 1991a). In the field, newly emerged pups quickly adopt adultlike elongate investigative postures and tail flag after detecting a rattlesnake, and they use caution in approaching a stick with a superficial resemblance to a snake (Poran and Coss 1990, Coss 1991a). During this period, laboratory-reared pups also spend more time investigating a caged rattlesnake than a caged gopher snake and use adultlike tactics of snake confrontation. This is potentially a period of high vulnerability in the field because adults rarely defend young above ground (Poran and Coss 1990).

Initial research on several geographically disparate populations of California ground squirrels suggested that differences in antipredator behavior across populations were associated with differences in predation regime (see Owings and Coss 1977, Coss et al. 1993). The differences were less pronounced than were those for venom resistance. Consequently, additional populations were studied in an effort to better understand the relationships between predation regime and antisnake behavior in the California ground squirrel. Before describing these differences, however, I present evidence that the behavioral phenotypes expressed in populations sympatric with both snakes are likely to represent the ancestral state of California ground squirrels.

Establishing Character Polarity

The relationship between ground squirrels and predatory snakes is ancient. Fossil deposits indicate temporal and regional contiguity of rattlesnakes, large colubrid snakes, and the *Otospermophilus* progenitor of modern ground-dwelling squirrels as early as the middle Miocene (Black 1963, Holman 1979). Ground squirrels and gopher snakes appear together in middle Pliocene deposits in Kansas (Hibbard 1941, Brattstrom 1967), and ancestral California ground squirrels and Pacific gopher snakes co-occur in nearly 1.8-million-year-old Irvingtonian fossil deposits in northern California (Firby 1968, Poran and Coss 1990). Rattlesnakes and California ground squirrels appear in the same Rancholabrean fossil assemblages of late Pleistocene age (Miller 1912, Stock 1918, Brattstrom 1953). Thus, the fossil record provides remarkably good evidence that the predator–prey relationship between ancestral California ground squirrels and gopher snakes pre-dates their divergence into Beechey (*S. b. beecheyi*) and Douglas (*S. b. douglasii*) ground squirrel subspecies in the middle Pleistocene (Smith and Coss 1984). Within each subspecies, population divergence has occurred mostly within the Rancholabrean time frame during sympatry with rattlesnakes and gopher snakes (Coss and Goldthwaite 1995).

Further insight is provided by comparing ground squirrel species that maintain taxonomic affinity with the stem *Otospermophilus* line with those whose ancestors diverged from it during the late Miocene (Shotwell 1956, Smith and Coss

1984). This initial surge of spermophile radiation was associated with brief cooling of the Northern Hemisphere (Hodell et al. 1986), a context that probably interrupted contact with predatory snakes in the Pacific Northwest and northern Great Basin. Among members of the cold-adapted *Spermophilus* subgenus, contact with snakes ended entirely for the ancestors of Arctic ground squirrels (*Spermophilus parryii*), who colonized northern Alaska earlier than 2.4 million years ago (Repenning et al. 1987) and are currently adapted to extreme cold (Barnes 1989). For other northern species, episodes of allopatry with snakes would have been most pronounced during renewed cooling of the Northern Hemisphere about 3.6 million years ago and during repeated episodes of glacial advances beginning in the late Pliocene (Hodell et al. 1986, Broecker and Denton 1989). As noted above, squirrel–snake sympatry continued throughout this time frame in warmer regions of the midwestern and southwestern United States.

Recognition of snakes is a highly specialized cognitive ability that encompasses contingencies and implies risks that lead to organized behaviors for dealing with them. Examination of ground squirrel species that maintained intermittent or continuous contact with snakes suggests that snake recognition was present in the common ancestor. Rock squirrels (*S. variegatus*), for example, retain a number of generalized morphological traits found in Miocene *Otospermophilus* (Black 1963). As apparent in their extensive use of trees for foraging and predator avoidance (Juelson 1970), rock squirrels also retain more generalized behaviors reflecting their tree squirrel heritage. Although sympatric with rattlesnakes and bullsnakes throughout much of their range, the antisnake behavior of rock squirrels has not been studied in detail. Our initial field research using a tethered bullsnake and a western diamondback rattlesnake (*Crotalus atrox*) indicates that rock squirrels will boldly pounce on the bullsnake and tail flag and harass the rattlesnake in a manner remarkably similar to that of California ground squirrels (compare Knickerbocker and Knickerbocker 1988, Coss and Owings 1989, Coss 1991a).

Richardson's ground squirrels (*S. richardsonii*) and Columbian ground squirrels (*S. columbianus*) are members of the *Spermophilus* subgenus that live in alpine and subalpine meadows and encounter rattlesnakes and gopher or bull snakes in some parts of their range (Hall 1981, Stebbins 1985). Both species recognize rattlesnakes and gopher snakes as predatory threats in a seminatural laboratory setting, and they confront these snakes by throwing substrate with their forepaws much like California ground squirrels (Towers 1990, Towers and Coss 1991). Columbian ground squirrels flag their long tails in a manner nearly identical to that of California ground squirrels. Because tail-flagging is uniquely associated with snakes (see Hennessy et al. 1981, Hennessy and Owings 1988, Hersek and Owings 1993, Coss and Biardi 1997), this behavior in taxa retaining long tails was probably a component of antisnake behavior before spermophile radiation in the late Miocene. Intriguing observations that eastern gray squirrels (*Sciurus carolinensis*) recognize snakes as dangerous and exhibit incipient tail flagging during close-range investigation (R. Sanchez personal communication) suggest that the raw material for conspicuous tail flagging might pre-date the early Miocene divergence of ground squirrels from tree squirrels.

Prairie dogs (*Cynomys*), another cold-adapted species, first appear in the late Blancan fossil record of Kansas about 2.5 million years ago (Eshelman 1975, L. Martin, personal communication), and the prairie rattlesnake (*C. v. viridis*), a major prairie dog predator, is thought to have originated in the western United States (Klauber 1972). Black-tailed prairie dogs (*Cynomys ludovicianus*) exhibit the ability to differentiate rattlesnakes from gopher or bullsnakes (Loughry 1989). They also share the common investigative behaviors, such as hesitant investigative approaches with elongate postures, elevated tails, and substrate throwing. Without long tails, they jump-yip as an alternative signaling analog to more specialized tail flagging (Owings and Owings 1979, Halpin 1983, Owings and Loughry 1985, Loughry 1988).

The study of Arctic ground squirrels from snake-free central Alaska provided the context for evaluating whether innate recognition of snakes presumed as the ancestral condition still persisted after 3–5 million years of relaxed selection. In above- and below-ground laboratory comparisons with laboratory-born California ground squirrels, laboratory-born Arctic ground squirrels failed to recognize a rattlesnake and a gopher snake as dangerous (Goldthwaite et al. 1990, Coss and Goldthwaite 1995). A caged rattlesnake or gopher snake, for example, was treated as a novel animate object despite its striking, rattling, or hissing. Caution was only apparent in the context of aversive feedback when an uncaged gopher snake was allowed to strike squirrels repeatedly. Following such strikes, several Arctic ground squirrels hesitated in their close-range investigation, adopted elongate postures, and threw substrate with their hind limbs in a manner analogous to that of other rodents that defensively bury a cylindrical prod that shocks them (see Heynen et al. 1989).

Comparison of Arctic and California ground squirrels revealed that the antisnake behavioral system is labile, with snake recognition and concomitant behavioral organization disintegrating entirely with prolonged relaxed selection in the million-year time frame. Snake recognition was not expected to decay in the thousand-year time frame because California ground squirrels from habitats where rattlesnakes are rare or absent maintain caution toward rattlesnakes, but not gopher snakes, like that of populations sympatric with both species of snakes (Owings and Coss 1977, Coss et al. 1993). Black-tailed prairie dogs from a site where rattlesnakes and bullsnakes are rarely encountered also continue to recognize snakes (Loughry 1988, 1989), again suggesting that the antisnake system is relatively robust. Unlike the precision required to recognize motionless snakes in heterogeneous settings, the organization of antisnake behavior is flexibly structured to accommodate stochastic changes in situational contexts (see Coss 1993). Some aspects of antisnake behavior have common components useful in other situations. Cautious investigation, for example, typically involves a high state of arousal and vigilance associated with regulation of distance and substrate throwing, features also exhibited by other rodents under threat (e.g., Heynen et al. 1989). Expression of these components of antisnake behavior was expected to vary in California ground squirrel populations with different histories of sympatry with snakes.

The Study Populations

The populations of California ground squirrels selected for study are members of the Douglas and Beechey subspecies distributed, respectively, north and south of the Sacramento and Feather rivers. Beechey ground squirrels were initially characterized as taxonomically distinct from the Fisher and Sierra subspecies on the basis of slight differences in body proportions and pelage coloration (Grinnell and Dixson 1918, Howell 1938, Hall 1981). These have been lumped into the Beechey subspecies because of their close genetic affinity, discussed in greater detail below (Goldthwaite 1989). Divergence of the two subspecies north and south of the contemporary Sacramento/San Joaquin Delta and San Francisco Bay is thought to have been initiated by the sudden onset of Great Valley drainage through this river system. Geological evidence indicates that rerouting of the Sacramento River drainage into San Francisco Bay occurred approximately 725,000 ± 15,000 years ago, providing a barrier to ground squirrel dispersal north and south in the Coast Range (Smith and Coss 1984).

Although northern Pacific rattlesnakes and Pacific gopher snakes are found throughout most of the range of California ground squirrels (Grinnell and Camp 1917, Nussbaum et al. 1983), some populations in each subspecies are found in areas from which both snakes are missing or rare (see fig. 9-1). Antisnake behavior was examined in four such populations of Douglas ground squirrels. Snakes are missing from the habitats of two Oregon populations (one at Logsden near Newport and one at the Finley National Wildlife Refuge) due to heavy coastal fog and winter rain. A third Douglas population at Mount Shasta does not encounter snakes due to altitude-based seasonal temperatures and a fourth, at Petaluma, rarely encounters either species of snake due to heavy summer fog (see Brattstrom 1965). Antisnake behavior was also examined in two populations of Beechey ground squirrels that experience relaxed selection from snakes. Snakes are absent from the Lake Tahoe Basin due to cool summer temperatures and are rare in the Sierra Valley Basin in the northern Sierra Nevada Range for the same reason. In all these populations, *in vivo* venom resistance and *in vitro* levels of venom resistance are very low (Poran et al. 1987, Towers and Coss 1990). Two other Beechey populations, Delta and Tracy, currently experience predation only by gopher snakes. The population called Delta is represented by animals collected from two adjacent islands in the Sacramento/San Joaquin River Delta that has probably lacked rattlesnakes since the late Pleistocene. Ground squirrels from the two islands have low venom resistance (Poran et al. 1987), although they proved genetically distinct in later study (Goldthwaite 1989). The Tracy population is a recent colonist of the rattlesnake-rare habitat on the floor of the Great Valley and exhibits moderate venom resistance. The remaining two Douglas study populations at Winters and Willows experience high levels of predation by gopher snakes and rattlesnakes, as do the two Beechey study populations at Folsom Lake and Walnut Creek. Venom resistance is moderate to high in these snake-sympatric populations.

Eleven of these populations were studied in a laboratory setting intended to

Figure 9-1 Geographic locations of 31 study populations of California ground squirrels (*Spermophilus beecheyi*). Asterisks indicate the five population sites north, and five population sites south of the San Francisco Bay-Sacramento River barrier used to calibrate Latter's ϕ^* genetic distance model to time (see text). Populations in italics are of the *beecheyi* subspecies, the remainder *douglasii*. Populations in bold text were selected for the study of antisnake behavior. Daggers (†) designate sympatry with rattlesnakes and carets (^) indicate sympatry with gopher snakes.

simulate conditions above ground in a squirrel colony. The Sierra Valley population was examined only under laboratory conditions simulating below-ground conditions in burrows. The Folsom Lake population from the Sierra Nevada foothills exhibited the highest level of venom resistance of any study populations and was studied in both contexts. Although snake density is generally high at Folsom Lake, this particular site was free of snakes because of human activity at a nearby boat-launching ramp.

Population Variation in Antisnake Behavior

Behavioral activity below ground was videotaped in low light using an artificial burrow with a sand floor and two nest chambers connected to two alleys that exited into a seminatural enclosure (see details in Coss and Owings 1978). After each squirrel habituated to the setting for 23 h, the burrow entrances were sealed and either a rattlesnake or gopher snake was introduced into the burrow for a 10-min trial (Towers and Coss 1990). Six male and six female adults each from Folsom Lake and Sierra Valley populations were tested once with each snake presented in a balanced order separated by 7–8 days.

Behavioral observations above ground were made in a small experimental room, the floor of which was covered with 5 cm of sand (Coss 1991a, Coss et al. 1993). A rattlesnake or gopher snake in a wire-mesh cage was placed in the center of the room and all squirrel–snake interactions were videotaped from overhead for 5 min following inspection of the snake and the ensuing tail piloerection thought to indicate the onset of snake recognition by the ground squirrel. Each snake was presented once in a balanced order separated by 5 days. The ground squirrels from all 11 populations ($n = 8$/population balanced for sex) were wild caught and maintained in the laboratory 3–9 months before testing. Several weeks before testing, squirrels in small groups were allowed to habituate to the experiment room using nest-boxes that resembled the wire-mesh cage containing the snake. Squirrels in these nest-boxes were removed from the room several hours before testing, leaving the subject in a wooden nest-box that could be raised remotely to initiate the trial.

Comparison of results across populations in both below- and above-ground contexts revealed only subtle changes in snake-recognition abilities and behavior with relaxed selection from snakes. In terms of substrate throwing, time spent within striking distance, and attempts to leave the artificial burrow, relax-selected Sierra Valley squirrels were generally as competent as snake-selected Folsom Lake squirrels in dealing with each species of snake (Towers and Coss 1990). Sierra Valley squirrels also discriminated the rattlesnake from the gopher snake using olfactory and acoustical cues, although they required more time to differentiate these snakes than equally inexperienced Folsom Lake squirrels that showed evidence of discrimination within the first minute of the encounter.

In the above-ground setting, all 88 squirrels exhibited the rapid shift to elongate postures, tail piloerection, and jumpiness after detecting the caged snake in the center of the laboratory room (see experiment protocol and behaviors described in Coss 1991a). Most squirrels in each population tail flagged, and nearly all squirrels harassed the snakes by throwing sand substrate, sometimes jumping sideways immediately afterward as if they anticipated the snake strikes that often followed. Analyses of variance with planned comparisons, grouping squirrels from the six snake-selected populations and five relax-selected populations, revealed that they differed appreciably ($p < .05$) in three of five behavioral measures. These were the magnitude of tail piloerection, the amount of time the squirrel faced the snake, and the percentage of time spent near the snake (fig. 9-2).

Averaged for both types of snakes, the relax-selected group exhibited significantly greater tail piloerection than did the snake-selected group. Indeed, none of the relax-selected populations exhibited tail piloerection scores that overlapped their snake-selected counterparts (fig. 9-2A). In other mammals, erection of body and tail hair reflects an increase in sympathetic nervous system arousal (see Siegel and Skog 1970, Fuchs et al. 1985), and tail piloerection previously has provided a reliable measure of population differences in snake-related contexts (Rowe et al. 1986, Coss 1991a).

The relax-selected group also spent more time than the snake-selected group facing the snakes (fig. 9-2B). Time-sampled proximity near the snakes was less sensitive in differentiating the two groups. Planned contrasts at the level of simple effects revealed that the relax-selected group spent more time near the gopher snake than did the snake-selected group (fig. 9-2E). The two groups did not differ appreciably in the number of substrate-throwing acts and tail-flagging cycles (see fig. 9-2C, D). As discussed above, these are two phylogenetically old motor patterns used to deal with snakes. Populations differed considerably in mean levels of substrate-throwing and tail-flagging activity, but these differences were not related to historical selection from snakes.

Pearson product-moment correlation analyses, with Bonferroni-adjusted probabilities, revealed significant correlations between individual responses to the two snakes that might reflect intra- and interpopulation differences in temperament. As a behavioral trait seen almost exclusively in snake-related circumstances (Hennessy et al. 1981, Hersek 1990, Hersek and Owings 1993), the propensity to tail flag during responses to the two snakes was more highly correlated ($r = .39$, $p < .01$) in squirrels from the snake-selected group than in squirrels from the relax-selected group ($r = .25$, $p =$ ns). The converse was true for substrate throwing, a motor behavior flexibly used to harass snakes above and below ground and occasionally used against conspecific (Levy 1977) and heterospecific burrow intruders. The amounts of substrate throwing in response to the two species of snakes were more highly correlated ($r = .45$, $p < .005$) in squirrels from the relax-selected group than in squirrels from the snake-selected group ($r = .27$, $p =$ ns).

Some of these group differences were affected by the order in which the snakes were presented. For the snake-selected group, the rattlesnake was distinguished ($p < .025$) from the gopher snake by arousal and vigilance measures irrespective of snake presentation order, although close proximity and substrate throwing were most pronounced ($p < .05$) when the rattlesnake was encountered first. The rattlesnake engendered greater ($p < .05$) arousal, substrate throwing, and tail flagging than the gopher snake in the relax-selected group, but only when the rattlesnake

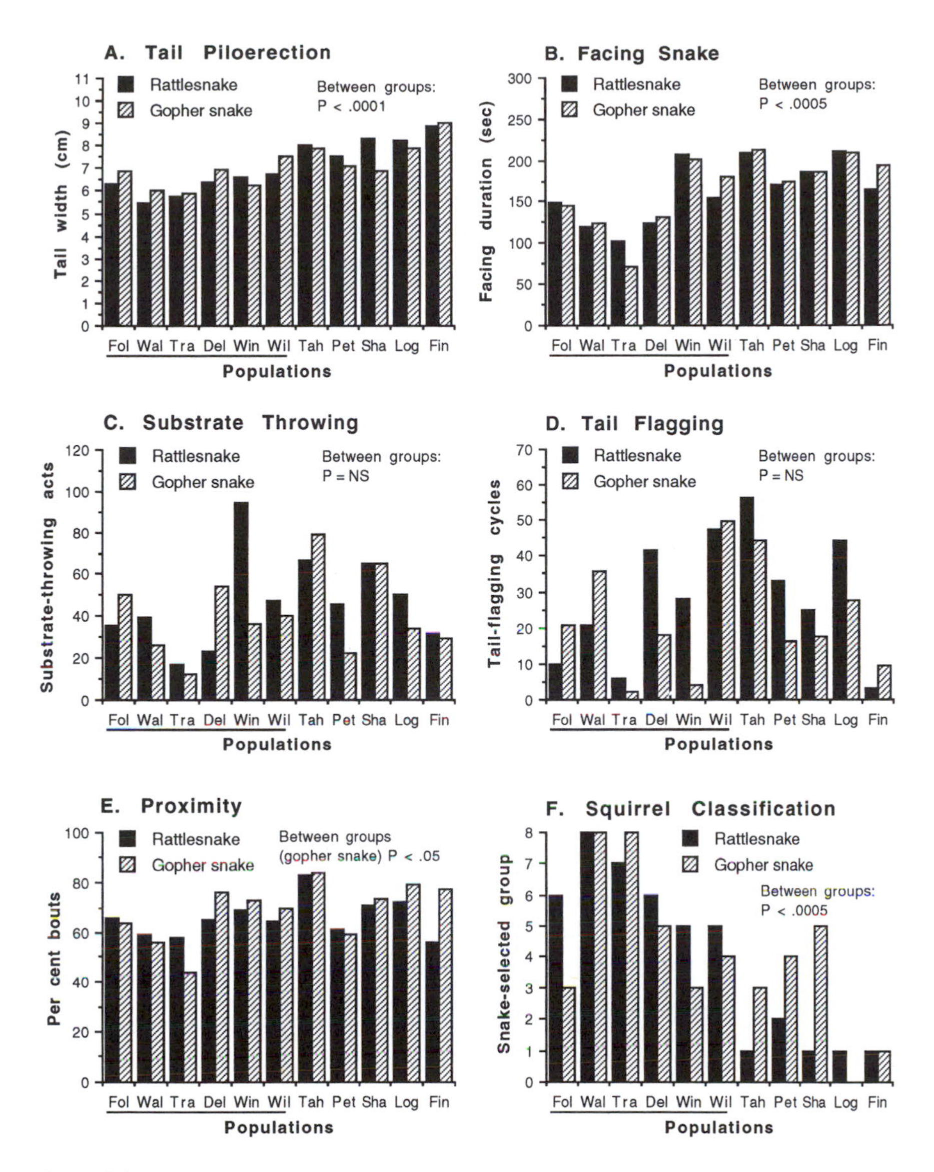

Figure 9-2 Comparisons of the antisnake behavior of 11 populations of adult wild-caught California ground squirrels. Population abbreviations refer to geographic sites. Underlined populations experience snake predation. Significance values characterize the results of planned comparisons of snake-selected and relax-selected groups averaged for both snakes (A–D). Groups differed significantly on proximity only for the gopher snake (E). These behavior patterns were examined by discriminant function and classification analyses to assign snake-selected group membership shown for each population (F).

was presented first, a finding suggestive of experiential refinement of the anti-snake system after exposure to the most dangerous snake.

Planned multivariate comparisons of the two groups using all five behavioral variables revealed significant group differences for each snake (multivariate $p < .0005$), with the tail piloerection and facing measures contributing the most to group separation. Subsequent discriminant function and classification analyses revealed that 85.0% and 77.1% of squirrels were correctly classified as members of their respective groups for the rattlesnake trial and 67.5% and 64.6% were correctly classified for the gopher snake trial (see fig. 9-2F).

In general, squirrels from populations whose ancestors experienced prolonged relaxed selection from snakes were more uniformly aroused and vigilant than squirrels from snake-selected populations, as represented by their visual monitoring of snake activity. However, arousal and vigilance were significantly correlated ($p < .02$) only in the snake-selected group of squirrels. In squirrels from the relax-selected group, greater physiological arousal was not reliably accompanied by markedly greater vigilance, possibly indicating disintegration in the linkage between these behaviors.

To describe the organization of these relationships, canonical correlation analysis was conducted on the two groups, excluding the eight Delta squirrels that encounter only gopher snakes to equilibrate for snake sympatry (Coss and Biardi 1997). The aim of this analysis was to determine the composition of unobserved traits common to two sets of measures by analyzing them as a linear composite of observed variables (Darlington et al. 1973). The five behavioral measures from encounters with the rattlesnake composed one set of variates, and their counterparts from encounters with the gopher snake composed the other set. Canonical correlation analysis revealed changes under prolonged relaxed selection in the cohesive properties of action patterns with potential adaptive properties. In this sense, cohesiveness refers to two or more variates loading on a single canonical factor. In snake-selected squirrels, the first three sets of canonical composites ordered by magnitude (see Thompson 1984) represented potentially adaptive composite groupings of vigilance, spacing, and confrontational behaviors. The fourth set reflected sympathetic nervous system arousal, completely uncoupled from any of the other variables (fig. 9-3). Squirrels whose ancestors had experienced prolonged relaxed selection did not exhibit any significant cohesive canonical correlations (fig. 9-4). Overall then, the relationships between the first three canonical functions observed in the responses of ground squirrels from locations with high snake densities seemed to reflect traits involved in successful confrontation of snake predators, whereas the mainly unidimensional structure of canonical composites in relax-selected populations may reflect deterioration of antisnake behavioral organization. High loadings of single measures (fig. 9-4) may simply represent the univariate Pearson product-moment correlations as is the case for substrate throwing ($r = .44$, $p < .005$).

The shift toward greater physiological arousal under conditions of prolonged relaxed selection was evident during the first seconds of dealing with the snake in which tail piloerection initially surged then stabilized well above baseline. Distinct arousal wave forms during piloerection also differentiated the two groups

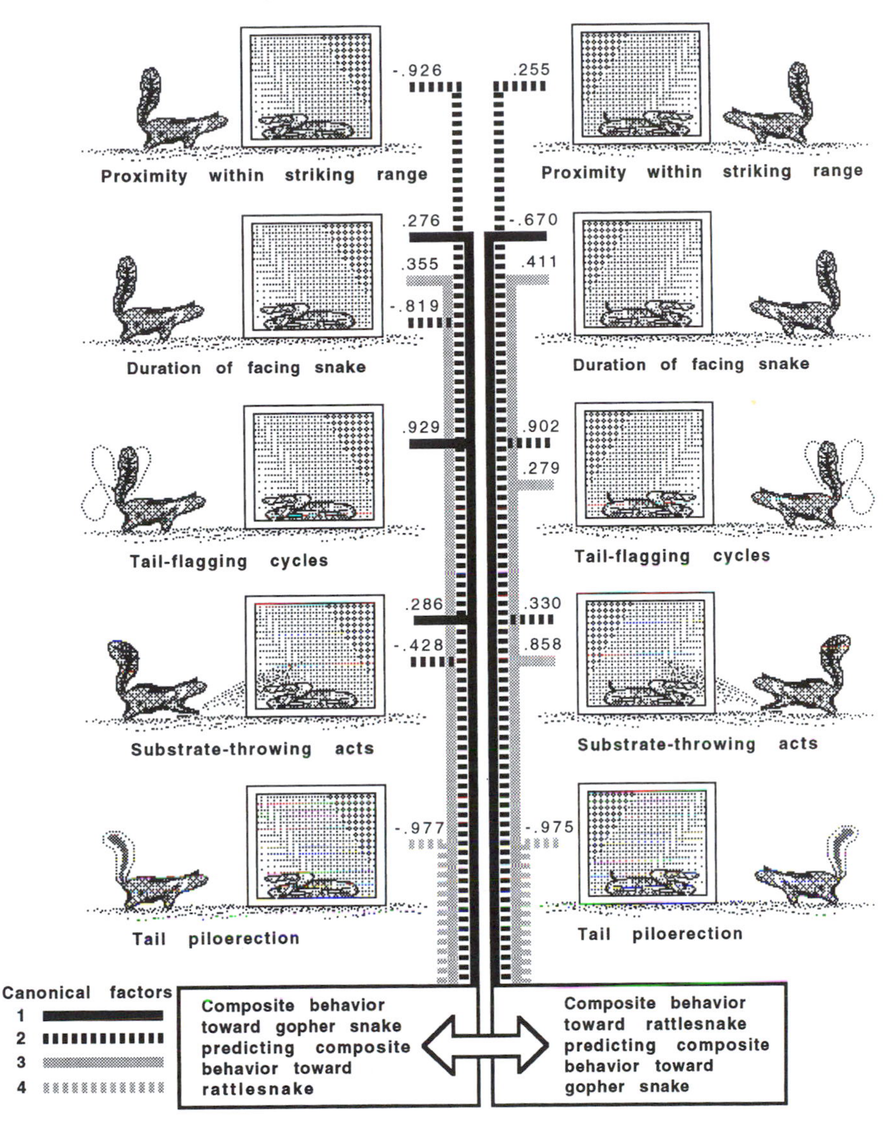

Figure 9-3 Factor structure of the first four canonical correlations for snake-selected ground squirrels during presentations of a rattlesnake or gopher snake (Bartlett's χ^2 approximation for correlations 1–4, respectively: $r^2 = .498, .368, .266, .188$; $p < .1$). Inset numbers represent varimax-rotated structure coefficients. Note that factor four is composed exclusively of the physiological arousal measure (tail piloerection), which does not contribute to the first three correlations.

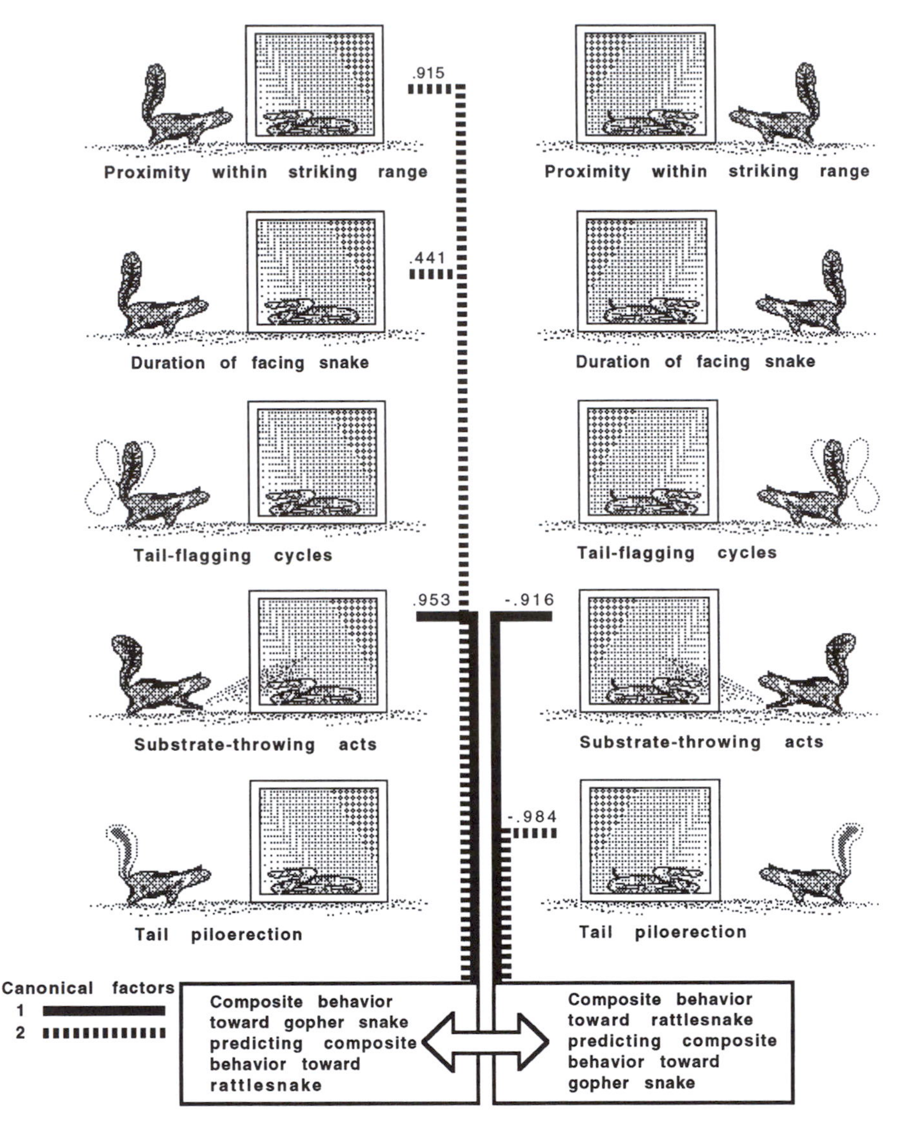

Figure 9-4 Factor structure of the first two canonical correlations for relax-selected ground squirrels during presentations of a rattlesnake or gopher snake. Inset numbers represent varimax-rotated structure coefficients. Only the first correlation is significant ($r^2 = .475$, $p < .1$) and is composed exclusively of substrate throwing. Compared with Figure 9-3, the canonical composites have much simpler factor structure, which might reflect deterioration of organized antisnake behavior with prolonged relaxed selection. In this group, tail piloerection is not isolated from investigative behavior (factor 2).

of squirrels (fig. 9-5A). Squirrels from the relax-selected group exhibited the same rise time as squirrels from the snake-selected group, but they typically displayed a more prolonged surge of arousal followed by stabilization at a higher amplitude (see fig. 9-5B–D). Discriminant analysis of four variables describing this arousal wave form, nearly balanced for snake species (multivariate $p < .005$), indicated that 16 of 18 squirrels were correctly classified as members of the relax-selected group and 16 of 20 squirrels were correctly classified as members of the snake-selected group.

These results indicate that although venom resistance and some behavior patterns have changed under relaxed selection, others have not. Those that have failed to change in expression as a consequence of relaxed selection appear to be traits that are phylogenetically old, as they are found in other species of ground-dwelling squirrels. Retention of their motoric properties linked to snake recogni-

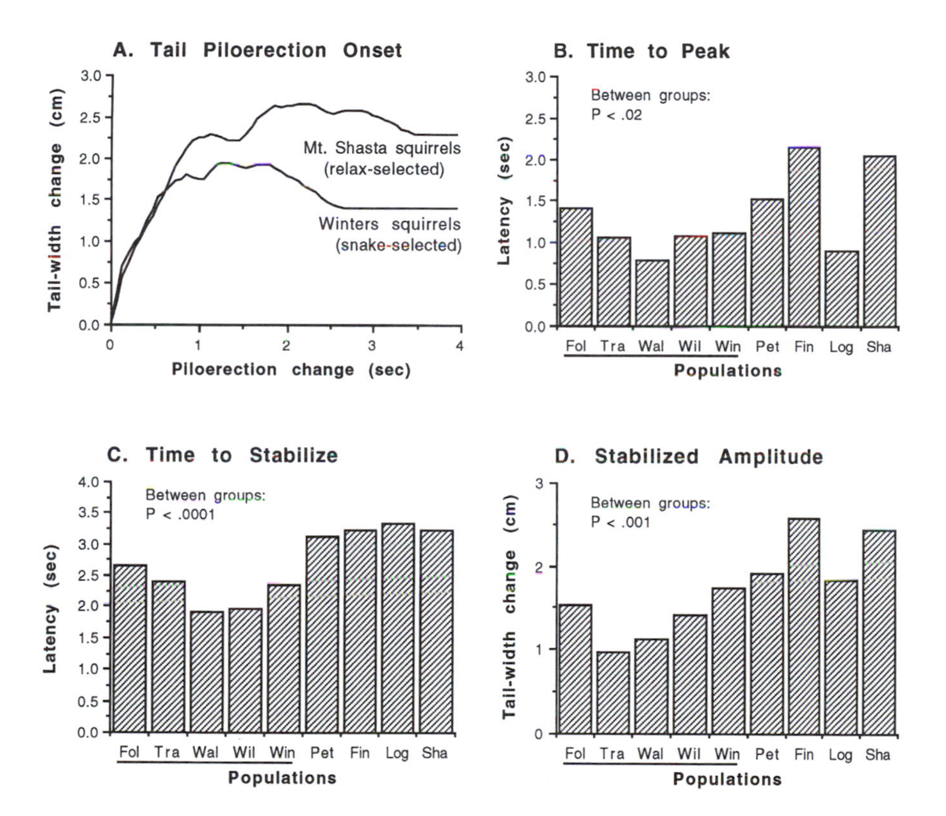

Figure 9-5 Tail piloerection onset after initial snake detection during the first snake trial characterizes the surge of sympathetic nervous system arousal among squirrels that remained immobile for 4 s. Average plot of four Winters and four Mount Shasta squirrels from California (A), approximately balanced for snake species, is smoothed in 133-ms increments. Underlined populations experience snake predation. Differences in arousal wave-form properties (B–D), shown for two or more squirrels from each population, contributed strongly to group membership in discriminant function and classification analyses.

tion characterizes the robustness of the lowest level of antisnake behavioral organization, irrespective of the specialization of tail flagging and general utility of substrate throwing. Higher levels of behavioral organization or cohesiveness showed marked disintegration under relaxed selection. These higher-order character states are particularly interesting because their evolutionary lability reflects their specialization for dealing with snakes. It is important to note that retention of a specialized recognition system under relaxed selection might also indicate the presence of constraints because it is linked to other sensory systems still under selection (see Basolo and Endler 1995). For example, snake recognition may be constrained because the recognition system is embedded in other perceptual systems. Before discussing this further, however, I provide a time scale within which maximum rates of disintegration of antisnake behavior patterns can be evaluated in populations of California ground squirrels.

A Time Frame for Disintegration of Antisnake Behavior in California Ground Squirrel Populations

Blood samples from 31 populations in California and Oregon (fig. 9-1) were screened electrophoretically using methods described in Smith and Coss (1984). Nine of 37 loci were found to be polymorphic. Latter's (1973) ϕ^*, a multilocus adaptation of F_{ST}, was used as the model for estimating times of divergence of nearby populations within the same subspecies (see Slatkin 1985). Pairwise genetic distances were calibrated to time using geological evidence for the rapid shift in drainage of a large Pleistocene lake that produced a major geographic barrier to gene flow about 725,000 years ago and is thought to have initiated Beechey and Douglas ground squirrel subspeciation (Smith and Coss 1984, Goldthwaite 1989). The resultant matrix of pairwise divergence times was used to construct a phenetic tree using an unweighted pair-group method of averages (UPGMA), a cluster method that assumes all branch lengths have equal evolutionary rates (fig. 9-6). The linear property of the entire phenetic tree was corroborated by the temporal coincidence of the youngest pairwise divergence age and radiocarbon ages bracketing a narrow interval of landscape stability leading to this divergence (see Coss et al. 1993).

Additional support for phenetic tree structure as related to paleoclimatic fluctuations that might have initiated population divergences and interrupted contact with snakes was obtained by superimposing the tree's branch points on the finely detailed $\delta^{18}O$ record from deep-sea core V19-30 raised near the Panama Basin (see Pisias and Shackleton 1984). A chi-square test revealed a significant ($p <$ .005) clustering of branch points in the colder part of the paleoclimatic record, interpreted as evidence that gene flow among ground squirrel populations in California and Oregon was markedly disrupted during repeated glacial periods. Time-series analyses of the same paleoclimatic record and fluctuations of pairwise population divergences at 5000-year intervals provided a similar interpretation (see Goldthwaite 1989).

The estimated maximum times since release from snake predation ranged from approximately 70,000–300,000 years. Gene flow from neighboring squirrel popu-

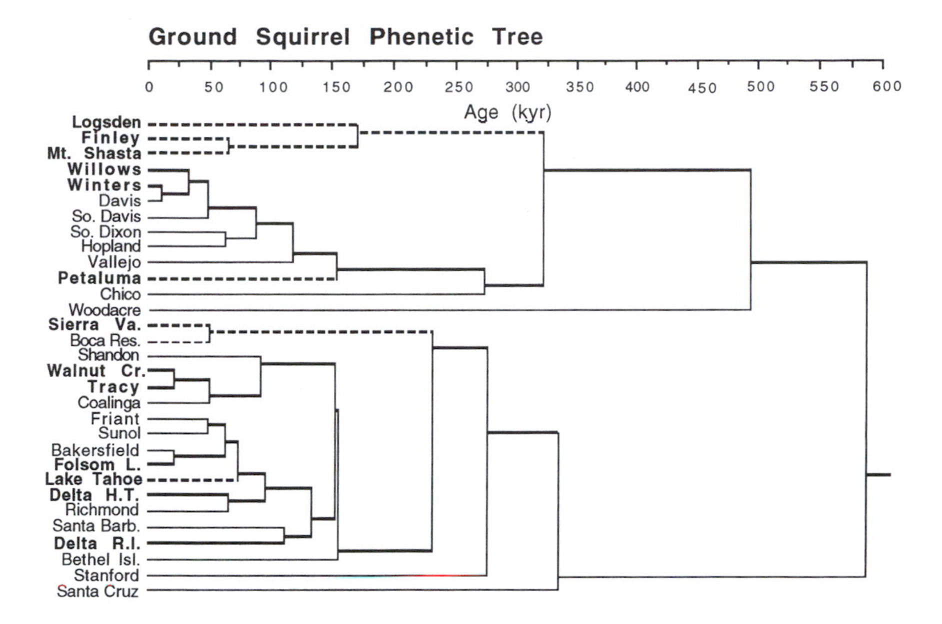

Figure 9-6 Phenetic tree of population divergences based on the UPGMA cluster method applied to a Latter's ϕ^* genetic distance matrix converted to time (Goldthwaite 1989). Populations in bold text were selected for the study of antisnake behavior. Dashed tree segments characterize the estimated duration of relaxed selection from rattlesnakes and gopher snakes.

lations during Ice Age times sustained the continuity of relaxed selection because rattlesnakes and gopher snakes were likely precluded from northerly habitats occupied today in low densities. The three populations with the longest time intervals of relaxed selection exhibited negligible *in vivo* and *in vitro* venom resistance (Poran et al. 1987), equivalent to that of Arctic ground squirrels. Thus, for Arctic ground squirrels whose ancestors experienced relaxed selection for 3–5 million years, the ability to recognize snakes and exhibit antisnake behavior disintegrates completely, yielding only disorganized caution after being bitten repeatedly (Goldthwaite et al. 1990). Data from California ground squirrels suggest, however, that disintegration of snake recognition is unlikely in a time frame of less than 300,000 years. In contrast, the ability to regulate physiological arousal while confronting snakes and the cohesiveness of action patterns used to investigate and harass them appear to be labile in the same time period. Tail flagging is phylogenetically old, yet its association with other confrontational behaviors is weakened considerably under relaxed selection (Coss and Biardi 1977). Other facets of higher-order behavioral organization illustrated by canonical correlation analyses might be recently derived. In either case, the ability to organize behavior appropriately requires the integration of cognition and decision making, which is less finely tuned than recognition systems that require much greater precision at perceptual integration.

Discussion and Conclusions

On the whole, relaxed selection from rattlesnakes and gopher snakes for a time period spanning an estimated 70,000–300,000 years does not compromise the snake-recognition abilities of California ground squirrels. All such populations treated the two snakes as dangerous adversaries, and both snake-selected and relax-selected groups were able to distinguish them, although above-ground experience with the rattlesnake first facilitated differentiation of the gopher snake by squirrels in the relax-selected group. Lack of individual variation in the recognition of snakes as dangerous is demonstrated in the published descriptions of 195 California ground squirrels studied in the laboratory since 1975 (see Coss 1993). That snake-recognition capabilities can eventually decay is shown by Arctic ground squirrels, whose ancestors were already present in a snake-free Alaskan habitat by the middle Pliocene (Goldthwaite et al. 1990, Coss and Goldthwaite 1995).

Retention of the ability to discriminate rattlesnakes and gopher snakes by members of relax-selected populations of California ground squirrels may reflect low heritability (Mousseau and Roff 1987, Hewitt 1990) for the trait in the ancestors of the relax-selected populations. More than 10 million years of selection by venomous rattlesnakes and constricting colubrids could have virtually exhausted previously existing genetic variation precluding rapid loss of this perceptual capability. Certainly, the specialized utility of snake-species recognition is sufficiently great that eventual erosion of variation would be expected. Whether this long history of snake selection could constrain disintegration for a period of 300,000 years is less clear.

There are two distinct levels of epigenetic constraints—the pleiotropic effects of numerous genes shaping brain development and the distributed, interactive nature of neural circuitry—that together might buffer rapid changes in behavioral organization under prolonged relaxed selection. Evidence for strong pleiotropic effects comes from the study of *Drosophila* mosaics, in which an estimated two-thirds of all vital genes play an essential role in shaping neural connectivity in the visual system (Thaker and Kankel 1992). Phenotypes with mutant gene products affecting deeply canalized developmental trajectories would be screened by selection much more readily than mutant phenotypes with highly specific anomalies permitting some recovery of function (e.g., Krishnan et al. 1993, Seeger et al. 1993, Haydon and Drapeau 1995, Miller and Niemeyer 1995). Under genetic drift, even subtle changes in patterns of neural circuitry shaped historically by selection might not be tolerated if they affect other functions currently under selection. For example, there is emerging evidence that neuronal columns in the primate inferior temporal cortex share partial information about specific perceptual features, a property that minimizes redundancy and likely increases processing capacity (Fujita et al. 1992, Gawne and Richmond 1993, Komatsu and Ideura 1993). Higher order cognitive systems responsible for pattern recognition involve larger brain regions and are usually described as modular or "informationally encapsulated." However, this view of localized cortical specialization has been challenged due to the difficulty of categorizing highly interdependent systems

(Farah 1994). The perceived regularity of overlapping seed heads in the spikelets of native California grasses (Crampton 1974) might employ the same visual-feature processing used to detect the scale patterns of snakes on heterogeneous substrates (e.g., Coss 1991a). Mutational disruption of essential pattern detection capabilities shared by more than one recognition system (Levine and Shefner 1991, Finkel and Sajda 1994) would therefore have profound effects on ground squirrel fitness despite relaxed selection operating in a single functional context.

Further elaboration of this view relies on an understanding of the organization of the innate perceptual abilities used to recognize specific threats. The innate ability to recognize specific visual schemata, such as different predators, involves the differentiation of unique perceptual features from background and the assignment of meaning to the invariant properties of these features. Together these provide the emergent context of perceived threat. Predators, such as owls and snakes, exhibit relatively unique silhouettes with complex surface patterns and textures that afford crypticity presumed to enhance stealth. Yet, the invariant properties of these cryptic patterns yield important recognition cues used by prey to defeat this crypticity (see Coss 1991a). In nest parasitism, egg recognition and rejection relies on the ability to detect differences in egg markings and color rather than indistinguishable egg contours. The propensity to learn egg markings during egg laying (Rothstein 1974, 1978) could provide a general solution for combating cowbird nest parasitism in North America. If applicable to cuckoo nest parasitism in Europe, Africa, and Asia (Lotem et al. 1995), such a motivational system might be more susceptible to disruption by relaxed selection over a few centuries than innate pattern-recognition systems that involve complex structuring of perceptual information. However, Rothstein (personal communication, 1998) has argued that persistence in the tendency of hosts to learn their own eggs under relaxed selection might be more common than evolutionary loss.

In the longer evolutionary time frame, reliance on a single recognition cue to distinguish a specific predator makes for a more fragile perceptual system than one that depends on multiple cues perceived and integrated by the same sensory modality. The general failure of Spanish pied flycatchers to recognize the redback shrike could indicate the fragility of a perceptual system that apparently relies on the conspicuous black eye mask as the predominant recognition cue to distinguish shrikes from novel passerines (Curio 1975). Similarly, threespine sticklebacks probably rely on the unique sculpin silhouette for predator recognition because the sculpin's eyes are cryptically colored, effectively deleting a feature known to be important in predator recognition (see Coss and Goldthwaite 1995). For less complex perceptual tasks, fewer genes might be involved in restructuring perceptual systems for innate predator recognition, a process that might make sculpin recognition relatively labile to the disruptive influences of founder events (see Pressley 1981, Foster and Ploch 1990). The greater complexity of perceptual systems required to integrate several cues during rapid decision making presumably share more essential feature-processing capabilities with other cognitive systems and involve many more genes yielding complex pleiotropic effects. Together these could buffer perceptual reorganization with relaxed selection. Owl and hawk recognition by Darwin's finches on Wenman Island and snake recognition by ground squirrels after thousands of years of relaxed selection provide some support for this argument because both perceptual systems rely on multiple-feature processing.

The studies of geographic variation in behavior described here can provide insight into how behavioral systems might vary in evolutionary plasticity. Perceptual systems attuned to specific circumstances may well require much more precision in their embryological development than do higher order systems that organize behavior for dealing with dynamic changes in circumstances. For example, tail flagging and substrate throwing are expressed under less urgent conditions than stereotyped pausing and evasive leaping associated with the immediacy of identifying something nearby as a snake; in snake-selected populations, tail flagging and substrate throwing are cohesively integrated with other antisnake behaviors, such as vigilance and distance regulation (Coss and Biardi 1997). Because snake encounters occur in a variety of microhabitats and spatial circumstances, the stochastic nature of these encounters and differential outcomes could preclude the evolution of stereotyped sequences of behavioral organization (e.g., Berridge 1990, 1994). Individual variation in the aspects of temperament that affect the willingness to engage snakes could similarly reflect unique genomic configurations not subject to selection because they result from the continuous recombination of unlinked genes (Lykken et al. 1992). Population differences in these behaviors, unrelated to snake selection, could involve other aspects of temperament involving timidity or aggressive behavior known in rodents to be sensitive to selection experiments (see Whitney 1970, Cairns et al. 1990).

It is reasonable to assert that individual variation in level of fearfulness might be associated with other behavioral attributes sensitive to changes in the selective regime. Tulley and Huntingford (1988) report that the fearfulness of threespine stickleback males, shown by recovery of normal swimming and feeding after a simulated attack from a model pike (*Esox lucius*), was positively correlated with their level of intraspecific fighting to defend territories. As might be expected, stickleback males from a site with low predation risk were much less fearful after the simulated pike attack than males from a site with high predation risk; yet males from both sites showed positive correlations of levels of antipredator fearfulness and intraspecific aggressiveness. In this context, degredation of antipredator behavior under relaxed selection might reflect changes in temperament rather than loss of predator recognition. Similarly, generalized temperament changes affecting risk perception might explain the rapid reorganization of schooling and predator inspection behavior of guppies (*Poecilia reticulata*) from Trinidad living for approximately 100 generations in sites with low predation (see Magurran et al. 1992, Magurran this volume).

In conclusion, population variation in predator-recognition abilities provides an ideal ecological context for studying the effects of prolonged relaxed selection. The value of examining predator–prey relationships in this context is found not only in the power to discriminate adaptive differentiation, but also to explore the cognitive and genetic causes of persistence under relaxed selection. Clearly, population differences provide the material for further elucidation of the causes of behavioral stasis and thus offer great potential for better understanding neural and genetic constraints on behavioral evolution.

Acknowledgments Research on population variation in ground squirrel antisnake defenses was supported by faculty research grant D-922, National Science Foundation grant BNS 84-06172, and National Science Foundation Animal Behavior Training grant DIR 92-7091. James E. Biardi conducted the canonical correlation analyses. I thank the editor, Dr. Susan A. Foster, for her valuable suggestions and Drs. Ronald O. Goldthwaite, Donald H. Owings, and Niels G. Waller for comments that improved the chapter.

References

Barnes, B. M. 1989. Freeze avoidance in a mammal: Body temperatures below 0°C in an Arctic hibernator. Science 244:1593–1595.

Basolo, A. L., and J. A. Endler. 1995. Sensory biases and the evolution of sensory systems. Trends in Ecology and Evolution 10:489.

Berridge, K. C. 1990. Comparative fine structure of action: Rules of form and sequence in the grooming patterns of six rodent species. Behaviour 113:21–56.

Berridge, K. C. 1994. The development of action patterns. *In* J. A. Hogan and J. J. Bolhuis, eds. Causal mechanisms of behavioural development, pp. 147–180. Cambridge University Press, Cambridge.

Black, C. 1963. A review of North American tertiary Sciuridae. Bulletin of the Museum of Comparative Zoology Harvard University 130:113–248.

Brattstrom, B. H. 1953. Records of Pleistocene reptiles from California. Copeia 1953 (3): 174–179.

Brattstrom, B. H. 1965. Body temperatures of reptiles. American Midland Naturalist 73: 376–422.

Brattstrom, B. H. 1967. A succession of Pliocene and Pleistocene snake faunas from the high plains of the United States. Copeia 1967 (1):188–202.

Broecker, W. S., and G. H. Denton. 1989. The role of ocean-atmosphere reorganization in glacial cycles. Geochimica et Cosmochimica Acta 53:2465–2501.

Cairns, R. B., J.-L. Gariépy, and K. E. Hood. 1990. Development, microevolution, and social behavior. Psychology Review 97:49–65.

Coss, R. G. 1991a. Context and animal behavior III: The relationship of early development and evolutionary persistence of ground squirrel antisnake behavior. Ecological Psychology 3:277–315.

Coss, R. G. 1991b. Evolutionary persistence of memory-like processes. Concepts in Neuroscience 2:129–168.

Coss, R. G. 1993. Evolutionary persistence of ground squirrel antisnake behavior: Reflections on Burton's commentary. Ecological Psychology 5:171–194.

Coss, R. G., and J. E. Biardi. 1997. Individual variation in the antisnake behavior of California ground squirrels (*Spermophilus beecheyi*). Journal of Mammalogy 73:294–310.

Coss, R. G., and R. O. Goldthwaite. 1995. The persistence of old designs for perception. Perspectives in Ethology 11:83–148.

Coss, R. G., K. L. Gusé, N. S. Poran, and D. G. Smith. 1993. Development of antisnake defenses in California ground squirrels (*Spermophilus beecheyi*): II. Microevolutionary effects of relaxed selection from rattlesnakes. Behaviour 120:137–164.

Coss, R. G., and D. H. Owings. 1978. Snake-directed behavior by snake-naive and -experienced California ground squirrels in a simulated burrow. Zeitschrift für Tierpsychologie 48:421–435.

Coss, R. G., and D. H. Owings. 1985. Restraints on ground squirrel antipredator behavior: Adjustments over multiple time scales. *In* T. D. Johnston and A. T. Pietrewicz, eds. Issues in the ecological study of learning, pp. 167–200. Lawrence Erlbaum Associates, Hillsdale, NJ.

Coss, R. G., and D. H. Owings. 1989. Rattler battlers. Natural History 5:30–35.

Crampton, B. 1974. Grasses in California. University of California Press, Berkeley.

Cruz, A., and J. W. Wiley. 1989. The decline of an adaptation in the absence of a presumed selection pressure. Evolution 43:55–62.

Curio, E. 1961. Rassenspezifisches Verhalten gegen einen Raubfeind. Experientia 17:188–189.

Curio, E. 1964. Zur geographischen variation des feinderkennens einiger Darwinfinken (Geospizidae). Verhandlungen der Deutschen Zoologischen Gesellschaft in Kiel (supplement) 28:466–492.

Curio, E. 1966. How finches react to predators. Animals 9:142–143.

Curio, E. 1969. Funktionsweise und Stammesgeschichte des Flugfeinderkennens einiger Darwinfinken (*Geospizinae*). Zeitschrift für Tierpsychologie 26:394–487.

Curio, E. 1975. The functional organization of anti-predator behaviour in the pied flycatcher: A study of avian visual perception. Animal Behaviour 23:1–115.

Curio, E. 1993. Proximate and developmental aspects of antipredator behavior. Advances in the Study of Behavior 22:135–238.

Darlington, R. B., S. L. Weinberg, and H. J. Walberg. 1973. Canonical variate analysis and related techniques. Review of Educational Research 43:433–454.

Davies, N. B., and M. de L. Brooke. 1989a. An experimental study of co-evolution between the cuckoo, *Cuculus canorus*, and its hosts. I. Host egg discrimination. Journal of Animal Ecology 58:207–224.

Davies, N. B., and M. de L. Brooke. 1989b. An experimental study of co-evolution between the cuckoo, *Cuculus canorus*, and its hosts. II. Host egg markings, chick discrimination and general discussion. Journal of Animal Ecology 58:225–236.

Eshelman, R. E. 1975. Geology and paleontology of the early Pleistocene (late Blancan) White Rock fauna from north central Kansas. University of Michigan Papers on Paleontology, Museum of Paleontology No. 13:1–60.

Farah, M. J. 1994. Neuropsychological inference with an interactive brain: A critique of the "locality" assumption. Behavioral and Brain Sciences 17:43–104.

Finkel, L. H., and P. Sajda. 1994. Constructing visual perception. American Scientist 82: 224–237.

Firby, J. B. 1968. Revision of the middle Pleistocene Irvington fauna of California (MA thesis). University of California, Berkeley.

Fitch, H. S. 1948. Ecology of the California ground squirrel on grazing lands. American Midland Naturalist 39:513–596.

Fitch, H. S. 1949. Study of snake populations in central California. American Midland Naturalist 41:513–579.

Foster, S. A. 1994. Inference of evolutionary pattern: Diversionary displays of three-spined sticklebacks. Behavioral Ecology 5:114–121.

Foster, S. A., and S. A. Cameron. 1996. Geographic variation in behavior: A phylogenetic framework for comparative studies. *In* E. Martins, ed. Phylogenies and the comparative method in animal behavior, pp. 138–165. Oxford University Press, New York.

Foster, S. A., and S. Ploch. 1990. Determinants of variation in antipredator behavior of territorial male threespine stickleback in the wild. Ethology 84:281–294.

Fuchs, S. A. G., E. M. Edinger, and A. Siegel. 1985. The role of the anterior hypothalamus in affective defense behavior from the ventromedial hypothalamus of the cat. Brain Research 330:93–107.

Fujita, I., K. Tanaka, M. Ito, and K. Cheng. 1992. Columns for visual features of objects in monkey inferotemporal cortex. Nature 360:343–346.

Gawne, T. J., and B. J. Richmond. 1993. How independent are the messages carried by adjacent inferior temporal cortical neurons? Journal of Neuroscience 13:2758–2771.

Goldthwaite, R. O. 1989. Ground squirrel antipredator behavior: Time, chance and divergence (PhD dissertation). University of California, Davis.

Goldthwaite, R. O., R. G. Coss, and D. H. Owings. 1990. Evolutionary dissipation of an antisnake system: Differential behavior by California and Arctic ground squirrels in above- and below-ground contexts. Behaviour 112:246–269.

Grinnell, J., and C. L. Camp. 1917. A distribution list of the amphibians and reptiles of California. University of California Publications in Zoology 17:127–208.

Grinnell, J., and J. Dixson. 1918. California ground squirrels. The Monthly Bulletin California State Commission of Horticulture (11–12) 7:597–708.

Hall, E. R. 1981. Mammals of North America, vol. 2, 2nd ed. John Wiley & Sons, New York.

Hall, M. L. 1983. Origin of Española island and the age of terrestrial life on the Galápagos islands. Science 221:545–547.

Halpin, Z. T. 1983. Naturally-occurring encounters between black-tailed prairie dogs (*Cynomys ludovicianus*) and snakes. American Midland Naturalist 109:50–55.

Haydon, P. G., and P. Drapeau. 1995. From contact to connection: Early events during synaptogenesis. Trends in Neuroscience 18:196–201.

Hennessy, D. F., and D. H. Owings. 1978. Snake species discrimination and the role of olfactory cues in the snake-directed behavior of the California ground squirrel. Behaviour 65:115–124.

Hennessy, D. F., and D. H. Owings. 1988. Rattlesnakes create a context for localizing their search for potential prey. Ethology 77:317–329.

Hennessy, D. F., D. H. Owings, M. P. Rowe, R. G. Coss, and D. W. Leger. 1981. The information afforded by a variable signal: Constraints on snake-elicited tail flagging by California ground squirrels. Behaviour 78:188–226.

Hersek, M. J. 1990. Behavior of predator and prey in a highly coevolved system: Northern Pacific rattlesnakes and California ground squirrels (PhD dissertation). University of California, Davis.

Hersek, M. J., and D. H. Owings. 1993. Tail flagging by adult California ground squirrels: A tonic signal that serves different functions for males and females. Animal Behaviour 46:129–138.

Hewitt, J. K. 1990. Changes in genetic control during learning, development, and aging. *In* M. E. Hahn, J. K. Hewitt, N. D. Henderson, and R. H. Benno, eds. Developmental behavior genetics: Neural, biometrical, and evolutionary approaches, pp. 217–235. Oxford University Press, New York.

Heynen, A. J., R. S. Sainsbury, and C. P. Montoya. 1989. Cross-species responses in the defensive burying paradigm: A comparison between Long-Evans rats (*Rattus norvegicus*), Richardson's ground squirrels (*Spermophilus richardsonii*), and thirteen-lined ground squirrels (*Catellus tridecemlineatus*). Journal of Comparative Psychology 103: 184–190.

Hibbard, C. W. 1941. New mammals from the Rexroad fauna, Upper Pliocene of Kansas. American Midland Naturalist 26:337–368.

Hickman, C. S., and J. H. Lipps. 1985. Geologic youth of Galápagos islands confirmed by marine stratigraphy and paleontology. Science 227:1578–1580.

Hirsh, S. M., and R. C. Bolles. 1980. On the ability of prey to recognize predators. Zeitshrift für Tierpsychologie 54:71–84.

Hodell, D. A., K. M. Elmstrom, and J. P. Kennett. 1986. Latest Miocene benthic $\delta^{18}O$ changes, global ice volume, sea level and the 'Messinian salinity crisis'. Nature 320: 411–414.

Holman, J. A. 1979. A review of North American tertiary snakes. Publication of the Museum, Michigan State University, Paleontological Series 1(6):203–260.

Howell, A. H. 1938. Revision of the North American ground squirrels, with a classification of the North American Sciuridae. North American Fauna 56:1–256.

Hughes, T. J. 1985. The great Cenozoic ice sheet. Palaeogeography, Palaeoclimatology, Palaeoecology 50:9–43.

Juelson, T. C. 1970. A study of the ecology and ethology of the rock squirrel, *Spermophilus variegatus* (Erxleben) in northern Utah (PhD dissertation). University of Utah, Salt Lake City.

Kaneshiro, K. Y. 1989. The dynamics of sexual selection and founder effects in species formation. *In* L. V. Giddings, K. Y. Kaneshiro, and W. W. Anderson, eds. Genetics, speciation, and the founder principle, pp. 279–296. Oxford University Press, Oxford.

Kavaliers, M. 1990. Responsiveness of deer mice to a predator, the short-tailed weasel: Population differences and neuromodulatory mechanisms. Physiological Zoology 63: 388–407.

Klauber, L. M. 1972. Rattlesnakes: Their habits, life histories and influence on mankind, 2 vols. University of California Press, Berkeley.

Komatsu, H., and Y. Ideura. 1993. Relationships between color, shape, and pattern selectivities of neurons in the inferior temporal cortex of the monkey. Journal of Neurophysiology 70:677–694.

Knickerbocker, M., and H. Knickerbocker. 1988. A Grand Canyon mini-adventure. Arizona Highways 65:12–15.

Korth, W. W. 1994. The Tertiary record of rodents in North America. Plenum Press, New York.

Krishnan, S. N., E. Frei, G. P. Swain, and R. J. Wyman. 1993. Passover: A gene required for synaptic connectivity in the giant fiber system of Drosophila. Cell 73:967–977.

Latter, B. D. H. 1973. Measures of genetic distance between individuals and populations. *In* N. E. Morton, ed. Genetic structure of populations, pp. 27–37. University of Hawaii Press, Honolulu.

Levine, M. W., and J. M. Shefner. 1991. Fundamentals of sensation and perception, 2nd ed. Brooks/Cole Publishing Company, Pacific Grove, CA.

Levy, N. 1977. Sound communication in the California ground squirrel. Masters thesis, California State University, Northridge.

Lotem, A., H. Nakamura, and A. Zahavi. 1995. Constraints on egg discrimination and cuckoo-host co-evolution. Animal Behaviour 49:1185–1209.

Loughry, W. J. 1987. The dynamics of snake harassment by black-tailed prairie dogs. Behaviour 103:27–48.

Loughry, W. J. 1988. Population differences in how black-tailed prairie dogs deal with snakes. Behavioral Ecology and Sociobiology 22:61–71.

Loughry, W. J. 1989. Discrimination of snakes by two populations of black-tailed prairie dogs. Journal of Mammalogy 70:627–630.

Lykken, D. T., J. McGue, A. Tellegen, and T. J. Bouchard Jr. 1992. Emergenesis, genetic traits that may not run in families. American Psychologist 47:1565–1577.

Magurran, A. E., B. H. Seghers, G. R. Carvalho, and P. W. Shaw. 1992. Behavioural consequences of an artificial introduction of guppies (*Poeciliar eticulata*) in N. Trinidad: Evidence for the evolution of anti-predator behaviour in the wild. Proceedings of the Royal Society of London Series B. Biological Sciences 248:117–122.

Miller, D. M., and C. J. Niemeyer. 1995. Expression of the unc-4 homeoprotein in *Caenorhabditis elegans* motor neurons specifies presynaptic input. Development 121:2877–2886.

Miller, L. H. 1912. Contributions to avian paleontology from the Pacific coast of North America. University of California Publications Bulletin of the Department of Geology 7:61–115.

Moksnes, A. E., and E. Røskaft. 1989. Adaptations of meadow pipits to parasitism by the common cuckoo. Behavioral Ecology and Sociobiology 24:25–30.

Mousseau, T. A., and D. A. Roff. 1987. Natural selection and the heritability of fitness components. Heredity 59:181–197.

Nussbaum, R. A., E. D. Brodie Jr., and R. M. Storm. 1983. Amphibians and reptiles of the Pacific northwest. Idaho University Press, Moscow, Idaho.

Øien, I. J., A. Moksnes, and E. Røskaft. 1995. Evolution of variation in egg color and marking pattern in European passerines: Adaptations in a coevolutionary arms race with the cuckoo, *Cuculus canorus*. Behavioral Ecology 6:106–174.

Owings, D. H., and R. G. Coss. 1977. Snake mobbing by California ground squirrels: Adaptive variation and ontogeny. Behaviour 62:50–69.

Owings, D. H., and W. J. Loughry. 1985. Variation in snake-elicited jump-yipping by black-tailed prairie dogs: Ontogeny and snake-specificity. Zeitschrift für Tierpsychologie 70:177–200.

Owings, D. H., and S. C. Owings. 1979. Snake-directed behavior by black-tailed prairie dogs (*Cynomys ludovicianus*). Zeitschrift für Tierpsychologie 49:35–54.

Pisias N. G., and N. J. Shackleton. 1984. Modelling the global climate response to orbital forcing and atmospheric carbon dioxide changes. Nature 310:757–759.

Polans, N. O. 1983. Enzyme polymorphisms in Galapagos finches. *In* R. I Bowman, M. Berson, and A. E. Leviton, eds. Patterns of evolution in Galapagos organisms, pp. 219–236. Pacific Division, American Association for the Advancement of Science, USA, San Francisco.

Poran, N. S., and R. G. Coss. 1990. Development of antisnake defenses in California ground squirrels (*Spermophilus beecheyi*): I. Behavioral and immunological relationships. Behaviour 112:222–245.

Poran, N. S., R. G. Coss, and E. Benjamini. 1987. Resistance of California ground squirrels (*Spermophilus beecheyi*) to the venom of the Northern Pacific rattlesnake (*Crotalus viridis oreganus*): A study of adaptive variation. Toxicon 25:767–777.

Pressley, P. H. 1981. Parental effort and the evolution of nest-guarding tactics in the threespine stickleback *Gasterosteus aculeatus* L. Evolution 35:282–295.

Repenning, C. A., E. M. Brouwers, L. D. Carter, L. Marincovich Jr., and T. A. Ager. 1987. The Beringian ancestry of *Phenacomys* (Rodentia: Cricetidae) and the beginning of the modern Arctic Ocean borderland biota. U.S. Geological Survey Bulletin 1687: 1–28.

Riechert, S. E. 1993. The evolution of behavioral phenotypes: Lessons learned from divergent spider populations. Advances in the Study of Behavior 22:103–134.

Rothstein, S. I. 1974. Mechanisms of avian egg recognition: Possible learned and innate factors. Auk 91:796–807.

Rothstein, S. I. 1978. Mechanisms of avian egg-recognition: Additional evidence for learned components. Animal Behaviour 26:671–677.

Rowe, M. P., R. G. Coss, and D. H. Owings. 1986. Rattlesnake rattles and burrowing owl hisses: A case of acoustic Batesian mimicry. Ethology 72:53–71.

Rowe, M. P., and D. H. Owings. 1990. Probing, assessment, and management during interactions between ground squirrels and rattlesnakes: Part 1: Risks related to rattlesnake size and body temperature. Ethology 86:237–249.

Schluter, D. 1984. Morphological and phylogenetic relations among Darwin's finches. Evolution 38:921–930.

Seeger, M., G. Tear, D. Ferres-Marco, and C. S. Goodman. 1993. Mutations affecting growth cone guidance in Drosophila: Genes necessary for guidance toward or away from the midline. Neuron 10:409–426.

Shotwell, J. A. 1956. Hemphillian assemblage from northeastern Oregon. Bulletin of the Geological Society of America 67:717–738.

Siegel, A., and D. Skog. 1970. Effects of electrical stimulation of the septum upon attack behavior elicited from the hypothalamus in the cat. Brain Research 23:371–380.

Slatkin, M. 1985. Gene flow in natural populations. Annual Review of Ecology and Systematics 16:393–430.

Smith, D. G., and R. G. Coss. 1984. Calibrating the molecular clock: Estimates of ground squirrel divergence made using fossil and geological time markers. Molecular Biology and Evolution 1:249–259.

Spencer, H. 1888. The principles of psychology, vol. 1, 3rd ed. D. Appleton, New York.

Stebbins, R. C. 1985. A field guide to western reptiles and amphibians, 2nd ed. Houghton Mifflin Company, Boston.

Stock, C. 1918. The Pleistocene fauna from Hawver cave. University of California Publications Bulletin of the Department of Geology 10:461–515.

Thaker, H. M., and D. R. Kankel. 1992. Mosaic analysis gives an estimate of the extent of genomic involvement in the development of the visual system in Drosophila melanogaster. Genetics 131:883–894.

Thompson, B. 1984. Canonical correlation analysis, uses and interpretation. Sage Publications, Beverly Hills, CA.

Towers, S. R. 1990. Ecological, evolutionary, and historical aspects of antipredator defenses in nearctic ground squirrels (PhD dissertation). University of California, Davis.

Towers, S. R., and R. G. Coss. 1990. Confronting snakes in the burrow: Snake species discrimination and antisnake tactics of two California ground squirrel populations. Ethology 84:177–192.

Towers, S. R., and R. G. Coss. 1991. Antisnake behavior of Columbian ground squirrels (*Spermophilus columbianus*). Journal of Mammalogy 72:776–783.

Tulley, J. J., and F. A. Huntingford. 1988. Additional information on the relationship between intra-specific aggression and anti-predator behaviour in the three-spined stickleback, *Gasterosteus aculeatus*. Ethology 78:219–222.

Tuomi, J. 1981. Structure and dynamics of Darwinian evolutionary theory. Systematic Zoology 30:22–31.

Victoria, J. K. 1972. Clutch characteristics and egg discrimination ability of the African village weaver *Ploceus cucullatus*. Ibis 114:367–376.

Whitney, G. 1970. Timidity and fearfulness of laboratory mice: An illustration of problems in animal temperament. Behavior Genetics 1:77–85.

Yang, S. Y., and J. L. Patton. 1981. Genic variability and differentiation in the Galapagos finches. Auk 98:230–242.

10

Variation in Advertisement Calls of Anurans across Zonal Interactions

The Evolution and Breakdown of Homogamy

MURRAY J. LITTLEJOHN

The allopatric mode of speciation has become a dominant paradigm (sensu Kuhn 1970) in evolutionary biology over the last 50 years (Mayr 1942, 1992). In this model, the geographic range of a species is fragmented, the previously dedifferentiating effect of gene flow is interrupted, and the now separated populations diverge. If there is enough genetic differentiation during this period of isolation, then the disjunct daughter populations may become separate biological species (sensu Mayr 1942, 1992). This level of divergence is achieved by the development of sufficient genetic incompatibility, as reflected in an absolute infertility or sterility of hybrids or by sufficient reduction in the absolute or relative levels of adaptedness of hybrids so that none survives to maturity when in competition with parental individuals. Full and complete allopatric speciation, then, is marked by the acquisition of those properties needed for extensive and continuing coexistence. Broad overlap of geographic ranges (sympatry) can then develop without any significant interactions between individuals of the derived populations.

In other situations, however, these essential properties may not have been acquired before the extrinsic barriers were removed. Thus, a critical stage is reached when the geographic ranges of previously separated daughter populations expand, and contact is established between individuals of the different genetic systems. Here the ecological compatibility, the specificity of mate choice, and the relative fitness of hybrids (if produced) are tested, and the following four outcomes may be envisaged. (1) If there is a cost to inbreeding, based on extrinsic and/or intrinsic factors, then the two lineages may diverge further in sympatry such that the attributes essential for stable coexistence arise, or are enhanced, through the direct action of natural selection within the context of the interaction (for references and recent commentaries, see Howard 1993, Littlejohn 1993, Butlin 1995). (2) If there is an ecological gradient and no cost to interbreeding because the hybrid progeny are as fit as, or fitter than, those of the parental taxa in part of the gradient, a geographically restricted and persistent hybrid zone may form (Moore 1977, Moore and Buchanan 1985, Littlejohn 1988, Hewitt 1989, Harrison 1993, Howard 1993). (3) If there is a cost to interbreeding and ecological compatibility does not arise, then contiguous and exclusive geographic ranges (parapatry) will be the result (Bull 1991). (4) The two taxa may fuse to form a single intermediate species.

Because of the underlying spatial pattern and the assumption that a relatively long time is required for allopatric speciation, a comparative approach seems to be the only tractable way to gain insights from natural situations. Such information can be obtained through the analysis of patterns of geographical variation in populations in situations presumed to represent different stages of this process of divergence. Investigation of the structure and dynamics of populations in zones of interaction may then provide insights into the evolution, maintenance, and breakdown of those properties (ecological, behavioral, and reproductive) essential for coexistence of individuals of the two derived systems.

It is important to state that I consider the effective partitioning of ecological resources to be an essential requirement for stable coexistence (Littlejohn 1993). Given this prerequisite, it is then of particular interest in the study of speciation to investigate patterns of geographical variation in behavioral characters that will promote positive assortative mating and consequent homogamy (sensu Littlejohn 1981, 1993). Analysis of behavioral interactions between closely related lineages in zones of overlap thus can provide insights into those processes that maintain or enhance homogamy or that result in its breakdown. Furthermore, the study of these interactions can increase our understanding of the ways in which behavior influences or affects the choice of a mate and the causes of evolutionary change in behavior that are a consequence of those interactions and so can provide insights into the processes of speciation.

The aim of this chapter is first to consider in general the evolutionary concepts, patterns, and processes associated with the origin, maintenance, and breakdown of homogamy and to discuss examples in which homogamy is presumed to have been improved in sympatry or to have broken down in hybrid zones. The conclusions drawn from these examples are reviewed within the broader context of general patterns of speciation. As the preliminary consideration, I deal with the following four topics: (1) the controversial issues of reproductive character displacement and reinforcement, (2) the stages in the development of sympatry, (3) the nature and dynamics of hybrid zones, as situations in which homogamy is absent, incomplete, or has broken down; and (4) the problematic topics of mate choice and sexual selection.

Reinforcement and Reproductive Character Displacement

Two terms, "reproductive character displacement" and "reinforcement" have been used to refer to the case in which closely related lineages appear to have diverged in sympatry in such a way that the level of homogamy is increased or maintained through the action of natural selection. These terms are sometimes used in different ways by different authors. The arguments for and against particular definitions and the controversy over the importance of reproductive character displacement and reinforcement of homogamy in speciation processes have been recently reviewed in several essays (e.g., Butlin 1989, Loftus-Hills and Littlejohn 1992, Howard 1993, Littlejohn 1993, Gerhardt and Schwartz 1995), and the appropriate references may be obtained from these works. Here I follow Loftus-Hills and Littlejohn (1992) and Howard (1993) and use "reinforcement" to refer to the selective processes by which homogamy is enhanced or maintained in sympatry, and "reproductive character displacement" to refer to the pattern of geographic variation in which differences in the systems of mate attraction and choice are greater in sympatry than in allopatry—without invoking any particular selective process. Assessment of the extent and pattern of interpopulational variation in components of homogamic systems, both in allopatry and sympatry, is thus a primary requirement for comparative studies of natural populations.

A cost to mismating, as measured by a relative reduction in reproductive fitness, is essential for the reinforcement of homogamy to occur during the speciation process (Littlejohn 1993). This cost may be based on intrinsic (endogenous or developmental) factors or on extrinsic (exogenous or ecological) factors. If there is no cost to mismating, then systems of actual or potential homogamy that evolved incidentally in allopatry are expected to break down with the establishment of secondary contact because of inevitable mistakes in mating; this situation would then lead to panmixia (Littlejohn 1988, 1993). Hence, an otherwise effective system of mate recognition is potentially reversible, and can only have a secondary role in the speciation process. Whether sympatry will develop depends primarily on ecological factors, such as the effective partitioning of limiting resources or the independent maintenance of the densities of both populations below the carrying capacity but above extinction (Littlejohn 1993). Accordingly, the transition from allopatry to sympatry represents a critical, but poorly understood, stage in geographic speciation (Otte 1989, Littlejohn 1993).

Investigations of the interactions between individuals of the daughter (i.e., cognate) populations are central to understanding the process of reinforcement. If there is a cost to mismating, then any allopatric divergence in systems of mate attraction and choice can set the directions in which subsequent reinforcing selection will proceed if parapatry is established. Otherwise, the pattern and direction of divergence may be derived from chance variation arising in disjunct allopatry before contact, or at the interface between the two interacting genetic systems. The nature and extent of divergence will be constrained by the properties of the communication channel used in attraction of mates and by the structure of coexisting signals of individuals of other syntopic (i.e., sharing the same habitat; Rivas 1964) and synchronic (breeding at the same time) populations that make up the reproductive and sensory environment.

Two patterns of spatial variation associated with reproductive character displacement are postulated to arise during reinforcement: (1) divergence, whereby the range of variation in the key attribute in each taxon remains about the same, but the means of one or both are displaced to produce a gap between the signals; and (2) partitioning of the original overlapping ranges of distribution by reduction in the ranges of variation of the key attribute in the signals of both taxa. But it seems that this latter condition could only apply to a fully sympatric distribution for the pair of taxa; otherwise, there would be an ongoing interaction at the leading edges of the invading phalanxes as they interacted with parapatric individuals that have similar values of the key attribute in their homogamic signals. Accordingly, only the former condition of divergence may be expected in systems with partly overlapping geographic distributions. Clearly, this aspect of zonal interactions warrants further consideration.

Overlaps of Geographic Distributions

The following three situations associated with the development of sympatry are visualized for members of contacting populations in which there is a cost to interpopulational mating. (1) They may be preadapted to coexist without any interactions at contact (the noninteractive case). (2) Divergence could occur in sympatric populations of both taxa (through bidirectional selection). (3) The displacement may be restricted to only one of the interacting taxa (through unidirectional selection); in this case, it is most likely that the invading taxon would undergo divergence (Littlejohn 1993).

In addition to the reduced reproductive success that would arise from the production of inferior hybrids as progeny, divergence could result from selection for the improvement of efficiency of communication in acoustic interactions between highly genetically dissimilar groups, such as crickets and frogs at the edge of a pond (Littlejohn 1977). The costs incurred in hybridization and reduced efficiency of communication in mate choice would be alleviated by the evolution and maintenance of distinctive signals and by the associated refinement of mechanisms of resolution and discrimination. Accordingly, no logical separation of these situations can be made, and all of the processes are considered as subsets of the more general process of reinforcement.

Hybrid Zones

In addition to investigation of the structure and dynamics of zonal interactions marked by extensive hybridization as evolutionary processes in their own right, much recent research has been directed at the significance of these interactions in the later stages of speciation as situations in which reinforcing selection might be expected to occur (see Otte and Endler 1989 and Harrison 1993 for reviews). As pointed out by Short (1969), and subsequently developed by Moore (1977), Littlejohn and Watson (1983, 1985, 1993), Moore and Buchanan (1985), Littlejohn (1988, 1993), and Moore and Price (1993), persistent hybrid zones may be arbitrarily classified into two categories on the basis of fitness of the hybrids compared with the progeny of pure parental crosses.

The first category involves universal hybrid inferiority in a zone of overlap and hybridization. In this situation, putative parental individuals are present in a proportion greater than that expected from chance recombination (usually set at 5%; Short 1969). The hybrid inferiority, and hence the cost to interbreeding, arises in either or a combination of two processes: intrinsically, through genetic incompatibility resulting from the basic discordance of the hybrid genomes, or extrinsically, where hybrids are poorly adapted in an ecological context relative to individuals of both parental taxa. Under this dynamic model of hybrid inferiority, the location of the hybrid zone is based on the chance position of the contact of expanding ranges, and the width of the zone is determined by the relative levels of dispersal and selective elimination (Barton and Hewitt 1989, Butlin 1989, Hewitt 1989, Ritchie et al. 1992, Szymura 1993). The presence of one or both of the above-mentioned costs is expected to result in the evolution or the maintenance, through reinforcing selection, of efficient homogamy in sympatry. However, the level of consequent reproductive character displacement may depend on the extent of geographic overlap and the rates of migration and gene flow from adjacent allopatric populations. See Howard (1993) for a recent summary.

The second category of zonal interaction involves hybrid superiority, wherein the recombination products are fitter than the progeny of the pure parental crosses and so displace them from the center of the hybrid zone (Moore 1977, Moore and Buchanan 1985, Moore and Price 1993). The result is a band of hybrids that separates pure parental individuals and prevents them from interacting directly. For this latter situation to be maintained, the greater relative fitness of the hybrids must be associated with factors in the extrinsic environment, and so the limits of the zone are usually determined by an ecological selective gradient (ecocline or ecotone) in which hybrids are superior to parental individuals. In this case there is no cost, in terms of Darwinian fitness, to interspecific mating within the ecological gradient. Any homogamy that is present at contact is expected to break down and eventually to disappear following chance cross-mating (Littlejohn 1993). Thus no reinforcement is expected under these conditions of hybrid superiority.

Harrison and Rand (1989) introduced the concept of mosaic hybrid zones, in which there are two discrete types of habitat distributed as a matrix of patches. Each taxon is assumed to be better adapted to one of the types of habitat, with the hybrids being restricted to patch boundaries. The primary determinant of the mosaic pattern is an exogenous factor that has a "coarse grain size" relative to the dispersal ability and mobility of the organisms (sensu Levins 1968, Pianka 1974). Sharp ecotones are assumed to exist, resulting in most hybrids being in close proximity to individuals of both parental taxa. Reinforcement is thus expected to occur, with different outcomes in different patches; this process is discussed by Gartside (1980) and Littlejohn (1981). The evolutionary dynamics and patterns of spatial variation in mosaic hybrid zones may be complicated and difficult to analyze by application of the comparative method.

Mate Choice

To understand fully the evolutionary dynamics of zonal interactions involving hybridization and reinforcement, mate attraction and choice must be investigated. The process of mate choice is sometimes divided into two stages: (1) species recognition, in which a class of appropriate mates (i.e., conspecific breeding individuals of the opposite sex) is first discriminated within the more diverse breeding array of potential mates; and (2) intersexual selection, in which a mate is chosen from within that subset (see discussions by Ryan and Rand 1993, Gerhardt and Schwartz 1995). I agree with Abt and Reyer (1993) that this dichotomy is neither justified nor useful. Rather than postulating that the receiving individual has a distinct concept of "same species" or "other species," a more appropriate postulate is to consider that the basic process of mate choice is the same at all levels of decision (Littlejohn 1993). The main purpose of mate choice is to find the best mate, not to find the "correct species"; species recognition emerges then as a consequence of finding the best mate (J. A. Endler, personal communication). The presence of persistent hybrid swarms comprising individuals exhibiting substantial variation in characters associated with mate attraction (e.g., between *Geocrinia laevis* and *G. victoriana*; Littlejohn 1988) implies that discrete characters for "species recognition" are unlikely to exist. This situation stands in contrast to that for sympatric populations, where there is extensive overlap in the timing and spatial patterns of breeding but no significant hybridization. In this instance, the divergences in criteria of mate choice may be so extreme that they serve to distinguish entire classes of individuals (species) as inappropriate mates.

In the present context, sexual selection is defined according to Arnold (1994, p. 9) as "selection that arises from differences in mating success (number of mates that bear or sire progeny over some standardized time interval)." This use of the term is close to that of Ryan and Rand (1993, p. 648), who defined sexual selection as: "variance in reproductive success that derives from variation in the ability to acquire mates." The distinction between intrasexual (competition for mates) and intersexual (epigamic) selection still applies. If the processes of mate choice, at both intrapopulational and interpopulational levels, have a common and unitary underlying selective mechanism, then reinforcement can be seen as an extension of the process of sexual selection. When mate choice is viewed in this way, the process of reinforcement is equivalent to unidirectional or to bidirectional divergent intersexual selection (sensu Littlejohn 1981), a term that warrants promotion, especially because of the ambiguity that arises through the long-standing use of the term "reinforcement" in psychology and behavior (Gerhardt and Schwartz 1995; see earlier discussion).

The reproductive biology of anuran amphibians provides a tractable system for investigating the operation and evolution of mechanisms of mate choice (Wells 1977). Vocal communication has a major role in mating, and geographical variation in this behavioral attribute can be objectively analyzed (Littlejohn 1977). Dense breeding aggregations can be formed, and individuals of several species may be present at the same time, so that mate choice involves both intraspecific and interspecific components. The advertisement call constitutes the principal homogamic mechanism in the courtship behavior of most synchronously breeding, syntopic species of anurans (Littlejohn 1977, 1988; Wells 1977, Gerhardt 1994, Gerhardt and Schwartz 1995), and is at the focus of subsequent discussion in this chapter.

The Advertisement Calls of Anuran Amphibians

Males broadcast a conspicuous acoustic signal, the advertisement call. Reproductively ripe conspecific females choose among the perceived signalers and then display positive phonotaxis to the source as the mate of their choice. There are two basic and contrasting patterns of anuran courtship (Wells 1977): breeding is short and explosive—the operational sex ratio approaches unity, males form dynamic breeding congresses, and most secure mates; or breeding is prolonged—the operational sex ratio is heavily biased toward males who generally establish and hold territories (sometimes including other resources of reproduction such as oviposition sites) from which they broadcast their advertisement calls and so compete for mates. Species with prolonged breeding have been best studied. The advertisement calls of most anuran species with prolonged breeding also have a second important function for an individual—namely, the announcement of an occupied territory. In this situation, the potential receivers are conspecific males and nearby males of other species that are using similar calling sites and broadcasting their advertisement calls in the same band of frequencies. These two advertising functions may be elicited by one structurally simple (monophasic) signal or may be partitioned between two discrete components of a diphasic signal (Narins and Capranica 1976, 1978; Littlejohn 1977, 1988; Wells and Schwartz 1984, Littlejohn and Harrison 1985).

Differentiation of Advertisement Calls

There are usually distinct gaps in the ranges of variation of key attributes of homogamic signals of members of different syntopic, synchronously breeding species. For example, mean values of pulse rate (pulse repetition rate or pulse repetition frequency) in the advertisement calls of several pairs of anuran species differ by a factor of about two or more (Littlejohn 1969), and ranges of variation do not overlap or abut but are separated by a distinct gap. The origin and maintenance of such gaps in the key factors of the signals, and the ranges of variation of them within the homogamic signals of each genetic system (i.e., species), are thus of considerable significance to understanding the process of mate choice in taxonomically complex reproductive environments.

A coefficient of call difference (CCD) can be used as an index of the difference between values of a key component of the advertisement call for pairs of taxa. The CCD has as its numerator the difference between the minimum value in the range of variation of the taxon with the highest value for the attribute of interest and the maximum value of the same attribute for the second taxon (gap). The denominator is the difference between their respective means ($\Delta\bar{x}$). Thus,

CCD = gap/$\Delta\bar{x}$. For temperature-sensitive call attributes, the value of the CCD depends on the effective temperature and the differences between the taxa in the regression coefficients. Methods of call analysis, sample sizes per individual, and the ways in which maximum and minimum values are inferred will also influence the value.

Spatial Variation in Advertisement Calls

Accurate descriptions of the spatial variation in the structure of advertisement calls are available for only a few pairs of similar, apparently closely related, anuran taxa with overlapping geographic distributions and similar temporal and spatial patterns of reproduction. Here I describe three examples in which the patterns of geographical variation in allopatric and sympatric populations can be interpreted as cases of reproductive character displacement resulting from reinforcing selection: the tree frogs *Litoria ewingii* and *L. verreauxii* in southeastern Australia (Littlejohn 1965); the chorus frogs *Pseudacris nigrita nigrita* and *P. feriarum* in the south-central United States (Fouquette 1975); and the narrow-mouthed toads *Gastrophryne carolinensis* and *G. olivacea* in the southern United States (Blair 1955, Loftus-Hills and Littlejohn 1992). I then describe two cases of zonal hybridization in which the advertisement calls of the hybridizing taxa are as distinctive as those of pairs of taxa that breed at the same times and places. Both examples are from southeastern Australia: the smooth froglets *Geocrinia laevis* and *G. victoriana* (Littlejohn 1988) and the three call races of spotted marsh frogs of the *Limnodynastes tasmaniensis* complex (Littlejohn and Watson 1993).

Examples of Reproductive Character Displacement in Advertisement Calls

Complex Character Displacement in Litoria ewingii *and* L. verreauxii

Litoria ewingii and *L. verreauxii* have wide-ranging peripheral geographic distributions with an extensive linear zone of sympatry which is about 50 km wide and extends parallel to the coastline for some 800 km (Littlejohn 1965, Littlejohn and Watson 1985). Because of the great length of the range, Littlejohn (1965) made an arbitrary division into western and eastern sympatry at 148° E latitude, so that "deep" and "shallow" regions of sympatry were recognized for each taxon. Reciprocal artificial (*in vitro*) crosses indicated that genetic compatibility is high, at least to metamorphosis (Watson and Martin 1968, Watson 1974). Although 3 of 66 amplexed pairs observed at one site in western sympatry were found to be heterospecific (all female *L. ewingii* × male *L. verreauxii*; Littlejohn and Loftus-Hills 1968), putative hybrids, based on intermediacy of advertisement calls, are rare (Nikolakopoulas 1985, Littlejohn and Watson unpublished observations).

The two taxa have similar advertisement calls, each consisting of a group of regularly repeated pulse trains (notes) of similar pulse rate and dominant frequency. There is diphasy in the advertisement calls of most individuals of *L. ewingii*, especially in allopatry and western sympatry, with the first (introductory) note being about twice the duration of the subsequent repeated notes (Littlejohn 1982, fig. 1). All notes in the call of *L. ewingii* are also of similar maximum amplitude. In contrast, the first few notes in the call of *L. verreauxii* are usually shorter and of lower amplitude than the subsequent repeated notes (Littlejohn and Watson unpublished observations). The functional significance of the differences in duration of notes has yet to be determined.

Whereas the repeated notes in the advertisement calls of allopatric populations of both species were found to be similar in all measured attributes, especially those of allopatric *L. verreauxii* from New South Wales and allopatric *L. ewingii* from Tasmania, notes of the sympatric populations differ strikingly in number of pulses, pulse rate, and depth of amplitude modulation (Littlejohn 1965). Values of means, coefficients of variation, and CCDs for pulse rates at 10°C for *L. ewingii* and *L. verreauxii* in mainland allopatric and in western and eastern sympatric populations are presented in table 10-1. There is a reduction of 16.6% in estimated means for pulse rates at 10°C between the allopatric mainland sample of *L. ewingii* and the eastern (i.e., deep) sympatric sample. For *L. verreauxii*, the estimated pulse rate at 10°C for the deep (i.e., western) sympatric sample is higher than that of the conspecific allopatric sample by 63.3%. It is here postulated that this asymmetry indicates that the latter taxon has undergone greater displacement as a consequence of invading more deeply into the range of *L. ewingii*. Coefficients of call difference for pulse rates at 10°C for *L. ewingii* and *L. verreauxii* in western and eastern sympatry, calculated from table 1 of Littlejohn (1965) and the original data set of Littlejohn (1965), are 0.50 ($\Delta\bar{x} = 69.7$; gap = 35) and 0.26 ($\Delta\bar{x} = 46.9$; gap = 12), respectively. Ratios of means for pulse rates (higher/lower) are 1.15 for the allopatric mainland samples, 2.02 for western sympatry, and 1.77 for eastern sympatry. For *L. verreauxii*, Gerhardt and Davis (1988) obtained mean values for pulse rates (calculated by a slightly different means from that of Littlejohn 1965) corrected to 10°C of 84.4 for an allopatric sample and 127.0 for a western sympatric sample (a difference of 50.5%). Numbers of pulses per note, on which effective temperature has little influence (Littlejohn 1965, Littlejohn and Watson unpublished observations), show similar trends to pulse rates.

Table 10-1 Estimated pulse rates (pulses/s) at 10°C for allopatric and sympatric populations of *Litoria ewingii* and *L. verreauxii*.

Taxonomic unit	Attribute	Allopatry	Western sympatry	Eastern sympatry
Litoria ewingii	Mean (range)	73.0 (52–104)	68.0 (48–84)	60.9 (48–76)
	CV% (*n*)	13.2 (74)	11.6 (39)	10.5 (30)
Litoria verreauxii	Mean (range)	84.3 (65–111)	137.7 (119–168)	107.8 (88–125)
	CV% (*n*)	14.0 (41)	9.3 (25)	10.6 (26)
Both taxa	Ratio of means	1.15	2.02	1.77
Both taxa	CCD	—	0.50	0.26

Data from Table 1 of Littlejohn (1965) and original files. CV%, coefficient of variation; CCD, coefficient of call differentiation (see text for explanation).

In contrast to high levels of amplitude modulation in calls of the allopatric samples (both close to 100%), the depth of modulation is reduced to a mean of about 56% (range about 10–90%; estimated from figure 4 of Gerhardt and Davis 1988) in calls of *L. verreauxii* from western (deep) sympatry (Gerhardt and Davis 1988). For *L. ewingii*, the depth of modulation remains close to 100% in all sympatric populations investigated (Littlejohn 1965, 1981; Littlejohn and Watson unpublished observations). Thus there is unilateral displacement of amplitude modulation in the advertisement calls of *L. verreauxii*. The geographic trend in decreased amplitude modulation in *L. verreauxii* with increasing depth of sympatry (see Littlejohn 1965, fig. 4), and the intraindividual variation in amplitude modulation described by Gerhardt and Davis (1988), reflect the progressive loss of distinctiveness of pulse coding as a key temporal factor in the western sympatric populations of this taxon (Littlejohn 1965). The reduction in depth of amplitude modulation in the western sympatric populations of *L. verreauxii* could, however, be an incidental consequence of reinforcing selection for higher pulse rates, coupled with the physical damping properties of the appropriate laryngeal structures. The opposite trend into deeper sympatry occurs in calls of *L. ewingii*, with decreased pulse rate, fewer pulses per call, and the formation of discrete pulses with distinct gaps between them (Littlejohn 1965).

Dominant frequencies of the calls of the two taxa are similar in eastern sympatry, with means differing by only 55 Hz and ranges of variation almost coinciding over a range of about 1950–2700 Hz; in western sympatry, however, the means differ by 568 Hz, with only about 250 Hz overlap in ranges of variation between about 2250 Hz and 2500 Hz (Littlejohn 1965). Thus there is divergent reproductive character displacement in dominant frequency in western sympatric populations of *L. verreauxii*. The similarity of dominant frequencies in the eastern sympatric samples indicates that there is the potential for acoustic interference because of synchronic and syntopic patterns of breeding, and this aspect is currently under investigation (Littlejohn and Watson, unpublished observations).

Littlejohn and Loftus-Hills (1968) reported that females of each taxon from western sympatry would respond only to a conspecific call from the same area. They found, however, that females of *L. ewingii* from western (shallow) sympatry did not discriminate between a conspecific call from the same area and a call of *L. verreauxii* from a remote allopatric population and that western sympatric females of *L. verreauxii* were not attracted to a conspecific call from remote allopatry. Although the natural stimuli used in these tests differed markedly in pulse rate (*L. ewingii* = 71, *L. verreauxii* = 150 Hz), they also differed in amplitude modulation, carrier frequency, and notes per call (Littlejohn and Loftus-Hills 1968). Accordingly, a second series of experiments was carried out, using as stimuli synthetic calls that differed only in pulse rate (*L. ewingii* = 80, *L. verreauxii* = 157) and, as a necessary consequence, number of pulses in a note (*L. ewingii* = 15, *L. verreauxii* = 31). Females of both species from western sympatry chose the signal that matched the conspecific advertisement call from their area of sympatry (Loftus-Hills and Littlejohn 1971). Because the envelopes of the notes of each species gradually increase in amplitude and have abrupt terminations, Loftus-Hills and Littlejohn (1971) argued that the perceived number of pulses (and duration of note) would vary depending on the received intensity of the call. Pulse rate, being based on the intervals between the peaks of the pulses, would not change with alterations in the amplitude of a note and was identified as a sufficient attribute for active discrimination (Loftus-Hills and Littlejohn 1971). Waage (1979) argued that the attributes affected by reproductive character displacement can only be those involved in specificity of mate choice and that the characters should be perceptible. Clearly, these criteria are met here.

In summary, the following patterns of geographical variation in different attributes of the advertisement calls of both species are consistent with reproductive character displacement: (1) bidirectional divergence in pulse rates in both species, (2) a unidirectional divergent trend of reduction in amplitude modulation and the associated loss of pulse distinctness of pulse coding in *L. verreauxii*, (3) greater overlap in dominant frequencies in eastern sympatry, (4) divergence in dominant frequency in western sympatric populations of *L. verreauxii*, and (5) the positive phonotactic responses of females of both species from western sympatry are consistent with the geographical variation in pulse rates, thus supporting the functional significance of the trends for mate choice and the associated homogamy. These preliminary observations, even though based on combined samples from extensive areas, indicate that the interaction deserves more detailed investigation so that geographic trends can be determined across the zone of sympatry.

Unidirectional Character Displacement in Pulse Rate in Pseudacris nigrita nigrita *and* P. triseriata feriarum

The two morphologically similar taxa, *Pseudacris nigrita nigrita* and *P. triseriata triseriata* (nomenclature of Conant and Collins 1991), hereafter referred to as *nigrita* and *feriarum*, have geographic ranges that are largely allopatric, except for a small area of sympatry along the Chattahoochee River drainage that marks the border of Alabama and Georgia, and extends into the Apalachicola River drainage in western Florida (Conant and Collins 1991). The diamond-shaped area of sympatry is about 150 km wide (from east to west) and extends for about 250 km from north to south (estimated from figure 4 of Fouquette 1975).

Given the small amount of experimental data (Mecham 1965) and the lack of natural hybrids in the area of sympatry (Fouquette 1975), little is known about the potential to hybridize and about the relative fitness of hybrids. Breeding seasons overlap considerably, but differences in habitat preference were noted by Crenshaw and Blair (1959) in southwestern Georgia, with *nigrita* tending to call and breed in upland pine flatlands and *feriarum* in floodplain pools of permanent or intermittently flowing streams. Fouquette (1975), however, found that in northern and western areas of sympatry, syntopic breeding assemblages were usual, and he placed greater emphasis on the distinctly different advertisement calls as the means whereby homogamic matings were achieved. The advertisement calls of these species are of similar basic structure, and each consists of a series of regularly spaced and discrete pulses (i.e., a pulse train) increasing in dominant frequency by about 500 Hz through the call; they differ strikingly, however, in pulse rate (Crenshaw and Blair 1959, Fouquette 1975). Spectrograms of the advertisement calls of each taxon from sympatry are presented by Crenshaw and Blair (1959).

Fouquette (1975) established a transect including 14 localities and obtained tape recordings of advertisement calls at each. The transect extended from allopatry (over about 300 km) for *nigrita* (localities 1–4) northward in peninsular Florida and westward to the zone of sympatry, then northward in sympatry (localities 5–8) for about 175 km, and allopatry for *feriarum* (localities 9–14) for about 650 km. Samples from extensive allopatry will indicate whether there are any trends of geographic variation independent of the interspecific interaction; if such trends are not present, then the observed shifts are unique to sympatry and thus represent character displacement (Waage 1979). Four attributes of the calls were measured by Fouquette (1975): duration of the pulse train, number of pulses in each pulse train, pulse rate, and mid-point of the dominant frequency band of the last pulse. For each taxon, the data, corrected to 14°C where appropriate, were combined for the three regions: allopatry for each taxon and sympatry. No intermediate calls or specimens were detected; nor were heterospecific amplexes observed. Thus, natural hybridization was unlikely to be frequent (Fouquette 1975).

There are considerable overlaps in dominant frequency and duration in the combined sympatric samples (table 10-2). The similarity in dominant frequencies could lead to serious interspecific acoustic interference at syntopic sites if calls are emitted simultaneously. Pulse numbers of the pooled sympatric samples have contiguous ranges of variation (table 10-2). Although values of the pooled allopatric samples overlap, pulse rates are distinctly different in sympatry, with no overlap in ranges (table 10-2). The ratio of the means for pulse rates of the sympatric samples at 14°C is 3.24. For the four sympatric localities, the gaps in ranges of pulse rates vary from 9 to 15 pulses/s. The values for CCD of pulse rate at 14°C at these four localities are 0.67, 0.65, 0.68, and 0.58 (mean = 0.64), but for the pooled data, the CCD is 0.43. Estimated coefficients of variation in pulse rate for the combined allopatric and sympatric samples, respectively (from figure 2 of Fouquette 1975), are lower in sympatry compared with allopatry for *nigrita*

Table 10-2 Means and ranges of four attributes of the advertisement calls of *Pseudacris nigrita nigrita* and *P. triseriata feriarum*, estimated from figure 2 of Fouquette (1975), in which values were corrected to 14°C where appropriate.

Feature of call	Geographic status of sample	*P. n. nigrita* Mean (range)	*P. n. nigrita* Number	*P. t. feriarum* Mean (range)	*P. t. feriarum* Number
Duration (s)	Allopatric	1.07 (0.72–1.35)	115	0.85 (0.35–1.25)	78
	Sympatric	1.12 (0.60–1.58)	72	1.00 (0.55–1.45)	117
Number of pulses	Allopatric	9.5 (6–12)	110	17.0 (10–24)	102
	Sympatric	9.5 (6–13)	85	22.0 (13–40)	137
Pulse rate (pulses/s)	Allopatric	8.3 (6.0–15.5)	118	16.5 (13.0–20.0)	109
	Sympatric	8.2 (6.0–11.0)	86	26.6 (19.0–35.0)	140
Dominant frequency (kHz)	Allopatric	3.40 (2.90–4.00)	116	3.25 (2.70–4.30)	110
	Sympatric	3.20 (2.70–4.05)	79	3.00 (2.65–3.60)	127

(12.5% vs. 21.7%), and higher in *feriarum* (16.9% vs. 9.1%). Following Otte (1989) and Littlejohn (1993), the unidirectional divergent reproductive character displacement in pulse rate for *feriarum* (from 16.5 pulses/s in allopatry to 26.6 pulses/s in sympatry) indicates that this taxon is the invading species.

On the Pearl River drainage, along the southern section of the border between Louisiana and Mississippi, *nigrita* forms a narrow hybrid zone (between 9 and 19 km wide, on the basis of four allozymic systems) with populations presumed to be of *feriarum* (Gartside 1980). Although not differentiated in adult morphology, more than 50% of the individuals within the hybrid zone were classified as hybrids on the basis of their allozymic profiles, and there was evidence of back-crossing. Even so, the taxa remain genetically distinct in areas adjacent to the narrow hybrid zone. Gartside (1980) also noted that there were no apparent differences between the advertisement calls of these two hybridizing taxa in the area of the transect (on the basis of subjective observations), and he suggested an ecological explanation for the high level of survival of hybrids in the Pearl River area, in contrast to the apparent absence of hybridization in the area of sympatry noted by Fouquette (1975). Gartside (1980, pp. 63–64) concluded: "In the present study, hybrids occur only in an ecologically discrete area—the Pearl River mixed hardwood bottom lands—separating areas of apparently similar parental habitats (piny woods)." Here, there is an implication of hybrid superiority, with no cost to interbreeding and no selective pressure for reinforcement and associated character displacement.

Unidirectional Character Displacement in Dominant Frequency in Gastrophryne carolinensis *and* G. olivacea

Gastrophryne carolinensis and *G. olivacea* have held a central position in the development of the concept of reinforcement since the publication by Blair (1955) of one of the first comprehensive and quantitative treatments of geographical variation in a behavioral attribute. There is a broad area of sympatry in central and eastern Texas and northward into Oklahoma and extensive areas of allopatry for *G. olivacea* in the west and *G. carolinensis* in the east (see Conant and Collins 1991). Blair investigated two attributes of the calls: call duration and mid-point of the emphasized band of frequencies. Subsequently, Awbrey (1965) obtained values for pulse rate from the same data set used by Blair (1955). He found that, for calls recorded between 24 and 26°C, there was slight overlap. The values for pulse rates of the sympatric populations were more distinctive than those of the allopatric populations, a finding consistent with reinforcement and reproductive character displacement.

Loftus-Hills and Littlejohn (1992) reexamined this interaction and investigated pulse rates, call duration, and spectral composition of the advertisement calls of the two species. The latter attribute was measured as dominant frequency (i.e., that frequency of greatest amplitude in an audiospectrographic section obtained near the middle of a call). These authors confirmed the presence of reproductive character displacement in spectral composition found by Blair (1955). In contrast to the findings of Blair (1955), there were no overlaps in the ranges of distribution of dominant frequency in calls from all areas, but a gap of 291 Hz between values for allopatric samples of *G. carolinensis* and allopatric/shallow sympatric samples of *G. olivacea* and of 728 Hz for the two species in the confirmed area of sympatry.

For the sympatric samples, the CCD for dominant frequency at 25°C is 0.51 ($\Delta\bar{x}$ = 1423 Hz, gap = 728 Hz; from original data set of Loftus-Hills and Littlejohn 1992); and the ratio of the means is 1.49. The displacement in dominant frequency is evident only in *G. carolinensis* (Loftus-Hills and Littlejohn 1992), so that the trend is unilateral. The ranges of variation in call duration for the two species overlap extensively between about 1000 and 1900 ms across the range of recording temperatures. Pulse rates of the two taxa overlap extensively below about 28°C, with increasing divergence at higher temperatures (Loftus-Hills and Littlejohn 1992). A high level of separation between the taxa was obtained in a scattergram of uncorrected paired values of pulse rate and duration for temperatures between 21.0 and 31.8°C, with overlap for only three individuals—two of *G. carolinensis* (one from allopatry) and one of *G. olivacea* (Loftus-Hills and Littlejohn 1992). While undoubtedly useful for taxonomic diagnosis, these differences may also reflect the action of reinforcing selection on processes of mate choice, and deserve further investigation.

There were no indications of any significant level of hybridization in the study of Loftus-Hills and Littlejohn (1992), with only one possible hybrid male being recognized on the basis of intermediacy of call in a sample of at least 83 and up to 110 males (Loftus-Hills and Littlejohn 1992). Blair (1955), however, identified 8 putative hybrids by this criterion in a total of 100 individuals recorded from the area of sympatry.

Volpe (1957) suggested that the variation in emphasized frequency in the advertisement calls observed by Blair (1955) might be explained as an incidental effect of natural selection acting on body size through ecological competitive interactions. As no consistent patterns of correlation were found between the three attributes of the advertisement calls and the snout-vent lengths of the emitters, pleiotropic effects of this variable on call structure were considered unlikely (Loftus-Hills and Littlejohn 1992). This finding thus counteracts the criticisms of Volpe (1957) and strengthens the original hypothesis of reinforcement as an explanation. But correlations with other morphological attributes, such as head width and the associated size of the supported vocal sac, which could influence the frequency composition of the signal (Loftus-Hills and Littlejohn 1992), should be explored.

Variation in Structure of Advertisement Calls along Transects across Regions of Zonal Hybridization

A Narrow Hybrid Zone between Geocrinia laevis *and* G. victoriana

There are two hybrid zones between the morphologically similar taxa *Geocrinia laevis* and *G. victoriana*: a long and irregular contact extending from the southern coastline of western Victoria northeastward for about 115 km (Littlejohn et al. 1971, Littlejohn and Watson 1973, Littlejohn 1988, Gollmann 1991, Littlejohn and Watson unpublished observations), and a shorter contact of about 10 km with an east-west orientation in the Grampians Ranges of central western Victoria (Littlejohn and Martin 1964, Gollmann 1991). (See Littlejohn and Watson [1985] for a map of the overall distributions of the two taxa.) The advertisement call of each species consists of a series of notes of similar dominant frequency within the range of 2300–3050 Hz (Littlejohn and Martin 1964, Littlejohn et al. 1971). The advertisement call of *G. victoriana* is strongly diphasic, with the first (introductory) note (and sometimes the second, and rarely the third) being much longer and of lower pulse rate than the subsequent series of similar repeated notes (Littlejohn and Harrison 1985). Although the advertisement calls of males of *G. laevis* are generally diphasic (Harrison and Littlejohn 1985), the ranges of variation of introductory notes and repeated notes in a population may overlap (e.g., by 7–9 pulses/s within the range of 23–35 pulses/s; Littlejohn et al. 1971). Within the recorded ranges, there were no significant effects of temperature on these attributes of the advertisement calls (Harrison and Littlejohn 1985).

The ranges of variation of the pulse rates of introductory and repeated notes allow diagnosis of the calls of putative parental and hybrid males, and the construction of hybrid (character) indices (Littlejohn et al. 1971, Littlejohn and Watson 1973, Littlejohn 1988). Typical values for introductory notes are ≤40 for *G. laevis* and ≥100 for *G. victoriana*; typical values for repeated notes are ≤100 for *G. laevis* and ≥350 for *G. victoriana*. The transition in structure of the advertisement calls across the southern hybrid zone is relatively steep, with most of the change occurring over 2–4 km (Littlejohn et al. 1971, Littlejohn and Watson 1973, 1985, unpublished observations; Gartside et al. 1979, Littlejohn and Gollmann, in preparation).

The CCD for pulse rates of repeated notes for two populations from areas adjacent to the southern hybrid zone, based on the data set used by Littlejohn (1988), within a temperature range of 10.7–12.5°C, is 0.82 ($\Delta\bar{x}$ = [436.5 – 43.1] = 393.4 pulses/s; gap = 321.2 pulses/s), and the ratio of the means is 10.12. For the same calls, the CCD for pulse rates of introductory notes is 0.75 ($\Delta\bar{x}$ = [151.3 – 20.4] = 130.9 pulses/s; gap = 98 pulses/s), and the ratio of the means is 7.40.

Thus, the calls of these parapatric and hybridizing taxa are more distinctive in pulse rates of both introductory and repeated notes than are the calls of the two pairs of extensively syntopic hylid taxa discussed earlier. In the calls of hybrids, the values of pulse rates of introductory notes are usually intermediate, while those of repeated notes are either of relatively low variability and intermediate or of high variability (i.e., mosaic; Littlejohn et al. 1971). These patterns of variability in repeated notes of parental and hybrid calls were depicted diagrammatically by Littlejohn et al. (1971). In the earlier studies, values for pulse rates of the first (introductory) note and of three successive (repeated) notes, from near the middle of the sequence and chosen to maximize variability, were used. The intermediate values of the hybrid index were then set around these criteria (Littlejohn et al. 1971, Littlejohn and Watson 1973, Gartside et al. 1979). Littlejohn (1988), however, considered values for all notes in one call of each individual.

Littlejohn and Watson (1974), through two-choice discrimination experiments carried out in the field, found that a normal advertisement call of each species with an introductory note followed by a series of repeated notes was effective in eliciting positive phonotaxis in gravid conspecific females from areas adjacent to the hybrid zone. They later demonstrated in two-choice discrimination trials (Littlejohn and Watson 1976) that a mosaic hybrid advertisement call (i.e., one containing elements of calls of both parental taxa), when offered as an alternative stimulus to the conspecific advertisement call, was effective in attracting gravid females of both parental taxa.

Through field playback experiments, Littlejohn and Harrison (1985) found that there was partitioning of function between the two components of the diphasic call of *G. victoriana.* When presented above a threshold of intensity (about 100 dB peak sound pressure level), the introductory note was shown to be involved in territorial interactions between males, with inhibition of advertisement calling and the production of conspicuous encounter calls (sensu Littlejohn 1977). No encounter calls were produced with repeated notes as the stimulus; rather, the effect was excitatory, with sustained and vigorous advertisement calling (Littlejohn and Harrison 1985). When an introductory note and a series of repeated notes were used as alternative stimuli, females of *G. victoriana* were attracted only to the repeated notes (Littlejohn and Harrison 1985), showing the mate-attracting function of the latter. During a similar playback regime, males of *G. laevis* showed no indications of acoustic territorial behavior, and no encounter calls were emitted; rather, the subjects ceased calling (Harrison and Littlejohn 1985). As no responsive gravid females of *G. laevis* were obtained during the field program, it is not known whether there is any partitioning of function between introductory and repeated notes in *G. laevis*; given the overlap in ranges of variation of pulse rate and the absence of an encounter call in males, such separation is considered unlikely (Harrison and Littlejohn 1985). The acoustic repertoires and associated behavior of males of the two taxa thus differ markedly.

Given the encounter call and associated aggressive and territorial behavior of males of *G. victoriana* (Littlejohn and Harrison 1985) and the apparent lack of such behavior and associated acoustic signals in males of *G. laevis* (Harrison and Littlejohn 1985), such an imbalance may be expressed to different extents in the hybrids. Hence, certain genotypes with high levels of aggression may be more successful in holding the preferred oviposition sites, and then of procuring mates, than are the less aggressive genotypes. Such interactions between males could thus have a significant influence on the structure of the hybrid populations (Gollmann 1991).

The region in which the southern hybrid zone is located has undergone considerable modification from clearing of the forest and conversion to farming since its occupancy by Europeans in the mid-nineteenth century (Bennett 1990). These changes may have favored the eastward expansion of the geographic range of *G. laevis* and the consequent displacement of *G. victoriana*, and it is highly likely that the zonal interaction is now in a dynamic phase (Littlejohn 1988).

Zonal Interactions of Different Width and Structure in the Limnodynastes tasmaniensis *Complex*

The wide-ranging southeastern Australian myobatrachid species *Limnodynastes tasmaniensis* includes three call races: northern, southern, and western (Littlejohn 1967, Loftus Hills 1973, Littlejohn and Roberts 1975, Roberts 1976, 1993, Littlejohn and Watson 1993). As indicated by the informal taxonomic notation, these three morphologically similar, largely allopatric taxa were initially resolved on the distinctive structure of the advertisement calls. The parapatric interactions between the northern and southern races, and between the northern and western races, are marked by hybrid zones (Littlejohn and Roberts 1975, Roberts 1976, 1993). The southern and western call races overlap in a narrow zone of sympatry with synchronous and syntopic breeding (see figure 2 of Littlejohn and Watson 1993), and hybridization is rare (Roberts 1976, Roberts and Littlejohn in preparation). The three call races thus constitute a ring species and provide a suitable system for assessing differentiation in advertisement calls with respect to the structure and dynamics of homogamy.

The interaction between the northern and western call races is located on the western side of the lower Murray River in southeastern South Australia, with two limited areas of hybridization about 6 and 20 km wide (Roberts 1993). The advertisement calls of the northern and western races are of similar basic structure, with each consisting of a short train of rapidly repeated notes which have amplitude-modulated envelopes, allowing the determination of pulse repetition rate (Littlejohn and Roberts, 1975, Roberts 1976, 1993). Only note repetition rate is significantly affected by temperature, with similar slopes but significantly different displacements for the races; accordingly, estimated values at 18°C were used by Roberts (1993) in subsequent comparisons. Roberts (1993, p. 108) commented that: "although values for dominant frequency, average number of notes per call, and note repetition rate show little or no overlap, no single call component is sufficiently different to allow recognition of intermediate phenotypes."

By constructing a character index for these three attributes from a large sample of calls from a remote allopatric population of each race, Roberts (1993) was able to classify males as either parental or intermediate individuals. He established one transect in each area; the larger northern contact (Morgan) included five localities over about 30 km, and the smaller, southern (Marne) included four localities over about 10 km. For each transect, the populations on the side of the western race included a few intermediates and only individuals of the one parental taxon. With the exception of one small sample ($n = 5$) on the Morgan transect, the samples from the zone of interaction included intermediates and putative parental males of both taxa. The interactions may thus be provisionally interpreted as zones of overlap with hybridization (see earlier discussion). On the basis of historical biogeographic information, Roberts (1993) suggested that the northern race may be expanding its geographic range to the south and west and displacing the western race, so that the interaction could thus be in a dynamic state. Because ranges of variation abut or overlap for all measured attributes, no values of CCD can be calculated.With the exception of an upward shift in pulse rate at one locality, Roberts (1993) found no evidence of positive assortative mating or reinforcement in the two zones of interaction between the northern and the western call races.

There are several areas of interaction between the northern and southern races (see figure 2 of Littlejohn and Watson 1993), with the main zone occurring in north-central Victoria. This zone has a length of about 215 km and a maximum width of between 90 and 135 km (Littlejohn and Roberts 1975). Calls of the northern and southern races are differentiated as much as in extensively synchronic, syntopic, closely related taxa (Littlejohn 1969) in notes per call (only one note in the southern race, and two or more in the northern race), note duration (<13 ms in the southern race and >20 ms in the northern race), and the absence of envelope modulation in the southern race; dominant frequencies, however, overlap broadly (ranges = 1832–2262 Hz for the southern call race, and 1566–1998 Hz for the northern call race; Littlejohn and Roberts 1975). Because of the relatively narrow range of temperatures over which the tape recordings were obtained (12.3–18.0°C), no attempt was made to determine the effects of this extrinsic variable on call structure. Populations containing hybrids were identified by the presence of individuals making calling sequences consisting of single notes and multiple-note calls, with the frequency of single notes in calling sequences of individuals decreasing with increasing distance across the zone away from the southern race. In their study, Littlejohn and Roberts (1975) considered note duration and the number of notes in the last five calls in a long and uninterrupted sequence to obtain an effective measure of the nature and extent of hybridization.

With the exclusion of one outlier (with extreme values) from the most distant population of the northern call race (locality 9), the data used by Littlejohn and Roberts (1975) were reexamined to obtain additional statistical information and graphical presentations. Because of the nature of the differences in notes per call (all single notes in the southern race and all two or more in the northern race), a meaningful CCD could only be calculated for the one distinctive attribute, note duration. Thus, there is a gap in ranges of 8.00 ms, and the difference between the means ($\Delta\bar{x}$) is 14.49 ms (northern call race = 25.13, southern call race = 10.64; from table 2 of Littlejohn and Roberts 1975); these values yield a CCD of 0.55, which is similar to those of extensively syntopic and synchronic taxa (see earlier discussion).

Although populations near the margins of the zone include hybrids and putative parental individuals of only one or other of the parental taxa, putative parental individuals of both races and recombination products are present at the central hybrid localities; hence, this interaction is also probably closer to an overlap with hybridization. Clearly, more samples are required from other transects in the main zone of interaction and from the other smaller interactions between the northern and the southern call race before a reliable assessment can be made. Because the northern and southern races hybridize freely wherever their ranges contact, well-differentiated, presumably homogamic, signals alone are insufficient to prevent interbreeding (Littlejohn 1993). The biogeographic patterns of distribution of these two call races are also consistent with an expansion of range by the northern race, with displacement of the southern race (Littlejohn and Roberts 1975, Littlejohn and Watson 1993). Thus, this interaction also appears to be in a dynamic state.

The calls of the southern race differ from those of the western race in all measured attributes, even more so than from the northern race because of the difference in dominant frequency. The extensive geographic ranges of the two taxa and the distinctiveness of call structure throughout their ranges indicate that they have become preadapted for effective homogamy where breeding in synchronic syntopy. Thus no reinforcement would be expected to occur in the small zone of sympatry, even if there were complete genetic incompatibility (but the putative hybrid suggests that there is at least some compatibility). Ecological partitioning, recency of contact, or marginal conditions for both taxa may provide explanations for their coexistence.

Summary and Conclusions

Values for the CCD for pulse rates in the sympatric populations discussed above range from 0.26 for the eastern sympatric populations of *Litoria ewingii* and *L. verreauxii* to 0.68 for one of the pair of sympatric samples of *Pseudacris n. nigrita* and *P. n. feriarum*. The low value of the former pair is about one-half that of the western sympatric populations (CCD = 0.50), and may be a consequence of the combining of several samples from a wide area. The ratios of the means of presumed key attributes in sympatry range from 1.49 for dominant frequency in *Gastrophyne carolinensis* and *G. olivacea* to 3.24 for pulse rates in *P. n. nigrita* and *P. n. feriarum*. Two other sets of values for pulse rates at 20°C were also estimated from the literature, using the regression equations provided: for *Hyla chrysoscelis* and *H. versicolor* (from data presented by Gerhardt 1982), CCD = 0.75 and ratio of means = 2.27; for *Pseudacris clarki* and *P. nigrita* (from data presented by Michaud 1964), CCD = 0.78 and ratio of means = 4.32. For the two cases of hybridizing taxa discussed earlier, the values are comparable to those of the pairs of sympatric populations. Thus, for pulse rates of repeated notes in *Geocrinia laevis* and *G. victoriana*, CCD = 0.82 and ratio of means = 10.12, and for note duration in the northern and southern call races of *Limnodynastes tasmaniensis*, CCD = 0.55, and ratio of means = 2.36.

For the three pairs of taxa with sympatric distributions, the following geographic trends emerge: (1) bidirectional displacement in pulse rate for *Litoria ewingii* and *L. verreauxii*; (2) unidirectional displacement in pulse rate *in Pseudacris triseriata feriarum* in sympatry with *P. nigrita nigrita*; (3) unidirectional displacement of dominant frequency in *L. verreauxii* in deep sympatry with *L. ewingii* and in *Gastrophryne carolinensis* in sympatry with *G. olivacea*; (4) decreased amplitude modulation with increasing depth of sympatry in *L. verreauxii* in sympatry with *L. ewingii*; and (5) the progressive loss of a key temporal attribute (pulse rate) in *L. verreauxii*, possibly as a consequence of the increase in pulse rate. As anticipated earlier, no clear case of partitioning of ranges of variation in the key attribute of a signal was found in these examples of reproductive character displacement. For the two pairs of hybridizing parapatric taxa, although advertisement calls of allopatric populations are as distinctive as those of extensively sympatric syntopic and synchronously breeding species, there have been extensive breakdowns in the potential homogamy. It was observations of this nature that led Littlejohn (1988, 1993) to conclude that the processes of mate choice subsumed under premating reproductive isolation in the biological species concept can only have a subsidiary role. For effective homogamy to evolve or be maintained, there must be a primary relative cost to interpopulational matings in terms of lower reproductive success (as compromised hybrid progeny if genetically compatible) and reduced efficiency of mate attraction and choice (for both compatible and incompatible systems) (Littlejohn 1993). Hence, ecological factors are considered to be of fundamental importance in the initial stages of establishment of zonal interactions in two ways: (1) the presence, adequacy, and effective partitioning of essential and limiting resources would allow coexistence, and then reinforcement can occur with the associated development of reproductive character displacement; and (2) available intermediate or novel habitats are required for occupancy by hybrids with superior survival because of their intermediacy of adaptation (Littlejohn 1993). But even less is currently known about the ecological aspects of these zonal interactions, particularly when associated with the evolution and maintenance of homogamy.

The examples of overlapping and zonally hybridizing pairs of populations carry the mark of recent and extensive environmental modification through human activities. The consequent inevitable destabilization implies that the zonal interactions have been shifted away from old equilibria and that new dynamic processes have been initiated—with all these changes taking place just at the time when sufficiently detailed and objective field studies have been started (Littlejohn and Watson 1985, 1993). Even so, it is important to obtain as extensive and full documentation as possible of the present spatial patterns, structure, and nature of such zonal interactions. Only then will it be possible to provide the appropriate baselines for the future research that will allow recognition of any new trends and the documentation of stability or, more likely, of displacements to new equilibria.

Descriptions of existing patterns of geographical variation in systems of homogamy and the application of a comparative approach may thus provide some insights into the critical steps in development of sympatry in pairs of closely related taxa. It is possible, of course, that an original zone of contact was subsequently displaced as a dynamic front along with regional climatic changes (as modeled in a general scenario by Littlejohn 1993), thus confounding the historical interpretation based on comparison of patterns gleaned from a comparison of extant populations. With allowance for this possibility, consideration of several examples of patterns of spatial variation in the factors associated with the origin and maintenance of homogamy (in this case the advertisement call of anuran amphibians) and the breakdown of homogamy in hybrid zones is the most reasonable source of information from natural systems that can allow the construction of realistic models for subsequent testing.

What general conclusions about processes of speciation in biparental organisms can be drawn from these examples? Only those situations explicable in terms of the classical allopatric model of speciation have been considered here. Are there similarities with the evolution and maintenance of homogamy in the parapatric and sympatric (i.e., nongeographic; Bush 1994) processes of speciation? Here the development or maintenance of homogamy most likely would be linked to the shift to the alternative set of ecological resources (see Bush 1993, 1994, for overviews and other references), at least where the distribution of the resources is coarse grained (sensu Levins 1968, Pianka 1974). The selective mating is then most likely to be incidental in origin, for it would arise indirectly as a consequence of the primary process of ecological divergence—where breeding occurs on a specific host or in a particular habitat (i.e., in allotopy; sensu Rivas 1964). The nature of the process of sympatric speciation, especially its initiation, in a fine-grained environment (sensu Levins 1968, Pianka 1974) is much harder to visualize and will not be addressed here.

How appropriate is the application of a term such as "homogamy" to processes of assortative mating in the context of hybrid zones and the incipient stages of the development of sympatry? The origins of the term have been discussed (Littlejohn 1993), and "homogamic mechanisms" were defined by Littlejohn (1981, p. 319) as "those factors or processes that directly increase an individual's probability of mating with a member of the same adaptive or genetically compatible set." Clearly the term homogamy, as a descriptor, can be applied to situations in which there is complete genetic incompatibility, as measured by full sterility, the failure to produce fertile hybrids, or where recombination products are so ill-adapted that they cannot survive to reproductive age because of their inability to compete with pure parental individuals for essential and limiting ecological resources. If effective homogamy is lacking or incompletely effective in these situations of "full speciation," then natural selection is expected to lead to the origin, improvement, and subsequent maintenance of assortative mating. Where potentially effective homogamy is present when individuals from each of two genetically compatible cognate populations come into contact and there is no cost to mating with an individual of the other population, the mating systems are expected to break down into panmixia because of inevitable mistakes in choosing a mate (Littlejohn 1988, 1993). If there is an ecological gradient and differential ecological adaptation, then the position of the contact would be displaced toward an area of competitive equilibrium. Here the fitness of hybrids would exceed those of the parental individuals (Littlejohn 1988, 1993). With no cost to choosing a mate from the alternative genetic system, the eventual outcome is a hybrid swarm in which any homogamy apparently is lacking (Littlejohn 1988, 1993). This outcome appears to be the case for *Geocrinia laevis* and *G. victoriana* and the northern and southern call races of *Limnodynastes tasmaniensis*, discussed earlier. In these situations, the use of terms associated with selective mating seems inappropriate. But these interactions may not have yet reached an equilibrium, or they have been recently reactivated because of extensive disturbances of habitat (Littlejohn and Watson 1985, 1993). The eventual outcome at equilibrium would then depend on factors such as the maximum distance that an individual moves during its lifetime, the rates of migration, and steepness and width of the ecological gradient on which the zone of interaction is situated. When equilibrium is reached, assortative mating may be favored by natural selection once more, with the associated development or restoration of homogamy.

References

Abt, G., and H.-U. Reyer. 1993. Mate choice and fitness in a hybrid frog: *Rana esculenta* females prefer *Rana lessonae* males over their own. Behavioral Ecology and Sociobiology 32:221–228.

Arnold, S. J. 1994. Is there a unifying concept of sexual selection that applies to both plants and animals? American Naturalist 144(supplement):S1–12.

Awbrey, F. T. 1965. An experimental investigation of the effectiveness of anuran mating calls as isolating mechanisms (PhD thesis). University of Texas, Austin.

Barton, N. H., and G. M. Hewitt. 1989. Adaptation, speciation and hybrid zones. Nature 341:497–503.

Bennett, A. F. 1990. Land use, forest fragmentation and the mammalian fauna at Naringal, south-western Victoria. Australian Wildlife Research 17:325–347.

Blair, W. F. 1955. Mating call and stage of speciation in the *Microhyla olivacea-M. carolinensis* complex. Evolution 9:469–480.

Bull, C. M. 1991. Ecology of parapartic distributions. Annual Review of Ecology and Systematics 22:19–36.

Bush, G. L. 1993. A reaffirmation of Santa Rosalia, or why are there so many kinds of *small* animals? *In* D. R. Lees and D. Edwards, eds. Evolutionary patterns and processes, pp. 229–249. Linnean Society of London, London, and Academic Press, London.

Bush, G. L. 1994. Sympatric speciation in animals: New wine in old bottles. Trends in Ecology and Evolution 9:285–288.

Butlin, R. K. 1989. Reinforcement of premating isolation. *In* D. Otte and J. A. Endler, eds. Speciation and its consequences, pp. 158–79. Sinauer Associates, Sunderland, MA.

Butlin, R. K. 1995. Reinforcement: An idea evolving. Trends in Ecology and Evolution 10:432–434.

Conant, R., and J. T. Collins. 1991. A field guide to reptiles and amphibians. Eastern and central North America, 3rd ed. Houghton Mifflin Company, Boston.

Crenshaw, J. W., and W. F. Blair. 1959. Relationships in the *Pseudacris nigrita* complex in southwestern Georgia. Copeia 1959:215–222.

Fouquette, M. J. 1975. Speciation in chorus frogs. I. Reproductive character displacement in the *Pseudacris nigrita* complex. Systematic Zoology 34:16–22.

Gartside, D. F. 1980. Analysis of a hybrid zone between chorus frogs of the *Pseudacris nigrita* complex in the southern United States. Copeia 1980:56–66.

Gartside, D. F., M. J. Littlejohn, and G. F. Watson. 1979. Structure and dynamics of a narrow hybrid zone between *Geocrinia laevis* and *G. victoriana* (Anura: Leptodactylidae) in south-eastern Australia. Heredity 43:165–177.

Gerhardt, H. C. 1982. Sound pattern recognition in some North American treefrogs (Anura: Hylidae): Implications for mate choice. American Zoologist 22:581–595.

Gerhardt, H. C. 1994. Reproductive character displacement of female mate choice in the grey treefrog, *Hyla chrysoscelis*. Animal Behaviour 47:959–969.

Gerhardt, H. C., and M. S. Davis. 1988. Variation in the coding of species identity in the advertisement calls of *Litoria verreauxi* (Anura: Hylidae). Evolution 42:556–563.

Gerhardt, H. C., and J. J. Schwartz. 1995. Interspecific interactions in anuran courtship. *In* H. Heatwole and B. K. Sullivan, eds. Amphibian biology, vol. 2. Social communication, pp. 603–632. Surrey Beatty and Sons, Chipping Norton, New South Wales, Australia.

Gollmann, G. 1991. Population structure of Australian frogs (*Geocrinia laevis* complex) in a hybrid zone. Copeia 1991:593–602.

Harrison, P. A., and M. J. Littlejohn. 1985. Diphasy in the advertisement calls of *Geocrinia laevis* (Anura: Leptodactylidae): Vocal responses of males during field playback experiments. Behavioral Ecology and Sociobiology 18:67–73.

Harrison, R. G. (ed). 1993. Hybrid zones and the evolutionary process. Oxford University Press, New York.

Harrison, R. G., and D. M. Rand. 1989. Mosaic hybrid zones and the nature of species boundaries. *In* D. Otte and J. A. Endler, eds. Speciation and its consequences, pp. 111–133. Sinauer Associates, Sunderland, MA.

Hewitt, G. M. 1989. The subdivision of species by hybrid zones. *In* D. Otte and J. A. Endler, eds. Speciation and its consequences, pp. 85–110. Sinauer Associates, Sunderland, MA.

Howard, D. J. 1993. Reinforcement: Origin, dynamics, and fate of an evolutionary hypothesis. *In* R. G. Harrison, ed. Hybrid zones and the evolutionary process, pp. 46–69. Oxford University Press, New York.

Levins, R. 1968. Evolution in changing environments: Some theoretical explorations. Princeton University Press, Princeton, NJ.

Kuhn, T. S. 1970. The structure of scientific revolutions, 2nd ed. University of Chicago Press, Chicago.

Littlejohn, M. J. 1965. Premating isolation in the *Hyla ewingi* complex (Anura: Hylidae). Evolution 19:234–243.

Littlejohn, M. J. 1967. Patterns of zoogeography and speciation in south-eastern Australian amphibia. *In* A. H. Weatherley, ed. Australian inland waters and their fauna. Eleven studies, pp. 150–174. Australian National University Press, Canberra.

Littlejohn, M. J. 1969. The systematic significance of isolating mechanisms. *In* Systematic biology. Proceedings of an international conference, pp. 459–482. National Academy of Sciences, Washington, DC.

Littlejohn, M. J. 1977. Long-range acoustic communication in anurans: an integrated and evolutionary approach. *In* D. H. Taylor and S. I. Guttman, eds. The reproductive biology of amphibians, pp. 263–294. Plenum Press, New York.

Littlejohn, M. J. 1981. Reproductive isolation: A critical review. *In* W. R. Atchley and D. S. Woodruff, eds. Evolution and speciation. Essays in honor of M. J. D. White, pp. 298–334. Cambridge University Press, Cambridge.

Littlejohn, M. J. 1982. *Litoria ewingi* in Australia: A consideration of indigenous populations, and their interactions with two closely related species. *In* D. G. Newman, ed., New Zealand herpetology. Proceedings of a symposium held at the Victoria University of Wellington 29–31 January 1980, pp. 113–135. New Zealand Wildlife Service, Wellington.

Littlejohn, M. J. 1988. Frog calls and speciation. The retrograde evolution of homogamic acoustic signaling systems in hybrid zones. *In* B. Fritzsch, M. J. Ryan, W. Wilczynski, T. E. Hetherington, and W. Walkowiak, eds. The evolution of the amphibian auditory system, pp. 613–635. John Wiley and Sons, New York.

Littlejohn, M. J. 1993. Homogamy and speciation: A reappraisal. *In* D. Futuyma and J. Antonovics, eds. Oxford surveys in evolutionary biology, vol. 9, pp. 135–165. Oxford University Press, New York.

Littlejohn, M. J., and P. A. Harrison. 1985. The functional significance of the diphasic advertisement call of *Geocrinia victoriana* (Anura: Leptodactylidae). Behavioral Ecology and Sociobiology 16:363–373.

Littlejohn, M. J., and J. J. Loftus-Hills. 1968. An experimental evaluation of premating isolation in the *Hyla ewingi* complex (Anura: Hylidae). Evolution 22:659–663.

Littlejohn, M. J., and A. A. Martin. 1964. The *Crinia laevis* complex (Anura: Leptodactylidae) in south-eastern Australia. Australian Journal of Zoology 12:70–83.

Littlejohn, M. J., and J. D. Roberts. 1975. Acoustic analysis of an intergrade zone between two call races of the *Limnodynastes tasmaniensis* complex (Anura: Leptodactylidae) in south-eastern Australia. Australian Journal of Zoology 23:113–122.

Littlejohn, M. J., and G. F. Watson. 1973. Mating-call variation across a narrow hybrid zone between *Crinia laevis* and *C. victoriana* (Anura: Leptodactylidae). Australian Journal of Zoology 21:277–284.

Littlejohn, M. J., and G. F. Watson. 1974. Mating call discrimination and phonotaxis by females of the *Crinia laevis* complex (Anura: Leptodactylidae). Copeia 1974:171–175.

Littlejohn, M. J., and G. F. Watson. 1976. Effectiveness of a hybrid mating call in eliciting phonotaxis by females of the *Geocrinia laevis* complex (Anura: Leptodactylidae). Copeia 1976:76–79.

Littlejohn, M. J., and G. F. Watson 1983. The *Litoria ewingi* complex (Anura: Hylidae) in south-eastern Australia VII. Mating-call structure and genetic compatibility across a narrow hybrid zone between *L. ewingi* and *L. paraewingi*. Australian Journal of Zoology 31:193–204.

Littlejohn, M. J., and G. F. Watson. 1985. Hybrid zones and homogamy in Australian frogs. Annual Review of Ecology and Systematics 16:85–112.

Littlejohn, M. J., and G. F. Watson. 1993. Hybrid zones in Australian frogs: Their significance for conservation. *In* D. Lunney and D. Ayers, eds., Herpetology in Australia: A diverse discipline, pp. 239–249. Royal Zoological Society of New South Wales, Mosman.

Littlejohn, M. J., G. F. Watson, and J. J. Loftus-Hills. 1971. Contact hybridization in the *Crinia laevis* complex (Anura: Leptodactylidae). Australian Journal of Zoology 19: 85–100.

Loftus-Hills, J. J. 1973. Comparative aspects of auditory function in Australian anurans. Australian Journal of Zoology 21:353–367.

Loftus-Hills, J. J., and M. J. Littlejohn. 1971. Pulse repetition rate as the basis for mating call discrimination by two sympatric species of *Hyla*. Copeia 1971:154–156.

Loftus-Hills, J. J., and M. J. Littlejohn. 1992. Reinforcement and reproductive character displacement in *Gastrophryne carolinensis* and *G. olivacea* (Anura: Microhylidae): A reexamination. Evolution 46:896–906.

Mayr, E. 1942. Systematics and the origin of species. Columbia University Press, New York.

Mayr, E. 1992. A local flora and the biological species concept. American Journal of Botany 79:222–238.

Mecham, J. S. 1965. Genetic relationships and reproductive isolation in southeastern frogs of the genera *Pseudacris* and *Hyla*. American Midland Naturalist 74:269–308.

Michaud, T. C. 1964. Vocal variation in two species of chorus frogs, *Pseudacris nigrita* and *Pseudacris clarki* in Texas. Evolution 18:498–506.

Moore, W. S. 1977. An evaluation of narrow hybrid zones in vertebrates. Quarterly Review of Biology 2:263–277.

Moore, W. S., and D. B. Buchanan 1985. Stability of the northern flicker hybrid zone in historical times: Implications for adaptive speciation theory. Evolution 39:135–151.

Moore, W. S., and J. T. Price. 1993. Nature of selection in the Northern Flicker hybrid zone and its implications for speciation theory. *In* R. G. Harrison, ed. Hybrid zones and the evolutionary process, pp. 196–225. Oxford University Press, New York.

Narins, P. M., and R. R. Capranica. 1976. Sexual differences in the auditory system of the tree frog *Eleutherodactylus coqui*. Science 192:378–380.

Narins, P. M., and R. R. Capranica. 1978. Communicative significance of the two-note call of the treefrog, *Eleutherodactylus coqui*. Journal of Comparative Physiology 127: 1–9.

Nikolakopoulos, N. 1985. Geographic isolation and mating call variation in the *Litoria ewingi* complex (Anura: Hylidae) (BSc honours thesis). Department of Zoology, University of Melbourne, Parkville.

Otte, D. 1989. Speciation in Hawaiian crickets. *In* D. Otte and J. A. Endler, eds. Speciation and its consequences, pp. 482–526. Sinauer Associates, Sunderland, MA.

Otte, D., and J. A. Endler (eds). 1989. Speciation and its consequences. Sinauer Associates, Sunderland, MA.

Pianka, E. R. 1974. Evolutionary ecology. Harper and Row, New York.

Ritchie, M. G., R. K. Butlin, and G. M. Hewitt. 1992. Fitness consequences of potential assortative mating inside and outside a hybrid zone in *Chorthippus parallelus* (Orthoptera: Acrididae): Implications for reinforcement and sexual selection theory. Biological Journal of the Linnean Society 45:219–234.

Rivas, L. R. 1964. A reinterpretation of the concepts "sympatric" and "allopatric" with proposal of the additional terms "syntopic" and "allotopic." Systematic Zoology 13: 42–44.

Roberts, J. D. 1976. Call differentiation in the *Limnodynastes tasmaniensis* complex (Anura: Leptodactylidae) (PhD thesis). University of Adelaide, Adelaide.

Roberts, J. D. 1993. Hybridisation between the western and northern call races of the *Limnodynastes tasmaniensis* complex (Anura: Myobatrachidae) on the Murray River in South Australia. Australian Journal of Zoology 41:101–122.

Ryan, M. J., and A. S. Rand 1993. Species recognition and sexual selection as a unitary problem in animal communication. Evolution 47:647–657.

Short, L. R. 1969. Taxonomic aspects of avian hybridization. The Auk 86:84–105.

Szymura, J. M. 1993. Analysis of hybrid zones with *Bombina*. *In* R. G. Harrison, ed. Hybrid zones and the evolutionary process, pp. 261–289. Oxford University Press, New York.

Volpe, E. P. 1957. Genetic aspects of anuran populations. American Naturalist 91:355–371.

Waage, J. K. 1979. Reproductive character displacement in *Calopteryx* (Odonata: Calopterygidae). Evolution 33:104–116.

Watson, G. F. 1974. The evolutionary significance of postmating isolation in anuran amphibians (PhD thesis). University of Melbourne, Parkville.

Watson, G. F., and A. A. Martin. 1968. Postmating isolation in the *Hyla ewingi* complex (Anura: Hylidae). Evolution 22:664–666.

Wells, K. D. 1977. The social behaviour of anuran amphibians. Animal Behaviour 25: 666–693.

Wells, K. D., and J. J. Schwartz. 1984. Vocal communication in a neotropical treefrog, *Hyla ebraccata*: Advertisement calls. Animal Behaviour 32:405–420.

11

Geographic Variation in Animal Communication Systems

WALTER WILCZYNSKI
MICHAEL J. RYAN

Intraspecific communication is fundamental to most social behavior. It is also a special problem in animal behavior because it necessarily involves the interaction of two systems within a species, a sender and a receiver (Walker 1957, Blair 1964, Capranica 1966, Schneider 1974, Hoy et al. 1977, Hopkins and Bass 1981, Gerhardt 1988, Brenowitz 1994). Sender and receiver components are almost always separable morphologically, physiologically, and behaviorally. Each may be under different mechanistic and developmental control, and, especially in those cases in which the senders and receivers are segregated by sex, the impact of selection pressures and constraints can be very different (Brenowitz 1986, Ryan 1986, 1988; Wilczynski 1986, Endler 1983, 1993). The presence of two different but necessarily interacting components make the evolution of communication systems a particularly challenging problem in behavioral biology.

In any communication system, the interaction between senders and receivers dictates some degree of matching such that the signal emitted by one member of the communicating pair is effectively received, recognized, and assessed by the other member (Blair 1964, Gerhardt 1982, 1988; Capranica and Moffat 1983, Littlejohn 1988, Ryan 1988, 1991; Endler 1993). Effective coupling of senders and receivers is crucial when communication underlies mate choice. Communication systems that accurately discriminate between heterospecifics and conspecifics, while effectively linking conspecifics to each other, are important for ensuring mating with genetically compatible conspecifics. As such, communication systems can be integral parts of speciation and the maintenance of species isolation (Blair 1958, Mayr 1963, Paterson 1985, 1993; Littlejohn 1981, 1988; Butlin 1987, Coyne and Orr 1989, Claridge 1993, Moore 1993, Wood 1993).

The natural variation among and within species in both signals and receivers provides a means for examining the factors contributing to the evolution of communication systems (Templeton 1981, Ryan and Keddy-Hector 1992, Paterson 1993). Among the different levels of variation observed, geographic variation provides the best material for disentangling the myriad factors shaping the evolution and divergence of communication systems and for testing fundamental ideas about the evolution of behavior (Endler 1983, Baker and Cunningham 1985, Nevo and Capranica 1985, Ryan and Wilczynski 1991, Loftus-Hills and Littlejohn 1992).

Heterospecific and Conspecific Variation

The obvious function of communication signals in separating conspecifics from heterospecifics has led to examination of species-specific characteristics of communication systems (e.g., Wells 1977, Hopkins 1980, Capranica and Moffat 1983, Walkowiak 1988, Penna et al. 1990, Wilczynski et al. 1993). For example, both sensory systems and signals used by frogs have been shown to be species-specific, and, on average, sensory systems have proven to have areas of expanded representation (Narins and Capranica 1976) or enhanced sensitivity (reviewed in Walkowiak 1988, Zakon and Wilczynski 1988) that match important features of the signal. The species-typical characteristics of signals and receivers, important for the recognition functions critical to reproduction, should tend to constrain the evolution of intraspecific diversity. Although the magnitude of this effect is difficult to quantify, phylogenetic relationships predict the maintenance of some degree of behavioral similarity (Ryan 1986, Ryan and Rand 1993a, 1995; Brenowitz 1994, Cocroft and Ryan 1995). Thus, intraspecific change should be constrained to occur within a species-typical framework.

Even above the species level, many features of vertebrate sensory systems are shared and may thus constrain or channel the evolution of call diversity. The discovery of such common sensory characteristics has led to the suggestion that some operations apparently specialized for communication may in fact be generalized neural processing operations coopted for recognizing conspecific signals. Rose (1986) suggested that midbrain feature detectors for amplitude modulation rates characteristic of many acoustic communication signals are no different from the neurons sensitive to temporal patterns in sound that are found in many vertebrate auditory systems, regardless of the use of such sounds in intraspecific communication. Similarly, Wilczynski and Capranica (1984) noted that the two-tone supression apparent in the peripheral auditory system of amphibians, while clearly important for bullfrog call recognition, is a common feature of all terrestrial auditory systems. Similarly, multiple syllables in bird song may have evolved to counteract habituation common to all sensory systems (Searcy 1992), and peripheral auditory system tuning characteristics common among species of *Physalaemus* may have channeled the evolution of calls toward features that better stimulate the ear's receptors (Ryan and Rand 1993a).

Despite potential constraints on diversity, the communication signals and receiver characteristics of species do vary geographically. The diversifying effects of variation in habitat, in pleiotropic effects brought on by evolutionary changes in noncommunication characters of organisms, and in selection due to interactions between species and among conspecifics may interact in different ways in different parts of a species' range, leading to significant differences among populations and among species occupying different habitats.

Environmental Effects and Variation in Communication Systems

Ecological factors can have direct effects on the evolution of geographic variation within a species and can account for some of the differences among species in different habitats. Signals must be transmitted through the environment from sender to receiver. Therefore, habitat characteristics can impose selection on the form of a communication signal (Lythgoe 1979, Brenowitz 1986, 1994; Endler 1991, Dusenbery 1992, Fleishman 1992, Narins 1995). Studies of acoustic signals have demonstrated adaptation to local environmental conditions enhancing transmission (Wiley and Richards, 1978, 1982; Gish and Morton 1981, Bowman 1983, Ryan et al. 1990a). In visual communication, variation in background clutter, ambient light, and, in aquatic environments, clarity of the transmission medium, can similarly influence a signal's effectiveness (Endler 1983, 1991, 1992; Fleishman 1992). Studies of both interspecific (Marchetti 1993) and intraspecific (McKenzie and Keenleyside 1970, Endler 1983, 1991; Reimchen 1989) signal variation have suggested that geographic variation in habitat characteristics can indeed affect the evolution of visual signals.

Environmental factors might also shape communication signals indirectly by acting on morphological traits correlated with aspects of the communication system (Ryan 1988). The most obvious of these is body size. For example, Nevo and Capranica (1985) suggested that in cricket frogs (*Acris crepitans*), dry conditions in western parts of their range favor larger body sizes that decrease desiccation (Nevo 1973). As call frequency and body size are negatively correlated in frogs (Ramer et al. 1983, Ryan 1985, Wagner 1989a, Keddy-Hector et al. 1992), western cricket frogs would have lower-frequency calls than eastern cricket frogs on this basis alone (Nevo and Capranica 1985). Narins and Smith (1986) made a similar suggestion to explain altitudinal variation in call frequencies in some tropical frogs. Tuning of the auditory system is also negatively correlated with body size in frogs (Wilczynski 1986, Zakon and Wilczynski 1988, Keddy-Hector et al. 1992), so environmental selection acting on body size might also affect the receiving portion of the communication system in these vertebrates.

Habitat differences in predation can also lead to geographic variation in communication systems. Endler's (1980, 1988) studies of guppies demonstrate that the presence of visually-hunting predatory fish in some areas provides a strong selection pressure on the color patterns male guppies use to attract females. Although there have been no studies of habitat differences in predation effects on acoustic communication systems as thorough as those of coloration in guppies, Ryan (1985, Ryan et al. 1982) demonstrated that bats prey on Túngara frogs by locating their calls and that the frogs' calling behavior was influenced by this predation. Presumably, geographic variation in bat predation could lead to geographic variation in calling in these frogs.

Social Behavior and Variation in Communication Systems

Heterospecific interactions and interference that occur as each species engages in its own mating behavior can influence the form of another species' communication behavior (Blair 1958, Walker 1974, Schwartz and Wells 1984, Butlin and Hewitt 1985, Gwynne and Morris 1986, Gerhardt 1988, Littlejohn 1988, Coyne and Orr 1989, Otte 1989, Loftus-Hills and Littlejohn 1992, Ryan and Rand 1993a,b). Where species breed together using acoustic signals, one often observes an apparent partitioning of communication channels (Drewry and Rand 1983, Duellman and Pyles 1983, Wilczynski et al. 1993). Such interactions could, in principle, lead to geographic variation if the mix of interacting species varies across a species' range, although there is little documentation of this.

One evolutionary issue that directly relates heterospecific interactions to geographic differences in a species' communication system is the phenomenon of "character displacement." In areas of its range where a species or population is sympatric with another having a similar communication system, there can be increased selection to limit mate choice "mistakes" (Brown and Wilson 1956, Littlejohn this volume). This can result in an accentuation of the differences in the courtship signals of the two groups. This idea has been applied to zones of overlap between subgroups within a species (i.e., pairs of subspecies or "incipient" species), as well as between pairs of genetically incompatible species (Nevo and Capranica 1985, Butlin 1987, 1989; Littlejohn 1988, Otte 1989). Butlin (1987) suggests that the term "character displacement" be used to describe this phenomenon where the sympatric groups are historically separate species producing infertile hybrids if mated, and that "reinforcement" be used where interacting species or sufficiently (genetically) different populations within a species may produce fertile hybrids with reduced fitness. In either case, the result in terms of geographic variation in the communication system within a species is the same: a shift in signal or receiver characteristics away from those of the interfering signal at points in the geographic range where groups interact. Because selection leading to this shift is absent at points in the range where only one population exists, differences between sympatric and allopatric populations within a species are expected.

Reproductive character displacement and reinforcement remain controversial (see also Littlejohn this volume, Verrell this volume). They have been challenged theoretically (Templeton 1981, Butlin 1987), and there have been few unequivocal empirical demonstrations of these phenomena. Most searches for character displacement have targeted the signals rather than the receiver portions of communication systems. The best examples occur in the calls of some frogs (Littlejohn 1965, Fouquette 1975, Ralin 1977, Loftus-Hills and Littlejohn 1992) and insects (Otte 1989, Benedix and Howard 1991). There is also some evidence that female discrimination can change geographically in ways that suggest character displacement (Wasserman and Koepfer 1977, Waage 1979, Gwynne and Morris 1986, Gerhardt 1994).

Intraspecific social interactions also drive the evolution of communication systems, although in a way less clearly predictive of particular patterns of geographic variation. Sexual selection induced by a bias among females for particular male signal characteristics is the primary example of a social factor that can drive the evolution of male communication signals. There is abundant evidence in many species that females not only prefer the signals of conspecifics to heterospecifics, but that they find some conspecific signals more attractive than others (Kirkpatrick 1982, Ryan 1985, Bradbury and Andersson 1987, Rand et al. 1992, Ryan and Keddy-Hector 1992, Endler and Houde 1995, Tokarz 1995, Wilczynski et al. 1995).

Disagreements exist about why female mating preferences are expressed (see reviews in Bradbury and Andersson 1987, Kirkpatrick and Ryan 1991, Andersson 1994, Tokarz 1995). Adaptive hypotheses posit that female preferences evolve either because females exerting those preferences produce more offspring due to immediate benefits provided by the male (e.g., parental care, nuptial gifts, or greater fertilization efficiency) or because the genes controlling signal characters become genetically correlated with a male's "good genes," and the preference then evolves via indirect selection. A hypothesis of "arbitrary" female mate preference is Fisher's theory of runaway selection (Fisher 1958), which suggests that female preferences evolve due to a genetic correlation with a male trait. A third hypothesis, "sensory exploitation" (and its more general form, "sensory drive"), suggests that there are preexisting biases in the female's sensory system, which may or may not be adaptive in the context of mate choice or other aspects of the animal's life such as foraging, and that males evolve traits that are more attractive to females given these sensory biases for particular stimulus configurations (Ryan 1990a, Ryan and Rand 1990, Endler 1992, Ryan and Keddy-Hector 1992, Enquist and Arak 1993).

Several authors (e.g., Fisher 1958, Ringo 1977, Lande 1981, West Eberhard 1983, Eberhard 1985) suggest that sexual selection driven by mate choice can have diversifying effects on communication systems, and others (Endler 1980, 1983; Eberhard 1985, Ryan and Keddy-Hector 1992) provide evidence that mate choice can provide strong directional selection on male signals. Sexual selection, when not mediated by "good genes," is unpredictable, which means that its expression in different conspecific populations could in principle lead to geographic variation in the characteristics of the communication system (Fisher 1958, Ringo 1977, West Eberhard 1983, Ryan 1990a,b). If mate choice has evolved under the influence of "good genes" or some instances of "sensory drive," the direction of evolutionary change might be more predictable but might still lead to geographic variation in communication signals if factors that influence fitness vary geographically (Endler 1993).

Ecological and social factors are obviously not mutually exclusive in their influence on communication systems. They, plus other factors such as genetic drift in isolated populations, and patterns of gene flow across a species' range, likely interact in complicated ways to yield the geographically changing profile of communication characteristics seen in many species. The interaction among various factors can be seen in one model system, the acoustic communication system of the cricket frog, *Acris crepitans*.

Geographic Variation in the Communication System of Cricket Frogs

Cricket frogs, *Acris crepitans*, are members of the family Hylidae. This species occupies much of eastern and central United States and is the only representative of the genus in the western part of its range across Texas and northeastern Mexico. Male cricket frogs produce a short, clicklike advertisement call, which they repeat in rapid bursts referred to as "call groups" (fig. 11-1, Nevo and Capranica 1985, Wagner 1989b, Ryan and Wilczynski 1991). The call serves as a mate recognition signal (Nevo and Capranica 1985, Ryan and Wilczynski 1988, Ryan et al. 1992) and also mediates aggressive interactions among males (Wagner 1989a,b,c). Therefore, the communication system in this species, as in most anurans, consists of a vocal signal (produced by males) and the auditory system (in females and males) receiving it.

Like other anurans, cricket frogs have two inner ear organs sensitive to sound (see Wilczynski and Capranica 1984, Zakon and Wilczynski 1988, Wilczynski 1992 for reviews of the amphibian auditory system). Each receptor structure, and each of the eighth nerve fibers connected to them, can be described in terms of its tuning. Tuning is the range of sound frequencies that will stimulate its receptors and the frequency to which it is most sensitive (its "best excitatory frequency"). In cricket frogs, as in many small anurans, the advertisement call stimulates only the basilar papilla (Capranica et al. 1973, Ryan and Wilczynski 1988). The amphibian papilla, which is larger and tuned to a wider range of lower frequencies, is not used for the reception of calls in cricket frogs.

The populations of *Acris crepitans* we examined (Ryan and Wilczynski 1991), occur along a transect from the Texas–Louisiana border to Lake Balmorrhea in west Texas (fig. 11-2). The transect passes through the ranges of two recognized subspecies of cricket frogs (Dessauer and Nevo 1969, Salthe and Nevo 1969), *A. c. crepitans*, which occupies the eastern portion of the range, and *A. c. blanchardi*, which occupies the western portion, as well as the zone of parapatry between them in east Texas. The eastern habitat of *A. c. crepitans* is piny woods characterized by wet, dense forests. The western areas occupied by *A. c. blanchardi* include post-oak savannah, blackland praries, Edwards Plateau, and Trans Pecos; all these habitats are drier and more open than those in the eastern areas (McMahon et al. 1984). Cricket frogs are also found in an isolated pine forest habitat in Bastrop County, an area of central Texas within the range of *A. c. blanchardi* and surrounded by the drier, open habitat characteristic of this subspecies. Preliminary allozyme analysis suggests that the Bastrop cricket frogs are more closely related to the *A. c. blanchardi* in the grasslands surrounding them than to the *A. c. crepitans* that live in similar forest habitat farther east.

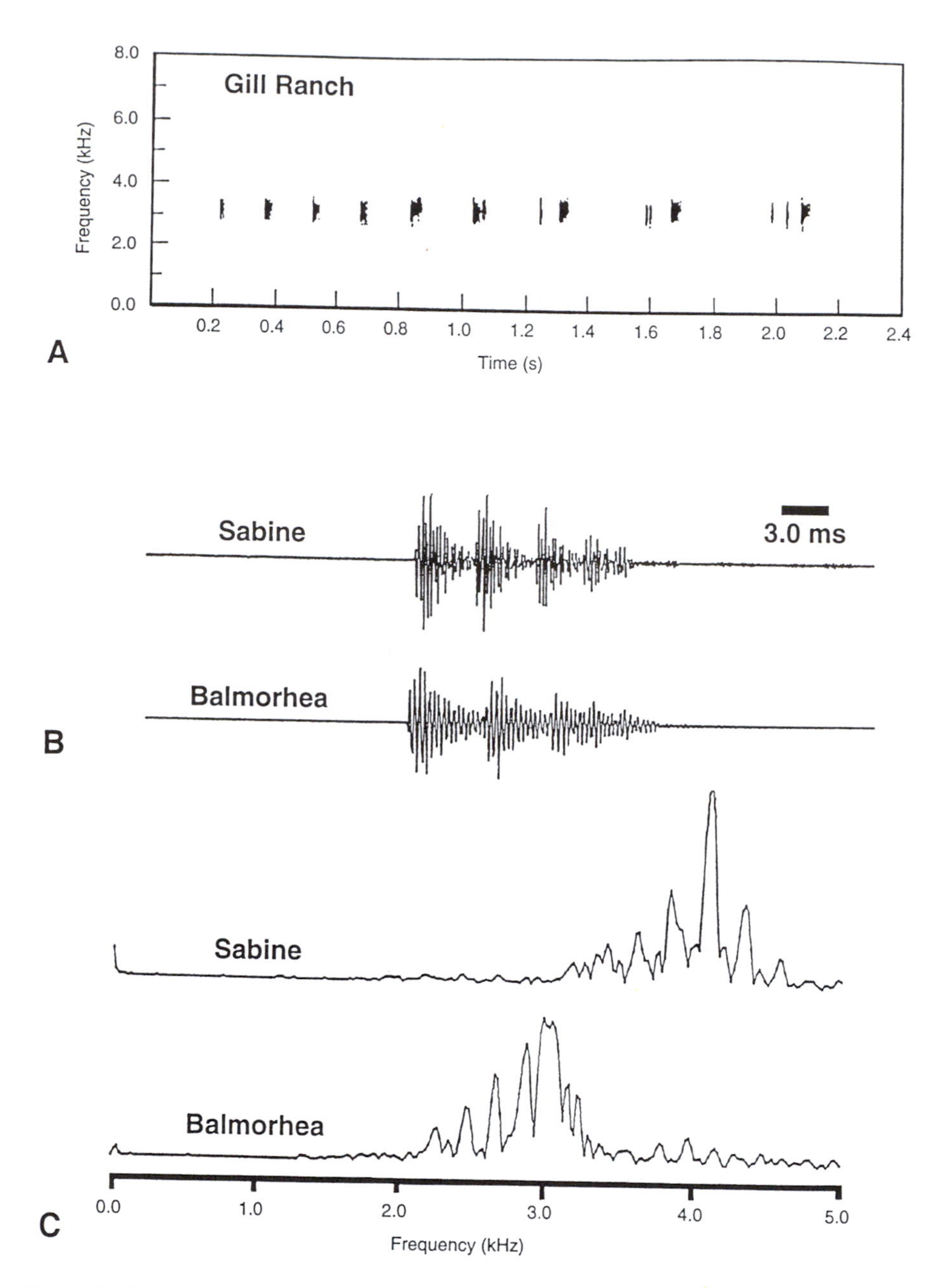

Figure 11-1 (A) Sonagram of a call group from a male cricket frog from the Gill Ranch population at the approximate center of our study transect (from Ryan and Wilczynski 1991). (B, C) Oscillograms and spectrograms of calls from populations on the eastern (Sabine) and western (Balmorhea) ends of our study transect. Note that the western calls and their component pulses are slightly longer (B) and much lower in frequency (C) than the eastern calls.

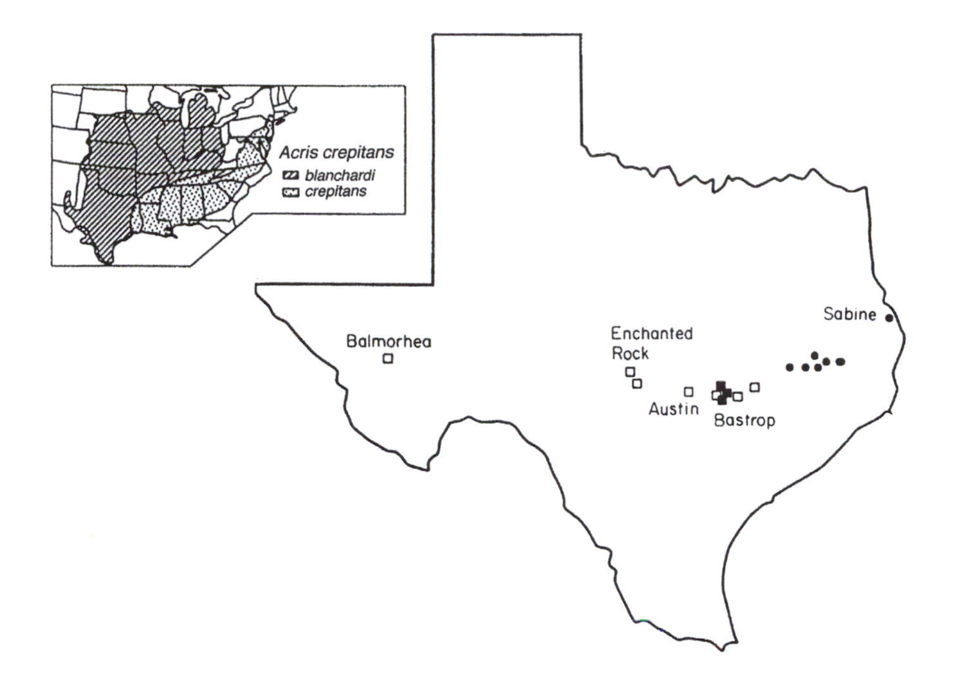

Figure 11-2 Location of cricket frog (*Acris crepitans*) populations studied to assess geographic variation in their communication system. (Circles) Populations of *A. c. crepitans* (all in forest habitats); (open squares) populations of *A. c. blanchardi* from open habitats; (filled squares) populations of *A. c. blanchardi* in forest habitats in the "Lost Pines" area of Bastrop, Texas. Inset at upper left shows the range of the two subspecies of *Acris crepitans*. (From Ryan and Wilczynski 1991.)

Geographic Variation in the Call

Temporal and spectral call characteristics show significant geographic variation among cricket frog populations along this transect (Ryan and Wilczynski 1991). Some temporal characteristics appear to vary randomly, but there is a strong clinal component to much of this variation. The dominant frequency of the call (the frequency with the most energy), call rate, and call group duration most reliably distinguish populations. In general, calls are higher in frequency, shorter, and produced at a faster rate in the eastern part of this range (fig. 11-1). Dominant frequency exhibits the strongest clinal variation of any call character as it descends from east to west.

Call variation is also significantly related to habitat and subspecies (Ryan and Wilczynski 1991). The results of a principal component analysis (PCA) of call variation among populations are shown in figure 11-3. There are two patterns of interest. First, the calls of *A. c. crepitans* tend to be more clumped on the PCA plot than the calls of open-habitat *A. c. blanchardi* populations. Also, the calls of *A. c. blanchardi* from the isolated forest habitats near Bastrop tend to segregate with the calls of *A. c. crepitans* on the PCA plot. However, two of the Bastrop

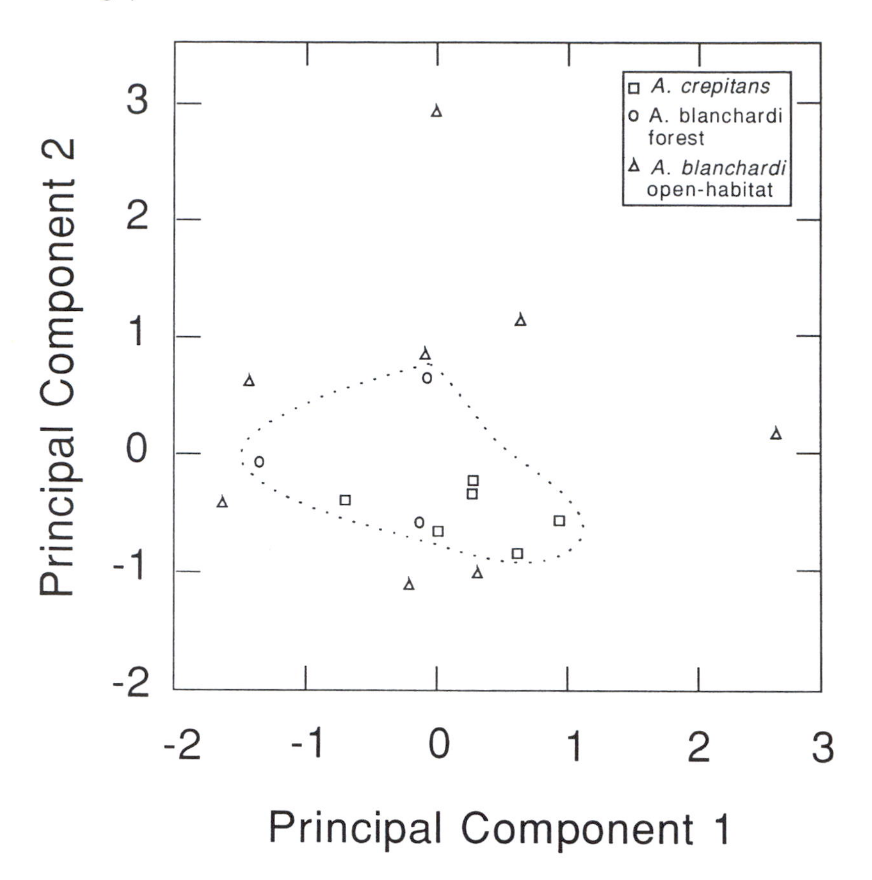

Figure 11-3 Results of a principal component analysis of cricket frog calls. Dotted line encloses populations from forest habitats. Temporal call characters provide the major loading on both components. Principal component 1 is determined mainly by the number of calls per call group, the number of pulses per call in the middle of a call group and at the beginning of a call group, and call group duration. Principal component 2 is determined mainly by the call durations at the end, middle, and beginning of a call group and by the number of pulse groups in calls at the beginning of a call group. Dominant frequency loads more heavily onto principal component 2 than 1, but is less a factor in specifying the components than any of the temporal features listed above. See Ryan and Wilczynski (1991) for a complete description of call characters.

populations are closer to an *A. c. blanchardi* population than to any *A. c. crepitans* population in the PCA-call space; in neither of these cases are the "nearest neighbors" on the PCA plot also the geographical nearest neighbors. The call analysis combined with the preliminary allozyme analysis suggest an evolutionary convergence in calls between *A. c. crepitans* and those of *A. c. blanchardi* that reside in the pine forests of Bastrop. The PCA analysis also shows that variation among populations in open habitats is much greater than variation among populations in the forest habitats.

Geographic Variation in the Auditory System

Across populations, basilar papilla tuning changes in the same general way as does call dominant frequency (fig. 11-4), leading to the maintenance of a rough match between calls and tuning at the population level (Wilczynski and Ryan 1988, Keddy-Hector et al. 1992, Wilczynski et al. 1992), just as has been seen in many frogs at the species level (Zakon and Wilczynski 1988). Complicating this relationship is the fact that basilar papillae of females are tuned to lower frequencies than those of males in all populations in which we sampled both sexes. Furthermore, average female basilar papilla tuning is lower than the average dominant frequency of the male calls in the same population.

In addition to the clinal variation in basilar papilla tuning, the degree of mismatch between female tuning and male calls differs among populations (fig. 11-5). In forest populations, the basilar papillae of males are tuned, on average, higher than the call, whereas those of females are tuned lower. Therefore, the call is pitched between the maximum sensitivities of the two sexes. In grassland populations, the papillae of both sexes are tuned lower than the population's call dominant frequency, and female papillae are tuned lower than those of males. Thus the difference between the dominant frequency of the average male call and the best excitatory frequency of the average female auditory system is much greater in grassland populations than in the forest populations.

Geographic Variation in Mate Choice

Two-choice phonotaxis experiments clearly indicate that females can discriminate call characters (fig. 11-6). When presented with calls that vary only in dominant

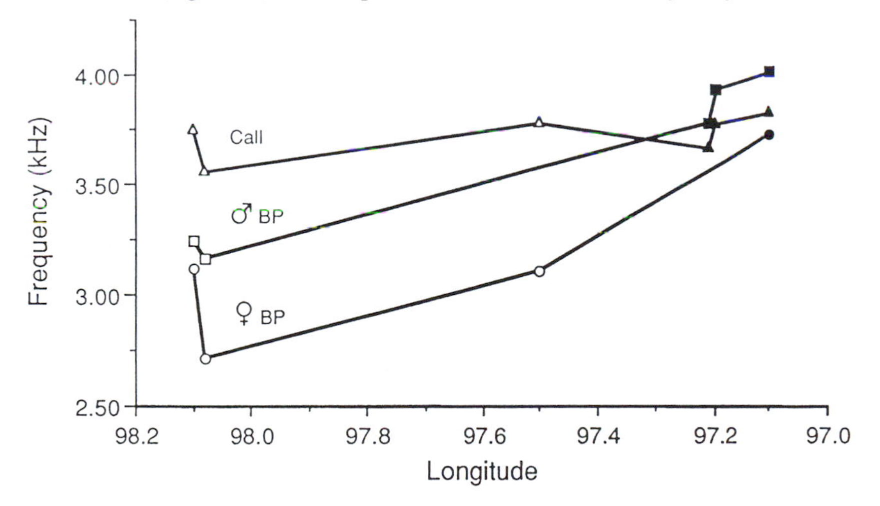

Figure 11-4 Mean advertisement call dominant frequency (triangles) and best frequency of the basilar papilla (BP) in males (squares) and females (circles) in six populations of cricket frogs (not all characters are available in all populations). Open symbols indicate populations from open habitats, filled symbols indicate populations from forest habitats. (From Wilczynski et al. 1992.)

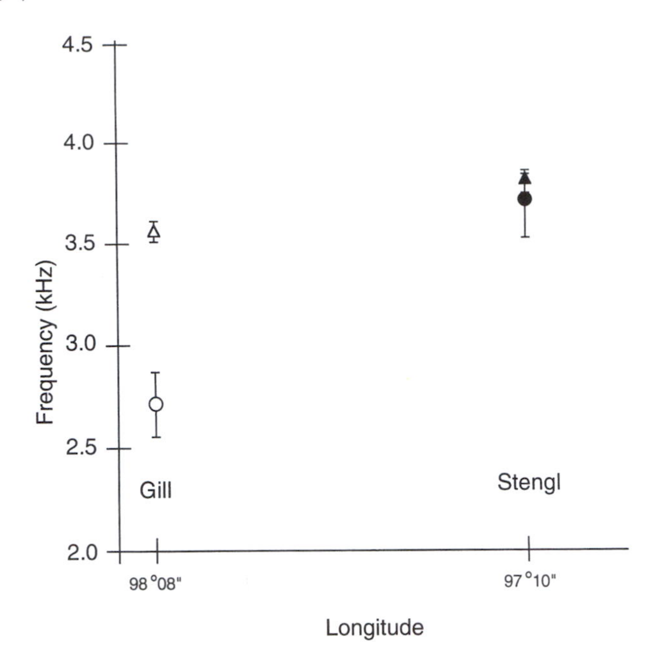

Figure 11-5. Mean (±SE) male call dominant frequency (triangles) and mean (±SE) female basilar papilla best excitatory frequency (circles) in two populations of cricket frogs, Gill Ranch and Stengl Ranch, from open and forested habitats, respectively. Note the difference in the degree of mismatch between male calls and female tuning in the two populations.

frequency, females in each of three populations prefer low-frequency calls to calls that are at the mean for their population or higher in frequency than that mean (Ryan et al. 1992). This suggests a within-population preference for males with low-frequency calls. These are, on average, calls of the larger males in the population (Wagner 1989a, Keddy-Hector et al. 1992). These preferences are predicted by the basilar papilla tuning in females, which is always lower than the average call dominant frequency in their home population. Additional confirmation comes from examining within-population variation in female tuning and mate choice. As for the calls, basilar papilla tuning is negatively correlated with body size in both sexes (Keddy-Hector et al. 1992). Consequently, larger females prefer lower call dominant frequencies than smaller females (Ryan et al. 1992).

The preference for lower-than-average dominant frequencies has implications for interpopulational mate choice as well. Given a choice between the average calls from their home population and a population with a higher call frequency, females should, and do, prefer calls from their home population (Ryan and Wilczynski 1988). Given a choice between the home call and one from a population with a lower frequency call they should, and do, choose the calls of the foreign population, which the phonotaxis experiments also show (Ryan et al. 1992). Mate

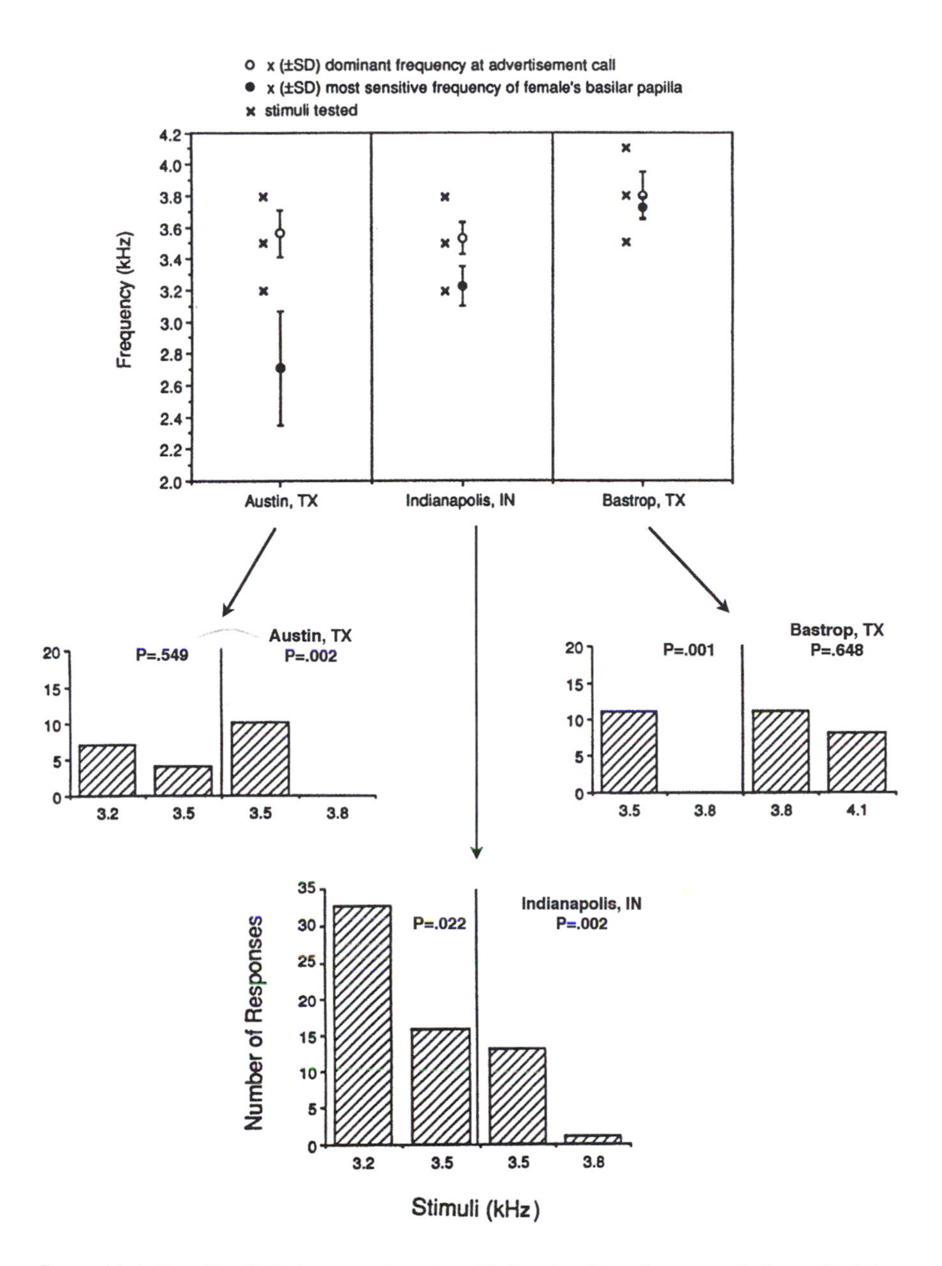

Figure 11-6 Results of choice experiments with females from three populations of cricket frogs. Top panel shows stimuli tested and their relationship to the dominant frequency of male calls and the best excitatory frequency of the basilar papilla of females in the tested population. Lower three panels show the choices made by females presented with calls at the population average versus calls lower than average (left side of graphs) or higher than average (right side of graphs). (From Ryan et al. 1992.)

choice is not based on population affinity per se, but on the relationship of male call dominant frequency and female basilar papilla tuning, causing females to potentially discriminate against foreign calls in some cases and against their own population's call in others. The preference is apparently derived simply from lower frequency calls being a better match with the tuning of the average female's basilar papilla, thereby stimulating her auditory system more (see also Ryan et al. 1990b).

Because female basilar papilla tuning is an important determinant of female mate choice, it is an important predictor of patterns of sexual selection in this species, suggesting that within all populations there is directional sexual selection for lower frequency calls. Furthermore, the mismatch between calls and tuning is greater in grassland populations than in forest populations, suggesting that selection for low-frequency calls may be greater in the more western, grassland populations.

The Evolution of Geographic Variation in Cricket Frogs

Analyzing the patterns of variation in signals and receivers allows an understanding of the interacting and competing forces shaping the evolution of this communication system. Our results, and those of others who have worked with this species (Nevo 1973, Nevo and Capranica 1985, Wagner 1989a,b), have indicated several factors contributing to this variation.

Body-Size Effects on the Communication System

Body size has an influence on both the signal (the call) and the receiver (basilar papilla tuning). Larger animals have lower frequency calls and have basilar papillae tuned to lower frequencies (Wagner 1989a,c, Keddy-Hector et al. 1992). Body-size differences no doubt also contribute to the sex difference in tuning, as females are larger, and have auditory systems tuned to lower frequencies, than males in this species (see also Wilczynski et al. 1984, Wilczynski 1986). Because western populations are on average larger than eastern populations, possibly as an adaptation to resist desiccation in drier western habitats (Nevo 1973), some of the clinal variation in call dominant frequency and basilar papilla tuning could be an indirect effect attributable to this selective influence.

Pleiotropic effects of body size may contribute to population differences, but they are not responsible for all the observed population variation in signal and receiver. Population and sex differences are still apparent when body size is statistically controlled (Ryan and Wilczynski 1988, 1991; Keddy-Hector et al. 1992). Morphological correlates of this can be seen in studies of the cricket frog vocal system. The size of a larynx and its component parts is an important determinant of its resonant properties and hence the frequency characteristics of the vocalizations it produces. Larger larynges are associated with lower call dominant frequencies. Male larynx size is significantly different among populations even after controlling for the effects of body-size differences (McClelland 1994, McClelland et al. 1996).

Selection by Environmental Acoustics

Habitat type (forest or open) is an important predictor of call features, independent of subspecies affiliation (Ryan and Wilczynski 1991). The isolated pine forest populations in the Bastrop area have calls more similar to those of the different subspecies in a similar east Texas habitat. One possible reason is that the calls have diverged in different habitats due to differences in habitat acoustics such that each habitat type, open and forest, contains populations with calls that minimize degradation, excess attenuation, or masking in that habitat.

An analysis of call degradation in open and forest habitats (Ryan et al. 1990a) shows that the forest habitat, as expected (Wiley and Richards 1978, 1982), causes much more call degradation than does the open grassland habitat (fig. 11-7). Furthermore, the calls from east Texas populations native to forest habitats are transmitted much more effectively than the slower, longer calls characterizing populations from the grassland habitat of central and west Texas (fig. 11-7). Somewhat suprisingly, our analysis provides little evidence that the acoustic features of the grassland habitats influence call evolution there. There is no significant difference in the degradation of calls from either habitat in the open grassland sites, and the forest call may even have a slight transmission advantage there.

Forest calls are also higher in frequency than open habitat calls, even though one might expect high frequencies to attenuate faster than low in acoustically cluttered environments. We believe that the higher frequency of the forest call may be an indirect consequence of the much more important selection on temporal features. Morphological studies of the larynx (McClelland 1994, McClelland

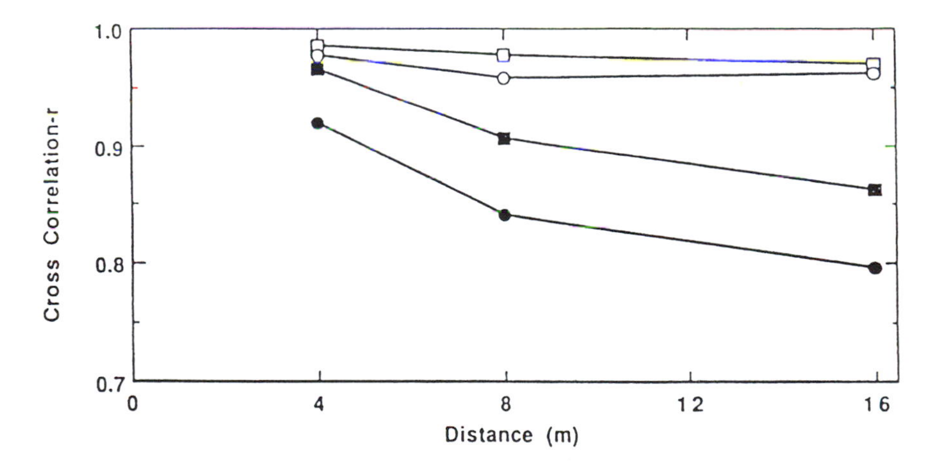

Figure 11-7 Call degradation over distance estimated by cross-correlation coefficients derived from correlating the call as recorded at the indicated distance with the call as originally broadcast. (Squares) Values for *A. c. crepitans* calls from east Texas forest population; (circles) Values for *A. c. blanchardi* calls from open habitat. Open symbols are results of broadcasts in an open habitat, filled symbols are results of broadcasts in a forest habitat. (From Ryan et al. 1990a.)

et al. 1998) indicate that, within a species, the various laryngeal structures are highly intercorrelated in size. These correlation studies also indicate that in larynges of the type found in cricket frogs, faster, shorter calls are predictive of smaller laryngeal muscles. As muscles decrease in size to protect calls from temporal degradation, allometric effects apparently reduce the size of other laryngeal components such as vocal cords and arytenoid cartilages, leading to higher dominant frequencies. Consequently, cricket frogs are forced to suffer a small disadvantage in call transmission in the frequency domain to achieve a larger advantage in call fidelity.

Environmental acoustics may also contribute directly to differences in auditory system characteristics. Signals should attenuate over distance more rapidly in the cluttered forest habitats. One way to achieve better sensitivity to the call is to achieve a better match between frequency peaks in the call (i.e., its dominant frequency) and the peak frequency sensitivity of the receiver (i.e., the best excitatory frequency of the basilar papilla). Indeed, one does observe that average female basilar papilla tuning is closer to average call dominant frequency in forest populations than in open, grassland populations. Other aspects of auditory sensitivity, such as absolute threshold and susceptiblity to masking by enviromental noise, which might also differ in the two habitat types, have yet to be explored.

Interaction of Social and Environmental Influences

No single factor accounts for the pattern of geographic variation seen in the cricket frog acoustic communication system. Pleiotropic effects of body size on calls and basilar papilla tuning, selection imposed by environmental acoustics, and the patterns of sexual selection based on tuning characteristics of the female basilar papilla interact to generate clinal and habitat-based variation. Moreover, the balance between the different influences may vary among populations.

Selection due to habitat acoustics and sexual selection appear to exert opposing forces on the cricket frog call. In all populations, mate choice selects for low-frequency calls. In forest habitats, however, the demands for preserving call fidelity over distance could provide a selective pressure in the opposite direction for calls that are faster, shorter, and, indirectly, higher in frequency. In these populations, therefore, the effects of sexual selection are mitigated. In open grassland habitats, sexual selection is relatively unopposed by habitat selection, resulting in lower frequency calls and, apparently, in calls that are also slower and longer due to the overall increase in the size of laryngeal components necessary to make such calls. Because forest habitats are more common at the eastern end of the range and dry, grassland habitats are more common at the western end, a clinal trend with shorter, faster, higher frequency calls to the east and longer, slower, lower frequency calls to the west results. Overlying these trends generated by different selection regimes is clinal variation in body size (Nevo and Capranica 1985, Ryan and Wilczynski 1991), which would pleiotropically affect calls and tuning, producing parallel clinal trends.

Migration of individuals between populations remains to be investigated. If populations interact, what might result is a pattern of gene flow from east to west, and from forest to grassland, driven by female mate choice for low-frequency calls. The net result of all these interacting factors is east–west clinal variation in call dominant frequency and basilar papilla tuning, overlain by habitat variation causing additional nonrandom call variation along the cline.

Regardless of the mix of evolutionary influences, a rough match between calls and basilar papilla tuning exists in each cricket frog population, just as it does in other cases in which geographic variation exists in communication systems. How the match between the signal and the receiver is maintained as a communication system evolves has been a question of considerable debate. In our populations, effects of body size, acting simultaneously on signal and receiver systems, may help maintain the match, but they are not solely responsible for it (Keddy-Hector et al. 1992). Strong genetic linkage between sender and receiver systems (e.g., Hoy et al. 1977, Kyriacou et al. 1992) is also unlikely in this species. The relationship between mean basilar papilla tuning and call dominant frequency varies among populations (Wilczynski et al. 1992), and the allometric relationship of each with body size also varies among populations (Keddy-Hector et al. 1992). This observation is important, as a genetic linkage has been proposed to play an important role in maintaining the congruence between signal and receiver in animal communication dyads (Alexander 1975, Hoy et al. 1977, Doherty and Gerhardt 1983, Boake 1991). The elimination of pleiotropic effects of body size and strong genetic linkage as most likely candidates for producing the relationship between calls and auditory tuning suggests some coevolutionary process underlying the call–tuning relationship.

Intraspecific Geographic Variation in Other Communication Systems

Geographic variation in communication signals and receivers has been documented in several vertebrate and invertebrate groups besides cricket frogs. Most studies of geographic variation in the communication systems of other frog species have examined patterns for evidence of character displacement in the call (Littlejohn 1965, Fouquette 1975, Ralin 1977, Nevo and Capranica 1985, Loftus-Hills and Littlejohn 1992). In addition, Littlejohn (1988; Littlejohn and Watson 1985) has used patterns of geographic variation to document the opposite phenomenon—the generation of stable hybrid zones in areas of sympatry where reproductive isolation breaks down. Both the investigations of reproductive isolation through character displacement and the breakdown of isolation with the subsequent formation of hybrid populations use patterns of geographic variation as a window into the dynamics of speciation (see Littlejohn this volume).

Visual communication systems in fish also show geographic variation. In these systems, attention has focused mainly on variation generated by habitat differences. In guppies (*Poecilia reticulata*), geographic variation in signals is generated mainly by differences in selection pressures among habitats (Endler 1983), although transmission characteristics of a population's habitat may also contribute to signal evolution (Endler and Houde 1995). Females base mate-choice decisions on male color patterns, particularly on the amount of orange coloration (Endler

1983, Houde 1987, 1988; Stoner and Breden 1988, Long and Houde 1989, Houde and Endler 1990, Endler and Houde 1995). Where predation is high, males lack bright color patterns, presumably because predatory fish use vision to find the guppies. Where predators are scarce or lacking, males are brightly colored.

As in frogs, the preferences of female guppies also vary geographically, but the pattern of choice is more complex than that seen in frogs to date. Endler and Houde (1995) report that female guppies generally prefer males from their home populations to alien males, but the male traits on which the preference is based vary among populations. Female preferences maintain a rough match with the expression of three male color traits across populations (amount of orange, amount of black, and degree of color contrast; see also Stoner and Breden 1988, Houde and Endler 1990), but not with many other visual features characteristic of males.

Among the most striking examples of geographic variation in communication systems is the presence of local song dialects in oscine birds. Indeed, the study of intraspecific variation in vocal communication leading to local "dialects" began with an investigation of the song bird *Zonotrichia leucophrys*, the white crowned sparrow (Marler and Tamura 1964; see Baker and Cunningham 1985 for a review of work in this species), although references to geographic variation in bird song do pre-date this work (e.g., Borror 1956, Marler and Isaac 1960, Armstrong 1963). Subsequent investigations have revealed local song dialects in many other passerines, including various sparrows (*Melospiza melodia*, Harris and Lemon 1972; *Zonotricha capensis*, Nottebohm and Selander 1972; *Melospiza georgiana*, Marler and Pickert 1984) and wrens (*Thrymomanes bewickii*, Kroodsma 1974; *Troglodytes troglodytes*, Kroodsma 1981; *Cistothorus palustris*, Kroodsma and Canady 1985), cardinals (*Richmondena cardinalis*, Lemon 1967, 1971), indigo buntings (*Passerina cyanea*, Shiovitz and Thompson 1970, Emlen 1971), cow birds (*Molothrus ater*, King et al. 1980), red-winged blackbirds (*Agelaius phoeniceus*, Searcy 1990), and rufous-sided towhees (*Pipilo erythrophthalmus*, Ewert and Kroodsma 1994).

Each bird species in which local dialects emerge has a species-typical song with discrete components or syllables common to the species. What varies among populations to yield dialects is the preponderance of certain syllables within the populations's song, variations in the sound of certain syllables, or the patterning of the syllables within the song, all within some species-specific limits (fig. 11-8). Most work has concentrated on describing signal variation, but there is now substantial behavioral evidence that population-level song preferences exist. In all cases studied so far, these preferences are for an individual's home dialect over foreign dialects. The ability to discriminate among dialects apparently is present in both sexes in at least some species (King et al. 1980, Baker 1983, Brenowitz 1983, Baker et al. 1987, Balaban 1988, Searcy 1990).

What makes the phenomenon of bird song dialects particularly interesting, and likely different from the situation in frogs and fish, is that many species with dialects learn their songs from conspecifics, and the generation of local dialects and preferences is thought to derive directly from the plasticity of the system that

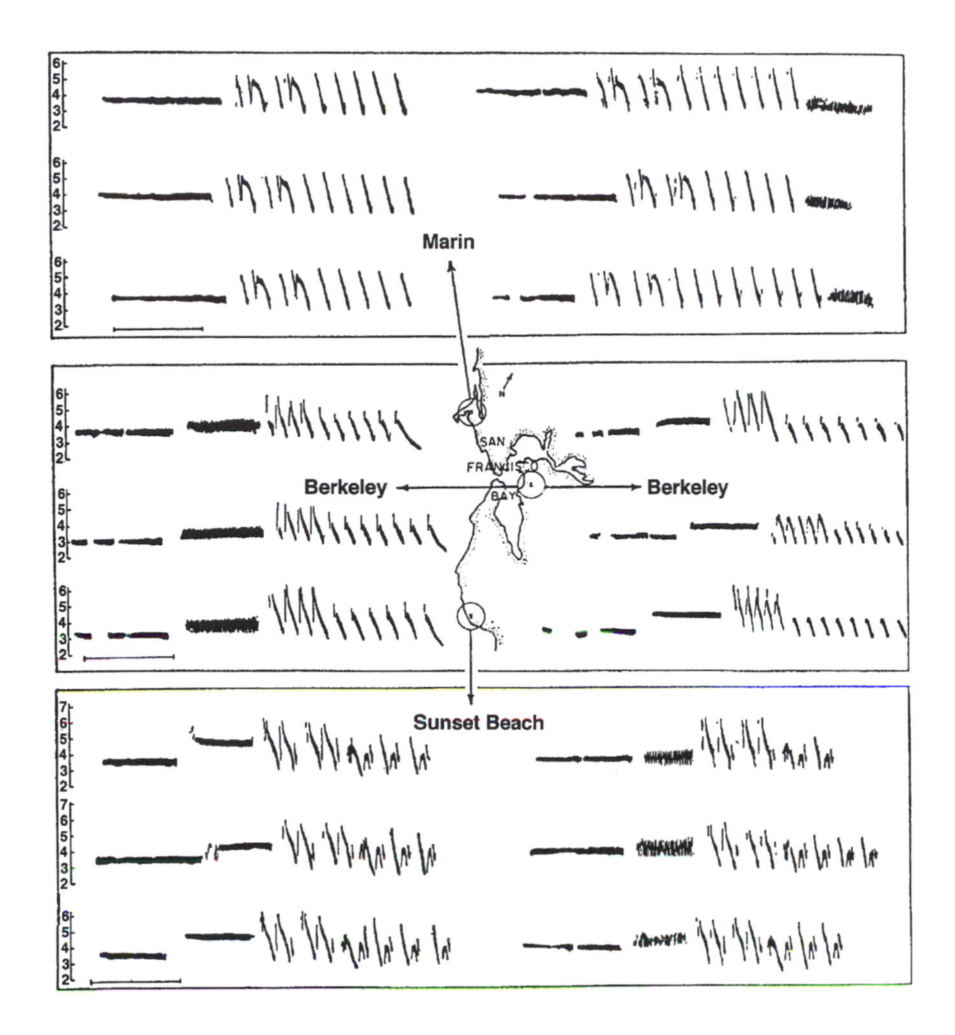

Figure 11-8 Sonagrams of songs from six male white-crowned sparrows from each of three locations noted on map in center of figure showing geographic dialects. Vertical scale is in kHz, horizontal scale, bar = 0.5 s. (Reprinted by permission from Marler and Tumara [1964]. Copyright 1964 by the AAAS.)

makes such learning possible (Marler and Tamura 1964, Kroodsma 1974, Baker et al. 1987, Baker and Cunningham 1985).

Geographic variation in bird song systems is not without genetic components, however. The plasticity in song acquisition leading to dialects is constrained by inherent biases in the acquisition process (Kroodsma and Canady 1985, Marler 1990, 1991). For example, Nelson et al. (1995) observed song acquisition in laboratory-reared white-crowned sparrows from sedentary and migratory populations supplied with a variety of tutoring tapes. Although there were no differences between populations in which song syllables they could learn, birds from migratory populations acquired song at an earlier age in a shorter time and sang more song variations while in their juvenile "plastic song" or acquisition stage. Nelson et al. (1995) suggest that the genetic differences in acquisition strategy are a reflection of the different challenges facing the populations. Migratory populations, faced with greater uncertainty about where individual birds will be during the breeding period, start learning faster and imitate more song types as juveniles, then crystalize one dialect from among them depending on the local dialect where they eventually settle. Sedentary populations, by contrast, have far more certainty about which dialect they will encounter and hence learn later, and with greater accuracy, while imitating fewer syllables.

The greater accuracy with which sedentary populations imitate song may also underlie the observation of Ewert and Kroodsma (1994) that individuals from nonmigratory populations of towhees (*Pipilo erythophthalmus*) share more songs with immediate neighbors than do individuals from migratory populations of this species. Ewert and Kroodsma (1994) also noted that, as for marsh wrens (Kroodsma and Canady 1985), nonmigratory individuals use a larger number of syllables than migratory individuals after their song has crystalized. The distinction remains in laboratory-reared wrens, suggesting a genetic basis for this difference.

In some bird species, female preference may be less plastic than male song. In cowbirds, regional preferences among females are genetically based, and these preferences channel song learning by the males into regional dialects matching those preferences (King and West 1983, 1987). Geographic variation in bird song is therefore a system that arises from the interaction of cultural evolution and learning and genetic constraints and consequences.

Diversity versus Stability in Communication Systems

The many examples of geographic variation in communication systems provide insight into, and raise additional questions about, the basic nature of animal communication systems. First and foremost among the lessons learned from such examples is that, although there are clearly phylogenetic and mechanistic constraints on both senders and receivers, environmental and social influences can exert strong diversifying effects. Regardless of the pattern of geographic variation, the mere presence of this variation indicates that mate-recognition systems need not be subject to strong stabilizing selection operating at the species level leading to narrow, species-specific characters stable across all populations.

Diversity appears in both sender and receiver portions of the communication dyad. This would almost certainly have to be the case, given the fundamental need of all communication systems to maintain some type of match between the sender's signal and the tuning of the receiver. Nevertheless, both systems are labile. The examples studied in detail to date provide no clear indication that one end of the dyad is more stable or severely constrains the characteristics of the other at the intraspecific level, although the issue of whether senders or receivers are more plastic deserves further attention (Brenowitz 1994).

Furthermore, the case of cricket frogs, in which both signal and receiver characteristics can be quantified, shows that a capacity for change in both parts of the communication dyad can lead to variation in the relationship between them. This variation indicates a looser coupling between the two parts of a species' communication system than has been suspected. In frogs, this has implications for the degree of sexual selection exerted on male calls by female preferences in different populations. It would be significant to determine if the match between signal and receiver varies in other communication systems, as the stability of this relationship at a species level has important mechanistic and evolutionary implications.

As they generate diversity, environmental and social influences can have different effects. West Eberhardt (1983) argued that environmental selection tends to drive traits toward optimization, whereas social, or sexual, selection has no optimal result and is therefore more unpredictable and diversifying (see also Ringo 1977, Lande 1982, Ryan 1990a,b). Indeed, we find some evidence of this. Where the call is under more strenuous environmental selection in the forest habitats, mean male and female basilar papilla tuning tend to lie consistently close to the mean call dominant frequency, whereas among grassland populations the relationship between tuning and the call is more variable (Wilczynski et al. 1992). Moreover, both principal component analysis and discriminant function analysis applied to measures of overall call structure in cricket frogs show that grassland populations are much more diverse and unpredictable than forest populations. Environmental diversity across the range of a species does lead to diversity in that species' communication system via selection due to habitat transmission or masking characteristics, predation differences, or, as may also be the case for cricket frogs, via selection on traits such as body size that pleiotropically affect the system. This diversity is, however, somewhat constrained so that common habitat types lead to similar communication adaptations within and across species.

Social or sexual selection—that is, selection due to the internal dynamics of the communication dyad—also leads to geographic differences within species, but has a more diversifying effect on the system. This is apparent in the more unpredictable patterns of cricket frog calls and tuning in open habitats, where environmental constraints are weak. Bird song dialects, which may be due primarily to social interactions rather than selection due to external forces, may be another example of the relatively unpredictable, highly diversifying nature of this type of change.

One final lesson is a methodological one. Dissecting the factors that influence the evolution of communication systems is aided greatly by examining patterns of geographic variation. All communication systems evolve amid complex social interactions between senders and receivers and a host of external factors such as habitat characteristics and predation pressures that can potentially drive the evolution of communication characters or constrain their expression. The fact that populations do vary in habitat and social interactions and thus in the constellation of potential influences on their behavior, allows one to test hypotheses about how specific factors, or particular patterns of interaction among them, contribute to the evolutionary and mechanistic processes that shape this fundamental component of animal behavior.

References

Alexander, R. D. 1975. Natural selection and specialized chorusing behavior in acoustical insects. *In* D. Pimentel, ed. Insects, society and science, pp. 35–77. Academic Press, New York.

Andersson, M. 1994. Sexual selection. Princeton University Press, Princeton, NJ.

Armstrong, E. A. 1963. A study of bird song. Oxford University Press, London.

Baker, M. C. 1983. The behavioral response of female Nuttall's white-crowned sparrows to male song of natal and alien dialects. Behavioral Ecology and Sociobiology 12: 309–315.

Baker, M. C., and M. A. Cunningham. 1985. The biology of bird-song dialects. Behavior and Brain Science 8:85–133.

Baker, M. C., K. J. Spitler-Nabors, A. D. Thompson, and M. A. Cunningham. 1987. Reproductive behaviour of female white-crowned sparrows: Effect of dialects and synthetic hybrid songs. Animal Behaviour 35:1766–1774.

Balaban, E. 1988. Bird song syntax: Learned intraspecific variation is meaningful. Proceedings of the National Academy of Science USA 85:3657–3660.

Benedix, J. H., and D. J. Howard. 1991. Calling song displacement in a zone of overlap and hybridization. Evolution 45:1751–1759.

Blair, W. F. 1958. Mating call in the speciation of anuran amphibians. American Naturalist 92:27–31.

Blair, W. F. 1964. Isolating mechanisms and interspecific interactions in anuran amphibians. Quarterly Review of Biology 39:334–344.

Boake, C. R. B. 1991. Coevolution of senders and receivers of sexual signals: Genetic coupling and genetic correlations. Trends in Ecology and Evolution 6:225–231.

Borror, D. J. 1956. Variation in Carolina wren songs. Auk 73:211–229.

Bowman, R. I. 1983. The evolution of song in Darwin's finches. *In* R. I. Bowman, M. Berson, and A. E. Leviton, eds. Patterns of evolution in the Galapagos, pp. 237–537. American Society for the Advancement of Science, San Francisco.

Bradbury, J. W., and M. B. Anderson (eds). 1987. Sexual selection: Testing the alternatives. John Wiley and Sons, Chichester, UK.

Brenowitz, E. A. 1983. The contribution of temporal song cues to species recognition in the red-winged blackbird. Animal Behaviour 31:1116–1127.

Brenowitz, E. A. 1986. Environmental influences on acoustic and electric animal communication. Brain Behavior and Evolution 28:32–42.

Brenowitz, E. A. 1994. Flexibility and constraint in the evolution of animal communication. *In* R. J. Greenspan and C. P. Kyriacou, eds. Flexibility and constraint in behavioral systems, pp. 247–258. Wiley, New York.

Brown, W. L., and E. O. Wilson. 1956. Character displacement. Systematic Zoology 5: 49–64.

Butlin, R. 1987. Speciation by reinforcement. Trends in Ecology and Evolution 2:8–13.

Butlin, R. 1989. Reinforcement of premating isolation. *In* D. Otte and J. A. Endler, eds. Speciation and its consequences, pp. 158–179. Sinauer and Associates, Sunderland MA.

Butlin, R., and G. K. Hewitt. 1985. A hybrid zone between *Chorthippus parallelus parallelus* and *Chorthippus parallelus erythropus* (Orthoptera: Acrididae): Behavioral characters. Biological Journal of the Linnean Society 26:287–299.

Capranica, R. R. 1966. Vocal response of bullfrog to natural and synthetic mating calls. Journal of the Acoustical Society of America 40:1131–1139.

Capranica, R. R., L. S. Frishkopf, and E. Nevo. 1973. Encoding of geographic dialects in the auditory system of the cricket frog. Science 182:1272–1275.

Capranica, R. R., and A. J. M. Moffat. 1983. Neurobehavioral correlates of sound communication in anurans. *In* J.-P. Ewert, R. R. Capranica, and D. J. Ingle, eds. Advances in vertebrate neuroethology, pp. 701–730. Plenum Press, New York.

Claridge, M. F. 1993. Speciation in insect herbivores: The role of acoustic signals in leafhoppers and planthoppers. *In* D. R. Lees and D. Edwards, eds. Evolutionary patterns and processes, pp. 285–297. Linnean Society Symposium Series No. 14. Academic Press, London.

Cocroft, R. B., and M. J. Ryan. 1995. Patterns of advertisement call evolution in toads and chorus frogs. Animal Behaviour 49:283–303.

Coyne, J. A., and H. A. Orr. 1989. Patterns of speciation in *Drosophila*. Evolution 43: 362–281.

Dessauer, H. C., and E. Nevo. 1969. Geographic variation of blood and liver proteins in cricket frogs. Biochemical Genetics 3:171–188.

Doherty, J. A., and H. C. Gerhardt. 1983. Hybrid frogs: Vocalizations of males and selective phonotaxis. Science 220:1078–1080.

Drewry, G. E., and A. S. Rand. 1983. Characteristics of an acoustic community: Puerto Rican frogs of the genus *Eleutherodactylus*. Copeia 1983:941–953.

Duellman, W. E., and R. A. Pyles. 1983. Acoustic resource partitioning in anuran communities. Copeia 1983:639–649.

Dusenbery, D. B. 1992. Sensory ecology. W. H. Freeman, New York.

Eberhard, W. 1985. Sexual selection and animal genitalia. Harvard University Press, Cambridge, MA.

Emlen, S. T. 1971. Geographic variation in indigo bunting song (*Passerina cyanea*). Animal Behaviour 19:407–408.

Endler, J. A. 1980. Natural selection on color patterns in *Poecilia reticulata*. Evolution 34:76–91.

Endler, J. A. 1983. Natural and sexual selection on color patterns in poeciliid fishes. Environmental Biology of Fishes 9:173–190.

Endler, J. A. 1988. Frequency-dependent predation, crypsis and aposematic coloration. Philosophical Transactions of the Royal Society of London B 319:505–523.

Endler, J. A. 1991. Variation in the appearance of guppy color patterns to guppies and their predators under different visual conditions. Vision Research 31:587–608.

Endler, J. A. 1992. Signals, signal conditions, and the direction of evolution. American Naturalist 139:S125–S153.

Endler, J. A. 1993. Some general comments on the evolution and design of animal communication systems. Philosophical Transactions of the Royal Society of London B 340: 215–225.

Endler, J. A., and A. E. Houde. 1995. Geographic variation in female preference for male traits in *Poecilia reticulata*. Evolution 49:456–468.

Enquist, M., and A. Arak. 1993. Selection for exaggerated male traits by female aesthetic senses. Nature 361:446–448.

Ewert, D. N., and D. E. Kroodsma. 1994. Song sharing and repertoires among migratory and resident rufous-sided towhees. Condor 96:190–196.

Fisher, R. A. 1958. The genetical theory of natural selection. Dover, New York.

Fleishman, L. J. 1992. The influence of the sensory system and the environment on motion patterns in the visual display of anoline lizards and other vertebrates. American Naturalist 139:S36–S61.

Fouquette, M. J. 1975. Speciation in chorus frogs. I. Reproductive character displacement in the *Pseudacris nigrita* complex. Systematic Zoology 24:16–23.

Gerhardt, H. C. 1982. Sound pattern recognition in some North American treefrogs (Anura: Hylidae): Implications for mate choice. American Zoologist 22:581–595.

Gerhardt, H. C. 1988. Acoustic properties used in call recognition by frogs and toads. *In* B. Fritzsch, M. J. Ryan, W. Wilczynski, T. E. Hetherington, and W. Walkowiak, eds. The evolution of the amphibian auditory system, pp. 455–483. John Wiley and Sons, New York.

Gerhardt, H. C. 1994. Reproductive character displacement of female mate choice in the grey treefrog, *Hyla chrysoscelis*. Animal Behaviour 47:959–969.

Gish, S. L., and E. S. Morton. 1981. Structural adaptations to local habitat acoustics in Carolina wren songs. Zeitschrift für Tierpsychologie 56:74–84.

Gwynne, D. T., and G. K. Morris. 1986. Heterospecific recognition and behavioral isolation in acoustic orthoptera. Evolutionary Theory 8:33–38.

Harris, M. A., and R. E. Lemon. 1972. Songs of song sparrows (*Melospiza melodia*): Individual variation and dialects. Canadian Journal of Zoology 50:301–309.

Hopkins, C. D. 1980. Evolution of electric communication channels in mormyrids. Behavioral Ecology and Sociobiology 7:1–13.

Hopkins, C. D., and A. Bass. 1981. Temporal coding of species recognition in an electric fish. Science 212:85–87.

Houde, A. E. 1987. Mate choice based upon naturally occuring color-pattern variation in a guppy population. Evolution 41:1–10.

Houde, A. E. 1988. Genetic differences in female choice between two guppy populations. Animal Behaviour 36:510–516.

Houde, A. E., and J. A. Endler. 1990. Correlated evolution of female mating preferences and male color patterns in the guppy *Poecilia reticulata*. Science 248:1405–1408.

Hoy, R. R., J. Hahn, and R. C. Paul. 1977. Hybrid cricket behavior: Evidence for genetic coupling in animal communication. Science 195:82–84.

Keddy-Hector, A. C., W. Wilczynski, and M. J. Ryan. 1992. Call patterns and basilar papilla tuning in cricket frogs. II. Intrapopulation variation and allometry. Brain, Behavior and Evolution 39:238–246.

King, A. P., and M. J. West. 1983. Female perception of cowbird song: A closed developmental program. Developmental Psychobiology 16:335–342.

King, A. P., and M. J. West. 1987. Different outcomes of synergy between song production and song perception in the same subspecies (*Molothrus ater ater*). Developmental Psychobiology 23:177–187.

King, A. P., M. J. West, and D. H. Eastzer. 1980. Song structure and song development as potential contributors to reproductive isolation in cowbirds (*Molothrus ater*). Journal of Comparative Physiology and Psychology 94:1028–1039.

Kirkpatrick, M. 1982. Sexual selection and the evolution of female choice. Evolution 36: 1–12.

Kirkpatrick, M., and M. J. Ryan. 1991. The paradox of the lek and the evolution of mating preferences. Nature 326:286–288.

Kroodsma, D. E. 1974. Song learning, dialects, and dispersal in the Bewick's wren. Zeitschrift fur Tierpsychologie 35:352–380.

Kroodsma, D. E. 1981. Winter wren singing behavior: A pinnacle of complexity. Condor 82:357–365.

Kroodsma, D. E., and R. A. Canady. 1985. Differences in repertoire size, singing behavior, and associated neuroanatomy among marsh wren populations have a genetic basis. Auk 102:439–446.

Kyriacou, C. P., M. L. Greenacre, M. G. Ritchie, B. C. Byrne, and J. C. Hall. 1992. Genetic and molecular analysis of the love song preferences of *Drosophila* females. American Zoologist 32:31–39.
Lande, R. 1981. Models of speciation by sexual selection on polygenic characters. Proceedings of the National Academy of Science USA 78:3721–3725.
Lande, R. 1982. Rapid origin of sexual isolation and character divergence in a cline. Evolution 36: 213–223.
Lemon, R. E. 1967. The response of cardinals to songs of different dialects. Animal Behaviour 15:538–545.
Lemon, R. E. 1971. Differentiation of song dialects in cardinals. Ibis 113:373–377.
Littlejohn, M. J. 1965. Premating isolation in the *Hyla ewingi* complex (Anura: Hylidae). Evolution 19:234–243.
Littlejohn, M. J. 1981. Reproductive isolation: A critical review. *In* W. R. Atchley and D. S. Woodruff, eds. Essays in honor of M. J. D. White, pp. 298–334. Cambridge University Press, Cambridge.
Littlejohn, M. J. 1988. Frog calls and speciation: The retrograde evolution of homogametic acoustic signaling systems in hybrid zones. *In* B. Fritzsch, M. J. Ryan, W. Wilczynski, T. E. Hetherington, and W. Walkowiak, eds. The evolution of the amphibian auditory system, pp. 613–635. John Wiley and Sons, New York.
Littlejohn, M. J., and G. F. Watson. 1985. Hybrid zones and homogamy in Australian frogs. Annual Review of Ecology and Systematics 16:85–112.
Loftus-Hills, J. J., and M. J. Littlejohn. 1992. Reinforcement and character displacement in *Gastrophryne carolinensis* and *G. olivacea* (Anura: Microhylidae): A reexamination. Evolution 46:896–906.
Long, K. D., and A. Houde. 1989. Orange spots as a visual cue for female mate choice in the guppy (*Poecilia reticulata*). Ethology 82:316–324.
Lythgoe, J. N. 1979. The ecology of vision. Oxford University Press, Oxford.
Marchetti, K. 1993. Dark habitats and bright birds illustrate the role of the environment in species divergence. Nature 362:149–152.
Marler, P. 1990. Innate learning preferences: Signals for communication. Developmental Psychobiology 23:557–568.
Marler, P. 1991. Song-learning behavior: The interface with neuroethology. Trends in Neuroscience 14:199–206.
Marler, P., and D. Isaac. 1960. Physical analysis of simple bird song as exemplified by the Chipping sparrow. Condor 62:124–135.
Marler, P., and R. Pickert. 1984. Species universal microstructure in the learned song of the swamp sparrow (*Melospiza georgiana*). Animal Behaviour 32:673–689.
Marler, P., and M. Tamura. 1964. Culturally transmitted patterns of vocal behavior in sparrows. Science 146:1483–1486.
Mayr, E. 1963. Animal species and evolution. Harvard University Press, Cambridge, MA.
McClelland, B. E. 1994. Population variation and sexual dimorphism in the larynx, ear, and forebrain nuclei of cricket frogs (*Acris crepitans*) (PhD thesis). University of Texas, Austin.
McClelland, B. E., W. Wilczynski, and M. J. Ryan. 1996. Correlations between call characteristics and morphology in male cricket frogs (*Acris crepitans*). Journal of Experimental Biology 199:1907–1919.
McClelland, B. E., W. Wilczynski, and M. J. Ryan. 1998. Intraspecific variation in larynx and ear morphology in male cricket frogs (*Acris crepitans*). Biological Journal of the Linnean Society 63:51–67.
McKenzie, J. A., and M. H. A. Keenleyside. 1970. Reproductive behavior of ninespine

sticklebacks (*Pungitius pungitius* L.) in South Bay, Manitoulin Island, Ontario. Canadian Journal of Zoology 48:55–61.

McMahon, C. A., R. G. Frye, and K. L. Brown. 1984. The vegetation types of Texas including cropland. Texas Parks and Wildlife, Austin, TX.

Moore, T. E. 1993. Acoustic signals and speciation in cicadas (Insecta: Homoptera). *In* D. R. Lees and D. Edwards, eds. Evolutionary patterns and processes, pp. 269–284. Linnean Society Symposium Series No. 14. Academic Press, London.

Narins, P. M. 1995. Comparative aspects of interactive communication. *In* A. Flock, ed. Active hearing, pp. 363–372. Elsevier, New York.

Narins, P. M., and R. R. Capranica. 1976. Sexual differences in the auditory system of the tree frog, *Eleutherodactylus coqui*. Science 192:378–380.

Narins, P. M., and S. L. Smith. 1986. Clinal variation in anuran advertisement calls: Basis for acoustic isolation? Behavioral Ecology and Sociobiology 19:135–141.

Nelson, D. A., P. M. Marler, and A. Palleroni. 1995. A comparative approach to vocal learning: Intraspecific variation in the learning process. Animal Behaviour 50:83–97.

Nevo, E. 1973. Adaptive variation in size in cricket frogs. Ecology 54:1271–1281.

Nevo, E., and R. R. Capranica. 1985. Evolutionary origin of ethological reproductive isolation in cricket frogs, *Acris*. Evolutionary Biology 19:147–214.

Nottebohm, F., and R. K. Selander. 1972. Vocal dialects and gene frequencies in the chingolo sparrow (*Zonotricha capensis*). Condor 74:137–143.

Otte, D. 1989. Speciation in Hawaiian crickets. *In* D. Otte and J. Endler, eds. Speciation and its consequences, pp. 482–526. Sinauer Associates, Sunderland, MA.

Paterson, H. E. H. 1985. The recognition concept of species. *In* E. Vrba, ed. Species and speciation, pp. 21–29. Transvaal Museum Monograph 4, Pretoria, South Africa.

Paterson, H. E. H. 1993. Animal species and sexual selection. *In* D. R. Lees and D. Edwards, eds. Evolutionary patterns and processes, pp. 209–228. Linnean Society Symposium Series No. 14. Academic Press, London.

Penna, M., C. Palazzi, P. Paolinelli, and R. Solis. 1990. Midbrain auditory sensitivity in toads of the genus Bufo (Amphibia—Bufonidae) with different vocal repertoires. Journal of Comparative Physiology A 167:673–681.

Ralin, D. B. 1977. Evolutionary aspects of mating call variation in a diploid-tetrapoid species complex of treefrogs (Anura). Evolution 31:721–736.

Ramer, J. D., T. A. Jenssen, and C. J. Hurst. 1983. Size related variation in the advertisement call of *Rana clamitans* (Anura: Ranidae), and its effect on conspecific males. Copeia 1983:141–155.

Rand, A. S., M. J. Ryan, and W. Wilczynski. 1992. Signal redundancy and receptor permissiveness in acoustic mate recognition by the Túngara frog, *Physalaemus pustulosus*. American Zoologist 32:81–90.

Reimchen, T. E. 1989. Loss of nuptial color in threespine sticklebacks (*Gasterosteus aculeatus*). Evolution 43:450–460.

Ringo, J. M. 1977. Why 300 species of Hawaiian *Drosophila*? The sexual selection hypothesis. Evolution 31:694–696.

Rose, G. J. 1986. A temporal-processing mechanism for all species? Brain Behavior and Evolution 28:134–144.

Ryan, M. J. 1985. The Túngara frog; a study in sexual selection. University of Chicago Press, Chicago.

Ryan, M. J. 1986. Factors influencing the evolution of acoustic communication: Biological constraints. Brain Behavior and Evolution 28:70–82.

Ryan, M. J. 1988. Constraints and patterns in the evolution of anuran acoustic communication. *In* B. Fritzsch, M. J. Ryan, W. Wilczynski, T. E. Hetherington, and W. Walkowiak, eds. The evolution of the amphibian auditory system, pp. 637–677. John Wiley and Sons, New York.

Ryan, M. J. 1990a. Sexual selection, sensory systems, and sensory exploitation. Oxford Surveys in Evolutionary Biology 7:157–195.

Ryan, M. J. 1990b. Signals, species, and sexual selection. American Scientist 78:46–52.

Ryan, M. J. 1991. Sexual selection and communication in frogs. Trends in Ecology and Evolution 6:351–354.

Ryan, M. J., R. B. Cocroft, and W. Wilczynski. 1990a. The role of environmental selection in intraspecific divergence of mate recognition signals in the cricket frog, *Acris crepitans*. Evolution 44:1869–1872.

Ryan, M. J., J. H. Fox, W. Wilczynski, and A. S. Rand. 1990b. Sexual selection for sensory exploitation in the frog, *Physalaemus pustulosus*. Nature 343:66–67.

Ryan, M. J., and A. C. Keddy-Hector. 1992. Directional patterns of female mate choice and the role of sensory biases. American Naturalist 139:S4–S35.

Ryan, M. J., S. A. Perrill, and W. Wilczynski. 1992. Auditory tuning and call frequency predict population-based mating preferences in the cricket frog, *Acris crepitans*. American Naturalist 139:1370–1383.

Ryan, M. J., and A. S. Rand. 1990. The sensory basis of sexual selection for complex calls in the Túngara frog, *Physalaemus pustulosus* (sexual selection for sensory exploitation). Evolution 44:305–314.

Ryan, M. J., and A. S. Rand. 1993a. Phylogenetic patterns of behavioral mate recognition systems in *Physalaemus pustulosus* species group (Anura: Leptodactylidae): The role of ancestral and derived characters and sensory exploitation. *In* D. R. Lees and D. Edwards, eds. Patterns and processes, pp. 251–267. Linnean Society Symposium Series No. 14. Academic Press, New York.

Ryan, M. J., and A. S. Rand. 1993b. Species recognition and sexual selection as a unitary problem in animal communication. Evolution 47:647–657.

Ryan, M. J., and A. S. Rand. 1995. Female responses to ancestral advertisement calls in Túngara frogs. Science 269:390–392.

Ryan, M. J., M. D. Tuttle, and A. S. Rand. 1982. Bat predation and sexual advertisement in a neotropical frog. American Naturalist 119:136–139.

Ryan, M. J., and W. Wilczynski. 1988. Coevolution of sender and receiver: Effect on local mate preference in cricket frogs. Science 240:1786–1788.

Ryan, M. J., and W. Wilczynski. 1991. Evolution of intraspecific variation in the advertisement call of cricket frog (*Acris crepitans*, Hylidae). Biology Journal of the Linnean Society 44:249–271.

Salthe, S. N., and E. Nevo. 1969. Geographic variation of lactate dehydrogenase in the cricket frog *Acris crepitans*. Biochemical Genetics 3:335–341.

Schneider, D. 1974. The sex attractant of moths. Scientific American 231:28–35.

Schwartz, J. J., and K. D. Wells. 1984. Interspecific acoustic interactions of the neotropical treefrog *Hyla ebraccata*. Behavioral Ecology and Sociobiology 14:211–224.

Searcy, W. A. 1990. Species recognition of song by female red-wing blackbirds. Animal Behaviour 40:1119–1127.

Searcy, W. A. 1992. Song repertoire and mate choice in birds. American Zoologist 32: 71–81.

Shiovitz, K. A., and W. L. Thompson. 1970. Geographic variation in song composition of the indigo bunting, *Passerina cyanea*. Animal Behaviour 18:151–158.

Stoner, G., and F. Breden. 1988. Phenotypic differentiation in female preference related to geographic variation in male predation risk in the Trinidad guppy (*Poecilia reticulata*). Behavioral Ecology and Sociobiology 22:285–291.

Tempelton, A. R. 1981. Mechanisms of speciation—a population genetic approach. Annual Review of Ecology and Systematics 12:23–48.

Tokarz, R. R. 1995. Mate choice in lizards: A review. Herpetological Monographs 9: 17–40.

Waage, J. K. 1979. Reproductive character displacement in *Calopteryx* (Odonata: Calopterygidae). Evolution 33:104–116.

Wagner, W. A. 1989a. Fighting, assessment, and frequency alteration in Blanchard's cricket frog. Behavioral Ecology and Sociobiology 25:429–436.

Wagner, W. A. 1989b. Graded aggressive signals in Blanchard's cricket frog: Vocal responses to opponent proximity and size. Animal Behaviour 38:1025–1038.

Wagner, W. A. 1989c. Social correlates of variation in male calling behavior in Blanchard's cricket frog. Ethology 82:27–45.

Walker, T. J. 1957. Specificity in the response of female tree crickets to calling songs of males. Annals of the Entomological Society of America 50:626–636.

Walker, T. J. 1974. Character displacement in acoustic insects. American Zoologist 14: 1137–1150.

Walkowiak, W. 1988. Neuroethology of anuran call recognition. *In* B. Fritzsch, M. J. Ryan, W. Wilczynski, T. E. Hetherington, and W. Walkowiak, eds. The evolution of the amphibian auditory system, pp. 485–510. John Wiley and Sons, New York.

Wasserman, M., and H. R. Koepfer. 1977. Character displacement for sexual isolation between *Drosophila mojavensis* and *D. arizonensis*. Evolution 31:812–823.

Wells, K. D. 1977. The social behavior of anuran amphibians. Animal Behaviour 25: 666–693.

West Eberhard, M. J. 1983. Sexual selection, social competition and speciation. Quarterly Review of Biology 58:155–183.

Wilczynski, W. 1986. Sexual differences in neural tuning and their effects on active space. Brain Behavior and Evolution 28:83–94.

Wilczynski, W. 1992. The nervous system. *In* M. Feder and W. Bruggren, eds. Environmental physiology of the amphibians, pp. 3–39. University of Chicago Press, Chicago.

Wilczynski, W., and R. R. Capranica. 1984. The auditory system of anuran amphibians. Progress in Neurobiology 22:1–38.

Wilczynski, W., A. C. Keddy-Hector, and M. J. Ryan. 1992. Call patterns and basilar papilla tuning in cricket frogs. I. Differences among populations and between sexes. Brain Behavior and Evolution 39:229–237.

Wilczynski, W., B. E. McClelland, and A. S. Rand. 1993. Acoustic, auditory, and morphological divergence in three species of neotropical frog. Journal of Comparative Physiology A 172:425–438.

Wilczynski, W., A. S. Rand, and M. J. Ryan. 1995. The processing of spectral cues by the call analysis system of the Túngara frog, *Physalaemus pustulosus*. Animal Behaviour 49:911–929.

Wilczynski, W., and M. J. Ryan. 1988. The amphibian auditory system as a model system for neurobiology, behavior and evolution. *In* B. Fritzsch, M. Ryan, W. Wilczynski, T. Hetherington, and W. Walkowiak, eds. The evolution of the amphibian auditory system, pp. 3–12. John Wiley and Sons, New York.

Wilczynski, W., H. H. Zakon, and E. A. Brenowitz. 1984. Acoustic communication in spring peepers. Call characteristics and neurophysiological aspects. Journal of Comparative Physiology A 155:577–584.

Wiley, R. H., and D. G. Richards. 1978. Physical constraints on acoustic communication in the atmosphere: Implications for the evolution of animal vocalizations. Behavioral Ecology and Sociobiology 3:69–94.

Wiley, R. H., and D. G. Richards. 1982. Adaptations for acoustic communication in birds: Sound transmission and signal detection. *In* D. E. Kroodsma and E. H. Miller, eds. Acoustic communication in birds, vol. 1, pp. 131–181. Academic Press, New York.

Wood, T. K. 1993. Speciation of the *Euchenopa binotata* complex (Insecta: Homoptera: Membracidae). *In* D. R. Lees and D. Edwards, eds. Evolutionary patterns and processes, pp. 299–317. Linnean Society Symposium Series No. 14. Academic Press, New York.

Zakon, H. H., and W. Wilczynski. 1988. The physiology of the VIIIth nerve. *In* B. Fritzsch, M. J. Ryan, W. Wilczynski, T. E. Hetherington, and W. Walkowiak, eds. The evolution of the amphibian auditory system, pp. 125–155. John Wiley and Sons, New York.

12

Geographic Variation in Sexual Behavior

Sex, Signals, and Speciation

PAUL A. VERRELL

The chapters in this volume share the theme that our understanding of pattern, process, and consequence in the study of behavioral evolution can be advanced by examining differences among conspecific populations. Traditionally, biologists have sought such understanding by comparing different species. Although differences among species are usually greater than differences among conspecific populations, so many factors can vary interspecifically that determining selection pressures driving behavioral divergence may be difficult. As Arnold (1992) has argued, confounding variables often are less prevalent in intraspecific studies; in addition, relatively small evolutionary changes may be perceptible.

From a historical perspective, there appear to be good reasons for believing that sexual behavior should show little variation among conspecific populations. First, early species concepts largely were typological, and accorded intraspecific variation with no reality, let alone importance (Mayr 1976). Partially due to such thinking, early ethologists argued that certain behavior patterns should be largely invariant. For Lorenz (1970), courtship behavior was a prime example of such a "fixed action pattern" (or FAP), a predictable and stereotyped sequence of actions that was elicited by a specific releasing stimulus. Later work revealed that such sequences are more variable than once thought, leading to the suggestion that the FAP be replaced by the MAP, or "modal action pattern." This stresses the average or modal nature of much behavior (Barlow 1977).

The second reason for expecting little intraspecific variation in sexual behavior derives in part from the modern synthesis. The "founding fathers" of modern evolutionary biology placed great emphasis on the role of species differences in preventing interspecific mating and wastage of reproductive effort in the production of unfit hybrid offspring (e.g., Dobzhansky 1937). Indeed, Tinbergen (1953) stated that one of the functions of mating behavior is to ensure such reproductive isolation among species. Presumably, selection would strongly favor the production of unambiguous signals, leading to invariance.

The earliest evidence demonstrating the existence of intraspecific variation in sexual behavior patterns came from studies that were firmly rooted in concepts that characterized the early stages of the modern synthesis. Under the influence of Dobzhansky, many workers sought to identify the earliest stages in the process of speciation, with flies of the genus *Drosophila* as popular animal subjects (see Chatterjee and Singh 1989 for a review). The typical approach was to collect samples from throughout the geographic range of a species, produce strains from these in the laboratory, and then stage both within- and between-strain encounters among all combinations. The extent of intermating among strains was used to make systematic decisions, assuming that sexually incompatible populations are at the earliest stages of species formation. This goal was fully in accord with the biological species concept then predominant in systematics. In this way, for example, incompatible populations of the South American fly *Drosophila paulistorum* were elevated to the status of semispecies (Carmody et al. 1962).

In more recent years there has been a surge of interest in the general issue of phenotypic differentiation among populations, at least in part due to renewed interest in the process of speciation. (Of course, research persists on interactions among fully formed species in areas of sympatry and in hybrid zones; e.g., Harrison 1993, Littlejohn this volume). Many authors have argued that only by examining variation among conspecific populations may we fully understand the earliest stages of the formation of new species (e.g., Lewontin 1974, Endler 1989). The extent to which sexual behavior has been studied in this context is disappointing. Some of this undoubtedly is due to remnants of typological thinking about species and their boundaries. But perhaps more important, the development of techniques in molecular biology has led increasing numbers of workers to address population variation at the genetic level. I find it discouraging that, in an influential discussion of geographic variation in birds, attention was focused on morphology and molecules; no mention was made of behavior, sexual or otherwise (Zink and Remsen 1986).

Aims

This chapter defines sexual behavior as all of the signals and responses that bring members of the same population together for the purpose of gamete union. These include the signals and responses which Tinbergen (1953) listed as orientation, synchronization, and persuasion, to which I add the fourth category of attraction. Variation in reproductive behavior patterns outside of this definition, such as direct competition for mates or parental care, are beyond the scope of this chapter (see Lott 1984 for a review).

My primary aims in this chapter are to (1) describe empirical approaches used to test for geographic variation among conspecific populations; (2) consider major evolutionary forces that may drive divergence; (3) review the consequences of divergence, especially as they relate to the formation of new species; and (4) suggest profitable areas for further study. Because data relevant to these topics often have been collected for other purposes, the literature is scattered and difficult to explore. I have chosen to focus on populations that *do* show divergence, although it should be borne in mind that there certainly are examples for which no divergence has been detected. The studies cited are not exhaustive, but they are representative. My hope is that they are sufficiently numerous and taxonomically diverse to encourage further research by ethologists and evolutionary biologists alike.

A Few Words on the Geography of Divergence

Before going further, brief mention must be made of the complexity of geographic aspects of divergence among conspecific populations. We are all familiar with populations that are allopatric to one another, and among which there is limited potential for migration and gene flow. However, as illustrated by Endler (1977), the spatial configuration of conspecific populations often is considerably more complex than the extreme of strict allopatry. For example, populations may be clinally distributed, with some potential for migration and gene flow between breeding units. My examples span much of the spatial variation discussed by Endler (1977).

Variation: What Data Are Available?

Variation in Signals

Intraspecific variation is most easily documented by studying signals, without paying attention to responses in the opposite sex. For this reason, the consequences of behavioral variation are uncertain. An example here would be the description of male courtship song across a broad geographic area in a species of acoustic insect. The assumption is that any variation "matters" in intersexual communication.

Discussion of a single study adopting this approach will suffice, chosen because, as I will discuss later, it provided the impetus for further work addressing variation in responsiveness among receivers. Duijm (1990) examined variation in male advertisement calls in French bushcrickets of the genus *Ephippiger*. In addition to finding differences between two subspecies of *E. ephippiger*, Duijm found clinal variation in the syllabic content of calls among populations of the subspecies *E. e. diurnus*.

Variation in Signals and Responses

Most studies of variation in signals and responses attempt to test the hypothesis that individual response is greatest toward signals produced in the "home" population. It is essential that there be a reliable bioassay for responsiveness; the ultimate bioassay would be preferential mating with individuals from the home population. Many studies, however, use differential proximity or approach as an index of choice, where the assumption must be that a proximity/approach preference would translate into actual mate choice.

As mentioned above, studies of signal variation may act as catalysts for further work addressing preferences. For example, the considerable geographic variation in syllabic content of male song described for *Ephippiger ephippiger* by Duijm (1990) led Ritchie (1991) to inquire if female preferences also vary geographically. Using phonotactic response to song playback as an index of preference, Ritchie found that females from a monosyllabic population preferred monosyllabic song over song from a population in which four syllables are produced.

A recent study of geographic variation in the courtship songs of planthoppers illustrates the importance of considering both signals and responses in constructing evolutionary scenarios. Although both male and female signals vary geographically in *Nilaparvata bakeri*, mate-choice experiments reveal no evidence of preferential mating with partners from the home population (Claridge and Morgan 1993). This study indicates that it may be unwise to assume, rather than demonstrate, covariation between signals and responses.

Surveys of Sexual Incompatibility

Surveys of sexual incompatibility are the oldest approach to studying geographic variation in sexual behavior. They involve staging heterosexual encounters to determine the extent of sexual incompatibility among an array of populations. Such incompatibility is assumed to result from variation in sexual behavior, although the nature of such variation may not be addressed. Where direct observations of behavior are not made, resources can be devoted to surveys of large numbers of populations across broad geographic areas.

The survey approach was pioneered by Dobzhansky in his efforts to understand the early stages of speciation in *Drosophila* (reviewed by Chatterjee and Singh 1989). Examples include studies of *D. paulistorum* (table 12-1, Carmody et al. 1962, Ehrman 1965), *D. prosaltans* (Dobzhansky and Streisinger 1944), *D. sturtevanti* (Dobzhansky 1944), and *D. silvestris* (Kaneshiro and Kurihara 1981). The survey approach has been less widely applied to vertebrates, although direct evidence for sexual incompatibility among conspecific populations is available for Asian newts of the genus *Cynops* (Kawamura and Sawada 1959, Sawada 1963) and North American salamanders of the genus *Desmognathus* (see table 12-2).

A number of the early survey studies attempted to relate degree of sexual incompatibility to the extent of geographic separation among populations. For example, Dobzhansky and Streisinger (1944) state that degree of incompatibility corresponds "in a general way" to the geographic origins of the populations studied in *D. prosaltans*. Testing for correlations among distance measures is fraught with statistical pitfalls because entries within each matrix of distance measures are not independent of one another. Techniques now exist to provide tests of significance for such correlations that were unavailable to earlier workers (see

Table 12-1 Levels of sexual incompatibility among laboratory strains derived from allopatric and sympatric "races" of the South American fly *Drosophila paulistorum* (0 = random mating, 1.00 = complete sexual incompatibility).

	Levels of Incompatibility	
Races Crossed	Allopatric	Sympatric
Amazonian × Andean	0.66	0.86
Amazonian × Guianan	0.76	0.94
Amazonian × Orinocan	0.61	0.75
Andean × Guianan	0.74	0.96
Orinocan × Andean	0.46	0.94
Orinocan × Guianan	0.72	0.85
Centro-American × Amazonian	0.71	0.78
Centro-American × Orinocan	0.73	0.85

Although levels of sexual incompatibility were higher in crosses involving sympatric races, substantial levels also were detected among allopatric races. Data from Ehrman (1965).

Oden and Sokal 1992). For example, Tilley et al. (1990) found significant levels of sexual incompatibility among salamander populations of the *Desmognathus ochrophaeus* complex. Using a matrix randomization technique, they showed that the degree of incompatibility is significantly correlated with geographic distance among populations, but not with genetic distance.

Some caution is necessary when interpreting the results of multiple "broad survey" studies if different authors use different protocols for measuring sexual incompatibility. Protocols developed largely for pairs of *Drosophila* populations (here denoted as *A* and *B*) include (1) mass-mating designs, where many individuals of both sexes of *A* and *B* are placed together, (2) simultaneous choice tests, where one (fe)male *A* is placed with one (fe)male *A* and one (fe)male *B*, and (3) sequential choice tests, where one (fe)male *A* is placed with one (fe)male *A* or one (fe)male *B*. At least in *Desmognathus* salamanders, levels of interspecific sexual incompatibility are higher in simultaneous than in sequential choice tests (Verrell 1990a).

Complete Case Studies

Case studies are the most comprehensive studies, and contain the following elements: (1) The extent of sexual incompatibility is determined among all of the constituents of an array of conspecific populations. (2) Detailed observational and manipulative analyses are used to determine the biological basis underlying incompatibility. Contributing factors may vary as a function of which populations in an array are considered in pairwise comparisons. (3) When combined with data on geographic distribution and genetic differentiation, hypotheses concerning the mode and tempo of intraspecific divergence may be addressed.

Table 12-2 Levels of sexual incompatibility in field-collected dusky salamanders of the genus *Desmognathus* (Plethodontidae) among conspecific populations, between populations of species in allopatry, and between populations of species in sympatry.

	% Insemination		
Cross	Within	Between	Note/reference
Conspecific Populations			
D. santeetlah	47	23	Crosses of three populations (Maksymovitch and Verrell 1993)
	42	13	
	37	25	
D. ochrophaeus complex	33–80	0–63	Minimum-maximum values for 35 crosses (Tilley et al. 1990)
D. imitator	57	15	Verrell and Tilley (1992)
D. fuscus fuscus	59	44	Verrell (unpublished data)
Between Species, Allopatric Populations			
D. santeetlah × *D. fuscus conanti*	66	36	Males presented with each type of female successively (Verrell 1990a)
	60	14	Males presented with each type of female simultaneously (Verrell 1990a)
D. apalachicolae × *D. f. conanti*	47	5	Verrell (1990b)
D. apalachicolae × *D. ochrophaeus*	63	0	Three populations of *D. ochrophaeus* (Verrell 1990b)
	47	33	
	47	5	
D. ochrophaeus × *D. imitator*	60	2	Two population of *D. ochrophaeus* (Verrell and Tilley 1992)
	42	0	
D. ochrophaeus × *D. f. conanti*	81	2	Two populations of *D. ochrophaeus* (Verrell unpublished data)
	70	22	
D. ochrophaeus × *D. f. fuscus*	53	3	Verrell (unpublished data)
Between Species, Sympatric Populations			
D. santeetlah × *D. ochrophaeus*	80	0	Verrell (1990c)
D. santeetlah × *D. imitator*	53	0	Verrell (1990c)
D. ochrophaeus × *D. imitator*	73	5	Verrell (1990c)
D. ochrophaeus × *D. f. fuscus*	53	0	Uzendoski and Verrell (1993)

Data are summed percentages of within- and between-population courtship trials resulting in insemination in laboratory tests.

Few species other than those in the genus *Drosophila* have been investigated in such a complete manner, and even then, most studies focus on interspecific comparisons. An exception is work on populations of *D. paulistorum*. Building on previous work that demonstrated significant levels of sexual incompatibility among certain "races" (Carmody et al. 1962), Koref-Santibanez (1972a) found quantitative variation among races in the performance of certain courtship elements. To determine whether these differences contributed to sexual incompatibility, Koref-Santibanez (1972b) observed behavioral interactions in interracial encounters staged between single males and females. The decision of whether to initiate courtship rested with males; in most cases, courtships that were initiated progressed to insemination. More recently, this type of experimental approach has been used to address sexual incompatibility in invertebrates such as planthoppers (Claridge et al. 1985, 1988), moths (Toth et al. 1992), and land snails (Baur and Baur 1992).

My own collaborative research on plethodontid salamanders has adopted a similar hierarchical approach to understanding the behavioral basis of sexual incompatibility among conspecific populations in a vertebrate genus (table 12-2). In *Desmognathus santeetlah*, significant levels of incompatibility exist among conspecific populations, although courtship behavior patterns appear to be identical (Maksymovitch and Verrell 1992, 1993). A more extensive study of the *D. ochrophaeus* complex also revealed significant levels of incompatibility among populations (Tilley et al. 1990). Ethological analyses of courtship indicate that most differences among populations are quantitative rather than qualitative (Herring and Verrell, 1996). The point in the courtship sequence at which heterotypic encounters fail is highly variable. In some pairs of populations, encounters are not initiated by males, perhaps due to variation in chemical cues produced by females (see Verrell 1989, Uzendoski and Verrell 1993). In others, courtship is initiated but fails later in the behavioral sequence (Verrell and Arnold 1989).

A Few Words on Field Experiments

Laboratory studies of signals, responses, and sexual incompatibility are valuable because they offer opportunities for precise measurement and experimental manipulation. However, they are open to the criticism that the laboratory environment is highly artificial compared to the natural context in which evolution occurs. What we see in the laboratory may not accurately portray what happens in nature. Recent work on the Australasian frog *Litoria ewingi* suggests the possibility of staging field experiments to study divergence of sexual behavior. Littlejohn et al. (1993; see also Littlejohn this volume) report divergence of male acoustic signals used in mate attraction between a "source population" and derived populations of frogs that were introduced to previously uninhabited areas as few as 38 years ago (about 19 "frog generations"). Studies of sexual incompatibility among translocated populations may prove most instructive in revealing the mode and tempo of intraspecific divergence.

Variation: What Are The Causes?

Correlated Responses

Differences in sexual signals among conspecific populations might evolve as apparent correlated responses to changes in other characters. Correlated responses arise in one of two ways. First, an association between the direct target of selection and correlated behavioral character(s) may be determined by different genes, with the correlation maintained by linkage disequilibrium. Selection may then directly maintain combinations of traits if their association is advantageous. Second, the same genes that are affected by selection also may be responsible for the production of sexual behavior (pleiotropy). Determining whether correlated responses are maintained by linkage disequilibrium or due to pleiotropy requires detailed genetic analysis. However, theory suggests that, in equilibrium populations under stabilizing selection, pleiotropy is the major cause of genetic correlations among characters (Lande 1980).

One way in which signals and/or responses may vary geographically is if they are correlated with another character that varies predictably, in a concordant manner, as a result of selection. For example, clinal variation in the advertisement call of male cricket frogs (*Acris crepitans*) was explained by Nevo and Capranica (1985) as a correlated response to differences in the intensity of selection for large body size (large, desiccation-resistant bodies might be selected in arid habitats). However, multivariate analyses have failed to support pleiotropy as the sole explanation for call variation (Ryan and Wilczynski 1991).

Stronger evidence supporting the view that sexual incompatibility may evolve as a correlated response comes from experimental studies in which individuals taken from the same parental stock are exposed to different types of nonsexual selection pressures. These different strains are then tested for sexual incompatibility after a number of generations. Partial, and sometimes transitory, sexual incompatibility has been detected in the laboratory among strains as a function of selection for such factors as phototactic and geotactic responses in *Drosophila pseudoobscura* (De Oliveira and Cordeiro 1980), *D. melanogaster* (Lofdahl et al. 1992), and the housefly *Musca domestica* (Soans et al. 1974), and for pH tolerance in *D. willistoni* (Del Solar 1966). In a fifth study, significant levels of sexual incompatibility evolved among populations of *D. pseudoobscura* reared on different food media (table 12-3, Dodd 1989).

Table 12-3 Levels of sexual incompatibility among laboratory populations of *Drosophila pseudoobscura* reared on either starch- or maltose-enriched media (−1.00 = complete dissortative mating, 0 = random mating, +1.00 = complete sexual incompatibility).

Range in Levels of Sexual Incompatibility	
Populations on Same Medium	Populations on Different Media
−0.06 to +0.15 (starch)	+0.30 to +0.49
−0.21 to +0.18 (maltose)	

No significant deviation from random mating was observed in 12 crosses staged among populations reared on the same medium. Significant levels of sexual incompatibility were found in 11 (69%) of 16 crosses staged among populations reared on different media. Data from Dodd (1989).

It is possible that selection for geotactic responses and pH tolerance might directly affect sexual interactions, say, by altering the microhabitats in which different strains court. For dietary factors, a less direct relationship may exist if the type of food ingested affects the composition of chemical cues used to attract and/or stimulate partners, as may be true for *Drosophila mojavensis* (Etges 1992, Brazner and Etges 1993). Partial sexual incompatibility among desert populations of this fly may also result from differences in the types of hydrocarbons that serve to waterproof the exoskeleton (Markow 1991). At least in the laboratory, hydrocarbon composition varies in response to ambient temperature. Because these hydrocarbons are also involved in close-range intersexual communication, sexual incompatibility may arise as an incidental consequence of thermal adaptation (Markow and Toolson 1990).

Founder Events

Another way in which sexual incompatibility among populations might arise is via founder events. Briefly, when a new population is formed from a small number of founder individuals, the genome of the population becomes reorganized by recombination and genetic drift. In the absence of extinction, the genome may attain a new equilibrium that expresses novel phenotypic characters. If these include elements of sexual behavior, then sexual incompatibility with the parental population may result (Mayr 1942, Carson and Templeton 1984). Although the concept of speciation by founder events has been criticized on theoretical grounds (Barton and Charlesworth 1984), it is a plausible mechanism for generating sexual incompatibility among laboratory populations (Giddings et al. 1989).

The best known experimental study, on *Drosophila pseudoobscura*, involved passing populations through a series of founder-flush cycles, in which populations founded by a few individuals were allowed to increase in size (Powell 1978). Partial sexual incompatibility arose after just four founder-flush cycles (table 12-4). Control populations failed to exhibit sexual incompatibility, although they had been maintained in isolation and were undoubtedly somewhat inbred (for a replication of Powell's original study, see Galiana et al. 1993). Similar results have been obtained for *D. sechellia* (Cobb et al. 1990) and for houseflies (Meffert and Bryant 1991). In the polychaete worm *Nereis acuminata*, sexual incompatibility was detected between field-caught worms and worms that had passed through just two founder-flush cycles in the laboratory (Weinberg et al. 1992).

The possibility that founder events may influence cultural evolution is suggested by comparison of the songs of mainland and island chaffinches (*Fringilla coelobs*) in New Zealand, in which aspects of vocal behavior are learned rather than transmitted genetically. Although broadly similar, the songs of island birds lack the elaborate end phrases produced by birds on the mainland. Baker and Jenkins (1987) suggest that the birds that founded the island population had already learned their songs from mainland conspecifics. An early bottleneck in population size may then have reduced the diversity of end phrases in the island birds. If learned end phrases are important in mate attraction and/or stimulation,

Table 12-4 The potential for founder-flush cycles to produce sexual incompatibility for laboratory populations of *Drosophila pseudoobscura* experiencing four such cycles (with appropriate controls also maintained): numbers of within- and between-population matings in all crosses involving population 1.

	Numbers of Matings	
Cross	Within	Between
1 × control	46	26
1 × 2	54	17
1 × 3	38	21
1 × 4	45	24
1 × 5	39	28
1 × 6	46	14
1 × 7	39	30
1 × 8	31	20

Levels of sexual incompatibility were measured in crosses involving all eight populations and a control. A significant excess of within-population matings was found in 6 (33%) out of 18 crosses, although only 3 populations were responsible for this level of sexual incompatibility. The data given are a fragment of the total data-set. Data from Powell (1978).

some degree of sexual incompatibility between mainland and island chaffinches might be expected (Baker and Cunningham 1985).

Local Adaptation: The Physical Environment

The concept of local adaptation to prevailing environmental conditions is a basic tenet of Neo-Darwinian evolutionary theory, and selection has long been invoked as the principle cause of geographic variation (Gould and Johnston 1972, Endler 1977, 1986). That selection may result in the local adaptation of animal communication systems is neither a novel nor recent idea (e.g., Morton 1975). For example, local adaptation appears to explain interspecific differences in the acoustic parameters of calls of birds inhabiting different habitat types (reviewed by Gerhardt 1983). In such studies, it is assumed (if not demonstrated) that signal "efficiency" varies as a function of how the physical structure of the environment facilitates or attenuates signal transmission. If conspecific populations inhabit an array of habitats that differ in their capacity for transmission, signals may diverge under the influence of local adaptation.

For some biologists, local adaptation of sexual signals is afforded supreme importance in the process of population divergence and, ultimately, speciation. This view is exemplified by Paterson (1993), for whom sexual behavior is *solely* the product of natural selection for signals that ensure effective mate recognition in the local habitat. He uses an example to illustrate his point. Imagine a parental population from which a number of individuals are taken to form a new population. The sexual signals by which individuals in the parental population locate and stimulate one another may not function as effectively if the derived population finds itself in a different habitat (e.g., grassland versus forest). Paterson states that there will follow a period of directional selection in the derived population, favoring the evolution of signals that most closely match the transmission properties of the new habitat. Signals and responses form what Paterson (1993) has termed the "mate-recognition system" of the population.

Such local adaptation is surely an important force driving geographic variation in sexual signals among conspecific populations. It is responsible, at least in part, for geographic variation in the male call of two subspecies of the frog *Acris creptitans*. Most of this variation is clinal on an east–west axis, but call characteristics also differ between open and forest habitats in a manner that matches the transmission properties of these habitat types (Ryan et al. 1990, Ryan and Wilczynski 1991, Wilczynski and Ryan, this volume).

Local Adaptation: The Biological Environment

Adaptation to the physical environment is not the only factor that may affect the evolution of signaling systems. In many animals, individuals of several species gather at the same location for breeding; such multispecies assemblages of males signaling simultaneously are common in frogs. Littlejohn (1977) has argued that signal interference may lead to interspecific competition for "air space" and that differences in male calls may have evolved in response to selection for minimizing acoustic interference among species. Similar pressures may be exerted on signals in other sensory modalities, too. If conspecific populations vary in the extent to which they occur in sympatry with other signaling species, competition for signaling space may lead to incidental signal divergence among conspecific populations.

Interspecific interactions also may affect signals used in intraspecific communication if they are exploited by "illegitimate receivers" using the signals to locate prey or hosts (as predators, parasites, or parasitoids). Exploitation by illegitimate receivers seems to have caused divergence of sexual signals among different species (reviewed by Otte 1974, Sakaluk 1990). For example, the presence of predatory bats has apparently shaped the evolution of male sexual signals in Neotropical katydids. Males of species occurring in the absence of bats produce complex, airborne calls of long duration. But in katydid species sympatric with bats, male airborne calls are simple, short, and infrequent. Sympatric species also produce unique substrate-borne tremulations, which presumably are not sensed or perceived by bats (Belwood and Morris 1987). Species that exhibit an evolutionary switch from one type of signal to another are said to use "private channels" of communication (Partridge and Endler 1987).

If different conspecific populations experience different types and/or intensities of illegitimate exploitation, sexual signals may diverge accordingly, and sexual incompatibility may be an incidental result of the use of population-specific private channels (Verrell 1991). Predation has apparently been of considerable importance in the evolution of geographic divergence of both male color patterns (displayed during courtship) and male mating behavior in the guppy, *Poecilia reticulata*. Depending at least in part on whether sympatric predators can perceive the color orange, males are either brightly colored and conspicuous or duller and more cryptic (Endler 1978, 1980, 1983). Orange spots are apparently a sexually selected character in guppies, for females prefer the brightest males in "orange populations" (Houde 1988). In a multipopulation comparison, Houde and Endler (1990) found that the degree of preference by female guppies for bright males is correlated with the average brightness of males in each population (table 12-5). Female preferences for males from their own populations may result in sexual incompatibility between partners from different localities (Endler and Houde 1995). It is also worth noting here that the extent to which males engage in conspicuous courtship or surreptitiously sneak matings varies among populations as a function of differences in predation pressure (table 12-6, Luyten and Liley 1985, Magurran and Seghers 1990).

The degree to which selection can enhance interspecific sexual incompatibility to reduce mismating among sympatric species is a subject of intense debate (see Howard 1993 for a recent review and Littlejohn this volume). Evidence that sexual incompatibility can evolve to high levels among conspecific populations and allopatric species (e.g., tables 12-1, 12-2) casts doubt on the necessity of invoking such selection as the sole mechanism for the creation of so-called sexual isolation among sympatric species (but see Coyne and Orr 1989, 1997).

Nevertheless, it is possible that *interspecific* interactions may influence the divergence of conspecific populations and so result incidentally in *intraspecific* sexual incompatibility. Consider two species, *A* and *B*. One population of *A* is in contact with a population of *B*, and these populations are sexually incompatible

Table 12-5 Geographic variation in male color patterns and female preferences in guppies (*Poecilia reticulata*).

Source of Males Tested	Source Populations of Females Tested	Preference-Color Correlation
A	A, B, C, D, E, F, G	.51*
B	A, B, G	.69*
C	A, C, D	.46
D	A, C, D	−.08
E	A, E, F	.30
F	A, E, F	−.18
G	A, B, G	−.05

Female sexual responsiveness was measured toward males from different populations differing in the amount of orange in their nuptial coloration (A = most orange, G = least orange). Females from populations with the most orange in males showed the greatest preference for orange males in these laboratory tests. *$p < .05$, Spearman rank-order correlation. Data from Houde and Endler (1990).

Table 12-6 Content and timing of male sexual behavior in guppies (*Poecilia reticulata*) from populations in headstream waters (low intensity of predation) and lowland waters (higher predation intensity).

	Location of Populations	
Behavior	Headstream	Lowland
Mean number of sigmoid displays	3.5, 6.5	1.5, 2.0
Mean display duration (s)	4.0, 4.0	2.5, 3.0
Gonopodial thrusts per 3 min	0.6, 0.8	1.3, 3.0

Two populations were studied from each type of location in the laboratory. Males displayed less frequently and more rapidly under increased predation risk. In addition, they engaged in higher rates of gonopodial thrusting, a less conspicuous means of insemination than display. Data from Luyten and Liley (1985).

as a result of selection for either reinforcement or reproductive character displacement (Butlin 1987). If selection arising from the presence of *B* alters the sexual behavior of *A* to a sufficient extent, the population of *A* that is in sympatry with *B* may become at least partially sexually incompatible with other conspecific *A* populations that are allopatric with *B*.

Weak evidence consistent with such a scenario comes from a single study. Crosses between populations of *Drosophila mojavensis* exhibit incomplete sexual incompatibility if one of the populations involved is sympatric with the sibling species *D. arizonensis*. Zouros and D'Entremont (1980) suggest that this intraspecific incompatibility is an incidental result of changes in the mate-recognition system of sympatric *D. mojavensis* selected to avoid mating with *D. arizonensis*. However, as noted earlier, differences in thermal adaptation and/or dietary composition also may be responsible for incidental sexual incompatibility among *D. mojavensis* populations (Markow and Toolson 1990, Etges 1992).

Sexual Selection

Darwin (1871) proposed his theory of sexual selection to account for the evolution of characters that provide a benefit to their bearer solely in relation to reproduction and so cannot be explained as products of natural selection. These characters, whatever form they take, provide such a benefit in one of two contexts: increasing the ability of their bearer to succeed in direct competition for mates, or increasing the attractiveness of their bearer to members of the opposite sex. Sexually selected characters are typically more developed in males and are found throughout the animal kingdom (Bradbury and Andersson 1987, Andersson 1994).

Considering signals used to attract and stimulate mates, there is ample evidence that the quality of such signals in senders and the responsiveness of receivers may vary among individuals within a population. Sexual selection will be generated if this variation translates into differential mating success among senders. Because sexual selection can drive the evolution of both sexual signals and preferences for signals, variation among conspecific populations in signals and responses may result in their divergence (for theory, see Lande 1981, West-Eberhard 1982, Nei et al. 1983, Wu 1985, Moller 1993, Schluter and Price 1993; for an empirical study see Gilburn and Day 1994). Sexual selection and species recognition can be regarded as different aspects of a unitary phenomenon in animal communication, with sexual selection resulting in species recognition (Ryan and Rand 1993).

Sexual selection also may be responsible for differences in male genital morphology among different species. These have long been explained as products of selection against interspecific mating (e.g., Mayr 1942). However, Eberhard (1985) has argued that such differences may be better explained as a consequence of sexual selection via intraspecific female mate choice. For example, he notes that complex and species-specific male genitalia exist even for species that do not occur in sympatry with other taxa, where there can be no threat of mismating. Further support comes from studies of male genitalia in primates, in which the most structurally complex penes are often found in those taxa in which the intensity of sexual selection likely is most intense (Dixson 1987, Verrell 1992, Harcourt and Gardiner 1994). I predict that there should also be variation in penile morphology among conspecific populations of certain primate species if sexual selection is an important force driving the evolution of male genitalia. Beyond some limited supporting evidence for lepidopterans (Shapiro 1978), this hypothesis has yet to be fully tested.

Multiple Factors: Complexity in the Real World

Clearly, multiple factors may drive the divergence of sexual signals and responses among conspecific populations. The evolution of population differences in nuptial coloration in guppies provides an excellent illustration of this (see Endler 1992, Houde 1997). Male coloration and female preferences represent an interacting compromise among such selection pressures as the local physical environment (background, quality of ambient illumination), the local biotic environment (especially the community of predators), sexual selection (favoring colorful males), and sensory systems in receivers (visual pigments).

The morphology and physiology of both sender and receiver in motor and sensory systems exert obvious influences on signal evolution. They act as constraints setting lower and upper limits on the physical properties of the signal. The physical structure of the local environment also influences signal evolution by facilitating or impairing effective transmission. Once again, selection will impose lower and upper bounds on the degree of variation that can be permitted if the signal is to be effectively transmitted and received. The local environment also may exert an influence via signal interference and illegitimate exploitation, again setting limits on signal variability.

Thus, sexual signals and responses are products of a complex, diverse, and dynamic array of influencing factors (Endler 1993). Overall, the amount of permissible variation that can exist will be limited by factors such as endogenous constraints, transmission properties of the environment, and interspecific interference/exploitation. However, within this range of permissible variation, sexual selection may cause evolutionary change, usually in the direction of elaboration or exaggeration beyond the mean (Ryan and Keddy-Hector 1992). If ranges of permissible variation in signals and responses differ among conspecific populations, variation in mate-recognition systems and sexual incompatibility may be incidental consequences.

A Few Words on the Genetics of Incompatibility, Signals, and Responses

A full understanding of the microevolution of population divergence and sexual incompatibility must include a consideration of genetics. For example, we may ask how many genes are involved (the number perhaps influencing rate of change) and how they are distributed across chromosomes (especially sex chromosomes versus autosomes). The standard experimental protocol is to cross closely related taxa and then examine the segregation of genes influencing reproductive incompatibility in backcross or F_2 hybrids. These genes are localized by observing their associations with mapped mutations. As of 1992, all work had focused on interspecific reproductive incompatibility, almost exclusively at the postmating level in *Drosophila* (Coyne 1992). However, studies are now underway that use similar methodologies to dissect the genetic basis of sexual incompatibility at the intraspecific level (Hollocher et al. 1997).

A genetic approach can also be used to investigate the phenotypic correlation that exists between signals and responses involved in mate recognition. An interesting issue is the manner in which signals and responses are coordinated between the sexes during evolutionary divergence: in "genetic coupling," signal and response are controlled by a common genetic substrate, and in "coevolution," signals and responses are controlled by different genetic substrates, and changes in one component lead to compensatory changes in the other (Boake 1991, Bakker and Pomiankowski 1995). In a review of nine studies examining signals and responses in interspecific hybrids, Butlin and Ritchie (1989) concluded that most data are insufficient to provide a clear distinction between the two alternatives. Where data are sufficient, the mechanism of genetic coupling is not indicated (see also Lofstedt 1990).

Finally, studies of the amount of additive genetic variance for sexual signals and responses are important if we are to determine whether there is the potential for these traits to undergo further evolutionary change. Empirical evidence of significant heritability of sexual signals is growing (e.g., Hedrick 1988, De Winter 1992, Houde 1992). This is despite the popular, but erroneous, assumption that traits closely associated with fitness should show little additive genetic variance (see Charlesworth 1987).

Variation: What Are the Consequences?

Examination of the literature on geographic variation in sexual behavior among conspecific populations reveals what ecologists, morphologists, physiologists, and molecular evolutionists have long known: many characters exhibit considerable intraspecific variation. As was made clear by the proponents of the modern synthesis in their move away from typological thinking, such variation (if heritable) forms the raw material for a response to selection. Variation is necessary for anagenetic (or "within-lineage") evolutionary change, as discussed in other chapters in this volume. Variation in sexual behavior among populations takes on a special significance if sexual incompatibility contributes to reduced gene flow. Sexual behavior then becomes important in terms of cladogenetic (or "among-lineage") change—that is, the process of speciation (Butlin and Ritchie 1994).

Whether information on sexual incompatibility can be used to delineate boundaries between species depends very much on one's concept of what constitutes a species. This subject has long been controversial, and a detailed treatment is beyond the scope of this chapter (for summaries of recent opinion, see Otte and Endler 1989, Ereshefsky 1992, Avise and Wollenburg 1997). For proponents of species concepts that stress genetic cohesion, either by isolation of one set of populations from others or by effective mate recognition within sets of populations, those sets that are sexually incompatible could be elevated to species status. The criterion of incompatibility is most readily demonstrated by the absence of gene flow between sympatric populations. For allopatric populations, no such "natural experiment" is available to us. Instead, we are forced to rely on demonstrations of signal–response variation and sexual incompatibility, often determined in laboratory settings. Examination of any volume of the *Journal of Chemical Ecology* (which mainly reports studies of arthropods) will reveal plenty of examples in which differences in sexual signals and/or responses and/or measures of sexual incompatibility are used to diagnose allopatric species or subspecies that are otherwise cryptic (see table 12-7; e.g., Foster and Roelofs 1987, Frerot and Foster 1991; for a well-argued alternative view, see Pinto et al. 1991).

Other species concepts place less emphasis on reproductive compatibility as a criterion of conspecificity (see Ereshefsky 1992). Most contemporary systematists stress that such compatibility represents an ancestral (or plesiomorphic) character

Table 12-7 Attraction of male moths to geographic-specific female sex pheromones.

	Mean Number of Males Captured	
Type of Lure	Christchurch	Aukland
(Z)-5-TAA and TDA	17.4	0
(Z)-5-TDA and (Z)-8-TDA	0.6	48.0
Blank controls	0.8	0

The chemistry of female sex pheromones in allopatric populations of the New Zealand leafroller *Ctenopseustis obliquana* varies geographically. Females in Christchurch produce a blend of (Z)-5-TAA (tetradeyl acetate) and TDA (tetradecenyl acetate), whereas those in Aukland produce a blend of (Z)-5-TDA and (Z)-8-TDA. In field trials at both localities, male moths were most attracted to the pheromone blend characteristic of females of their own population. Data from Foster and Roelofs (1987).

state. Given their insistence on considering only derived (or apomorphic) character-states in making systematic decisions, these biologists accord little, if any, role to reproductive compatibility as a criterion for conspecificity. The fact that reproductive incompatibility must then represent a derived state (or an emergent property of many derived states) seems often to be either unnoticed or considered irrelevant. Especially (but not only) in the absence of derived characters of other kinds (such as morphology and molecules), I see no reason systematists may not separate populations that share derived reproductive characters from those that do not.

It should be noted that there are biologists who are quite content to diagnose species purely on the basis of behavioral differences, sometimes in the absence of information on sexual incompatibility. Such taxa have been termed "ethospecies." For example, the wolf spiders *Schizocosa ocreata* and *S. rovneri* are morphologically similar but completely sexually incompatible due to differences in their courtship behavior. These ethospecies are capable of forming viable F_1 hybrids when forcibly mated (Stratton and Uetz 1986). The reception of the ethospecies concept by biologists at-large can be gauged from an exchange of opinions regarding its significance for arachnids (see Vlijm 1986). These range from positive ("we should search for characters that allow spider species to distinguish one another": W. Shear) to profoundly negative ("As a plesiomorphic character, the ability to interbreed is not a valid criterion for species membership": J. Coddington), stressing an obvious lack of consensus. I believe that the study of geographic variation of sexual behavior among conspecific populations is interesting and valuable in its own right, regardless of whether it is has any significance in systematics. However, I note that by elevating divergent populations to a higher systematic status, we may effectively destroy what we set out to investigate: variation among conspecific populations.

Avenues for Further Work

Most studies of phenotypic variation involve comparisons of different species. Examples abound in which workers have sought to explain interspecific variation in one character as a function of variation in other parameters, such as secondary sexual characters as a function of social organization (reviewed in Harvey and Pagel 1991, Martins 1996). Developments in comparative biology cast doubt on the value of such studies if they fail to account for the underlying phylogenetic relationships among the species studied. Only by considering such relationships can we disentangle the independent evolutionary origins of adaptations from historical effects arising from common ancestry (Brooks and McLennan 1991, Harvey and Pagel 1991, Greene 1994). In addition, phylogenetic approaches enable us to better identify important selective pressures. For example, McLennan (1996) asked whether mate choice, intermale competition, or paternal care was primarily responsible for the evolution of nuptial coloration in male sticklebacks. Based on phylogenetic relationships among species in male color and breeding behavior (and some experimental manipulations), she concluded that male coloration was most strongly associated with intersexual interactions, indicating an important role for mate choice.

The study of phenotypic variation among conspecific populations also will surely benefit from this improvement in comparative biology. Just as different species are phylogenetically connected with one another via ancestor-descendant relationships, so are conspecific populations. Avise (1994) argues cogently that populations have evolutionary histories in the same way as do species. He suggests that molecular techniques enabling us to map gene lineages will permit the reconstruction of historical relationships among both species (as "branches") and conspecific populations (as terminal "twigs" on phylogenetic trees). Mapping transitions among behavioral states and/or the distribution of multiple characters onto independently derived phylogenies will enable us to provide better tests of hypotheses concerning likely selection pressures, the direction of evolution and the nature of correlations among suites of characters.

A second area that will surely prove fruitful for study concerns the manner in which divergence in mate-recognition systems is correlated with divergence in other aspects of the total reproductive system. This approach was recently applied in an extensive survey of various types of reproductive incompatibility among species of *Drosophila*. Using a statistical technique to control for phylogenetic relatedness, Coyne and Orr (1989, 1997) found that interspecific sexual incompatibility evolves more rapidly than other genetic incompatibilities that result in hybrid dysfunction. Further, this pattern was largely a result of the faster evolution of sexual incompatibility among species occurring in sympatry, as predicted if selection acts against the wastage of reproductive effort between species in secondary contact (Butlin 1987).

Studies of relationships among different forms of reproductive incompatibility among conspecific populations have yet to be conducted in such a systematic manner as undertaken by Coyne and Orr (1989, 1997) for different species. Such studies undoubtedly will prove crucial to a complete understanding of the mode and tempo of the evolution of reproductive closure among populations, and so to the very process of speciation.

Acknowledgments The ideas presented in this chapter were formed during discussions and correspondence with many people, but most especially my *Desmognathus* co-workers and various colleagues now or once at the University of Chicago. These include Steve Arnold, Brian Charlesworth, Jerry Coyne, Lynne Houck, Esther Maksymovitch, Allen Orr, Nancy Reagan, Steve Tilley, and Kerry Uzendoski. I also thank Chris Boake, Bill Cresko, John Endler, Susan Foster, Carl Gerhardt, Tim Halliday, Kimberley Herring, Norah McCabe, Hugh Paterson, Mike Ritchie, Mike Ryan, John Thompson, and Dave Wake for discussion, ideas, and comments. All errors and misinterpretations are my own. Work on *Desmognathus* salamanders has been generously supported by grants from the National Science Foundation (BSR 85-06766 to S. Arnold, BSR 85-08363 to S. Tilley, and BSR 89-06703 to S. Arnold and P. Verrell) and by a research minigrant from Washington State University (to P. Verrell). I dedicate this paper to the memory of my father, Ronald Verrell.

References

Andersson, M. 1994. Sexual selection. Princeton University Press, Princeton, NJ.
Arnold, S. J. 1992. Behavioural variation in natural populations. VI. Prey responses by two species of garter snakes in three regions of sympatry. Animal Behaviour 44:705–719.
Avise, J. C. 1994. Molecular markers, natural history and evolution. Chapman and Hall, New York.
Avise, J. C., and K. Wollenburg. 1997. Phylogenetics and the origin of species. Proceedings of the National Academy of Science USA 94:7748–7755.
Baker, A. J., and P. F. Jenkins. 1987. Founder effect and cultural evolution of songs in an isolated population of chaffinches, *Fringilla coelobs*, in the Chatham Islands. Animal Behaviour 35:1793–1803.
Baker, M. C., and M. A. Cunningham. 1985. The biology of bird-song dialects. Behavior and Brain Science 8:85–133.
Bakker, T. C. M., and A. Pomiankowski. 1995. The genetic basis of femate mate preferences. Journal of Evolutionary Biology 8:129–171.
Barlow, G. W. 1977. Modal action patterns. *In* T. A. Sebeok, ed. How animals communicate, pp. 98–134. Indiana University Press, Bloomington, IN.
Barton, N. H., and B. Charlesworth. 1984. Genetic revolutions, founder effects, and speciation. Annual Review of Ecology and Systematics 15:133–164.
Baur, B., and A. Baur. 1992. Reduced reproductive compatibility in *Arianta arbustorum* (Gastropoda) from distant populations. Heredity 69:65–72.
Belwood, J. J., and G. K. Morris. 1987. Bat predation and its influence on calling behavior in Neotropical katydids. Science 238:64–67.
Boake, C. R. B. 1991. Coevolution of senders and receivers of sexual signals: Genetic coupling and genetic correlations. Trends in Ecology and Evolution 6:225–227.
Bradbury, J. W., and M. B. Andersson (eds). 1987. Sexual selection: Testing the alternatives. Wiley, New York.
Brazner, J. C., and W. J. Etges. 1993. Premating isolation is determined by larval rearing substrates in *Drosophila mojavensis*. II. Effects of larval substrates on time to copulation, mate choice and mating propensity. Evolutionary Ecology 7:605–624.
Brooks, D. R., and D. A. McLennan. 1991. Phylogeny, ecology, and behavior. University of Chicago Press, Chicago.
Butlin, R. K. 1987. Speciation by reinforcement. Trends in Ecology and Evolution 2:8–13.
Butlin, R. K., and M. G. Ritchie. 1989. Genetic coupling in mate recognition systems: What is the evidence? Biological Journal of the Linnean Society 37:237–246.
Butlin, R. K., and M. G. Ritchie. 1994. Behaviour and speciation. *In* P. J. B. Slater and T. R. Halliday, eds. Behaviour and evolution, pp. 43–79. Cambridge University Press, Cambridge.
Carmody, G., A. Diaz Collazo, T. Dobzhansky, L. Ehrman, I. S. Jaffrey, S. Kimball, S. Obrebski, S. Silagi, T. Tidwell, and R. Ullrich. 1962. Mating preferences and sexual isolation within and between the incipient species of *Drosophila paulistorum*. American Midland Naturalist 68:67–82.
Carson, H. L., and A. R. Templeton. 1984. Genetic revolutions in relation to speciation phenomena: The founding of new populations. Annual Review of Ecology and Systematics 15:97–132.
Charlesworth, B. 1987. The heritability of fitness. *In* J. W. Bradbury and M. B. Andersson, eds. Sexual selection: Testing the alternatives, pp. 21–40. Wiley, New York.

Chatterjee, S., and B. N. Singh. 1989. Sexual isolation in *Drosophila*. Indian Review of Life Sciences 9:101–135.

Claridge, M. F., J. Den Hollander, and J. C. Morgan. 1985. Variation in courtship signals and hybridization between geographically definable populations of the rice brown planthopper, *Nilaparvata lugens* (Stal). Biological Journal of the Linnean Society 24: 35–49.

Claridge, M. F., J. Den Hollander, and J. C. Morgan. 1988. Variation in host plant relations and courtship signals of weed-associated populations of the brown planthopper, *Nilaparvata lugens* (Stal), from Australia and Asia: A test of the recognition species concept. Biological Journal of the Linnean Society 35:79–93.

Claridge, M. F., and J. C. Morgan. 1993. Geographical variation in acoustic signals of the planthopper, *Nilaparvata bakeri* (Muir), in Asia: species recognition and sexual selection. Biological Journal of the Linnean Society 48:267–281.

Cobb, M., B. Burnet, R. Blizard, and J.-M. Jallon. 1990. Altered mating behavior in a Carsonian population of *Drosophila sechellia*. Evolution 44:2057–2068.

Coyne, J. A. 1992. Genetics and speciation. Nature 355:511–515.

Coyne, J. A., and H. A. Orr. 1989. Patterns of speciation in *Drosophila*. Evolution 43: 362–381.

Coyne, J. A., and H. A. Orr. 1997. "Patterns of speciation in *Drosophila*" revisited. Evolution 51:295–303.

Darwin, C. 1871. The descent of man and selection in relation to sex. John Murray, London.

Del Solar, E. 1966. Sexual isolation caused by selection for positive and negative phototaxis and geotaxis in *Drosophila pseudoobscura*. Proceedings of the National Academy of Science USA 56:484–487.

De Oliveira, A. K., and A. R. Cordeiro. 1980. Adaptation of *Drosophila willistoni* experimental populations to extreme pH medium. II. Development of incipient reproductive isolation. Heredity 44:123–130.

De Winter, A. J. 1992. The genetic basis and evolution of acoustic mate recognition signals in a *Ribautodelphax* planthopper (Homoptera, Delphacidae). 1. The female call. Journal of Evolutionary Biology 5:249–265.

Dixson, A. F. 1987. Observations on the evolution of the genitalia and copulatory behaviour in male primates. Journal of Zoology 213:423–443.

Dobzhansky, T. 1937. Genetics and the origin of species. Columbia University Press, New York.

Dobzhansky, T. 1944. Experiments on sexual isolation in *Drosophila*. III. Geographic strains of *D. sturtevanti*. Proceedings of the National Academy of Science USA 30: 335–339.

Dobzhansky, T., and G. Streisinger. 1944. Experiments on sexual isolation in *Drosophila*. II. Geographic strains of *Drosophila prosaltans*. Proceedings of the National Academy of Science USA 30:340–345.

Dodd, D. M. B. 1989. Reproductive isolation as a consequence of adaptive divergence in *Drosophila pseudoobscura*. Evolution 43:1308–1311.

Duijm, M. 1990. On some song characteristics in *Ephippiger* (Orthoptera: Tettigoniidae) and their geographic variation. Netherlands Journal of Zoology 40:428–453.

Eberhard, W. G. 1985. Sexual selection and animal genitalia. Harvard University Press, Cambridge, MA.

Ehrman, L. 1965. Direct observation of sexual isolation between allopatric and between sympatric strains of the different *Drosophila paulistorum* races. Evolution 19:459–464.

Endler, J. A. 1977. Geographic variation, speciation and clines. Princeton University Press, Princeton, NJ.

Endler, J. A. 1978. A predator's view of animal color patterns. Evolutionary Biology 11: 319–364.

Endler, J. A. 1980. Natural selection on color patterns in *Poecilia reticulata*. Evolution 34:76–91.

Endler, J. A. 1983. Natural and sexual selection on color patterns in poeciliid fishes. Environmental Biology of Fishes 9:173–190.

Endler, J. A. 1986. Natural selection in the wild. Princeton University Press, Princeton, NJ.

Endler, J. A. 1989. Conceptual and other problems in speciation. *In* D. Otte and J. A. Endler, eds. Speciation and its consequences, pp. 625–648. Sinauer Associates, Sunderland, MA.

Endler, J. A. 1992. Signals, signal conditions, and the direction of evolution. American Naturalist 139(supplement):S125–S153.

Endler, J. A. 1993. Some general comments on the evolution and design of animal communication systems. Philosophical Transactions of the Royal Society of London B 340: 215–225.

Endler, J. A., and A. E. Houde. 1995. Geographic variation in female preferences for male traits in *Poecilia reticulata*. Evolution 49:456–468.

Ereshefsky, M. (ed). 1992. The units of evolution: Essays on the nature of species. MIT Press, Cambridge, MA.

Etges, W. J. 1992. Premating isolation is determined by larval substrates in cactophilic *Drosophila mojavensis*. Evolution 46:1945–1950.

Foster, S. P., and W. L. Roelofs. 1987. Sex pheromone differences in populations of the brownheaded leafroller, *Ctenopseustis obliquana*. Journal of Chemical Ecology 13: 623–629.

Frerot, B., and S. P. Foster. 1991. Sex pheromone evidence for two distinct taxa within *Graphania mutans* (Walker). Journal of Chemical Ecology 17:2077–2093.

Galiana, A., A. Moya, and F. J. Ayala. 1993. Founder-flush speciation in *Drosophila pseudoobscura*: A large-scale experiment. Evolution 47:432–444.

Gerhardt, H. C. 1983. Communication and the environment. *In* T. R. Halliday and P. J. B. Slater, eds. Animal behaviour, vol. 2. Communication, pp. 82–113. Blackwell Scientific, Oxford.

Giddings, L. V., K. Y. Kaneshiro, and W. W. Anderson (eds). 1989. Genetics, speciation and the founder principle. Oxford University Press, New York.

Gilburn, A. S., and T. H. Day. 1994. Evolution of female choice in seaweed flies: Fisherian and good genes mechanisms operate in different populations. Proceedings of the Royal Society of London B 255:159–165.

Gould, S. J., and R. F. Johnston. 1972. Geographic variation. Annual Review of Ecology and Systematics 3:457–498.

Greene, H. W. 1994. Homology and behavioral repertoires. *In* B. K. Hall, eds. Homology: The hierarchical basis of comparative biology, pp. 369–391. Academic Press, New York.

Harcourt, A. H., and J. Gardiner. 1994. Sexual selection and genital anatomy of male primates. Proceedings of the Royal Society of London B 255:47–53.

Harrison, R. G. (ed). 1993. Hybrid zones and the evolutionary process. Oxford University Press, New York.

Harvey, P. H., and M. D. Pagel. 1991. The comparative method in evolutionary biology. Oxford University Press, New York.

Hedrick, A. V. 1988. Female choice and the heritability of attractive male traits: An empirical study. American Naturalist 132:267–276.

Herring, K., and P. Verrell. 1996. Sexual incompatibility and geographical variation in mate recognition systems: Tests in the salamander *Desmognathus ochrophaeus*. Animal Behaviour 52:279–287.

Hollocher, H., C.-T. Ting, F. Pollack, and C.-I. Wu. 1997. Incipient speciation by sexual isolation in *Drosophila melanogaster*: variation in mating preference and correlation between the sexes. Evolution 51:1175–1181.

Houde, A. E. 1988. Genetic difference in female choice between two guppy populations. Animal Behaviour 36:510–516.

Houde, A. E. 1992. Sex-linked heritability of a sexually selected character in a natural population of *Poecilia reticulata* (Pisces: Poeciliidae) (guppies). Heredity 69:229–235.

Houde, A. E. 1997. Sex, color, and mate choice in guppies. Princeton University Press, Princeton, NJ.

Houde, A. E., and J. A. Endler. 1990. Correlated evolution of female mating preference and male color patterns in the guppy *Poecilia reticulata*. Science 248:1405–1408.

Howard, D. J. 1993. Reinforcement: origin, dynamics and fate of an evolutionary hypothesis. *In* R. G. Harrison, ed. Hybrid zones and the evolutionary process, pp. 46–69. Oxford University Press, New York.

Kaneshiro, K. Y., and J. S. Kurihara. 1981. Sequential differentiation of sexual behavior in populations of *Drosophila silvestris*. Pacific Science 35:177–183.

Kawamura, T., and S. Sawada. 1959. On the sexual isolation among different species and local races of Japanese newts. Journal of Science of Hiroshima University Series B 18:17–31.

Koref-Santibanez, S. 1972a. Courtship behavior in the semispecies of the superspecies *Drosophila paulistorum*. Evolution 26:108–115.

Koref-Santibanez, S. 1972b. Courtship interaction in the semispecies of *Drosophila paulistorum*. Evolution 26:326–333.

Lande, R. 1980. The genetic covariance between characters maintained by pleiotropic mutations. Genetics 94:203–215.

Lande, R. 1981. Modes of speciation by sexual selection on polygenic characters. Proceedings of the National Academy of Science USA 78:3721–3725.

Lewontin, R. C. 1974. The genetic basis of evolutionary change. Columbia University Press, New York.

Littlejohn, M. J. 1977. Long range acoustic communication in anurans: an integrated and evolutionary approach. *In* D. H. Taylor and S. I Guttman, eds. The reproductive biology of amphibians, pp. 263–294. Plenum Press, New York.

Littlejohn, M. J., G. F. Watson, and J. R. Wright. 1993. Structure of advertisement calls of *Litoria ewingi* (Anura: Hylidae) introduced into New Zealand from Tasmania. Copeia 1993:60–67.

Lofdahl, K. L., D. Hu, L. Ehrman, J. Hirsch, and L. Skoog. 1992. Incipient reproductive isolation and evolution in laboratory *Drosophila melanogaster* selected for geotaxis. Animal Behaviour 44:783–786.

Lofstedt, C. 1990. Population variation and genetic control of pheromone communication systems in moths. Entomologia Experimentata et Applicata 54:199–218.

Lorenz, K. 1970. Studies in animal and human behavior. Methuen, London.

Lott, D. F. 1984. Intraspecific variation in the social systems of wild vertebrates. Behaviour 88:266–325.

Luyten, P. H., and N. R. Liley. 1985. Geographic variation in the sexual behavior of the guppy, *Poecilia reticulata* (Peters). Behaviour 95:164–179.

Magurran, A. E., and B. H. Seghers. 1990. Risk sensitive courtship in the guppy (*Poecilia reticulata*). Behaviour 112:194–201.

Maksymovitch, E., and P. Verrell. 1992. Courtship behavior of the Santeetlah dusky salamander, *Desmognathus santeetlah* Tilley (Amphibia: Caudata: Plethodontidae). Ethology 90:236–246.

Maksymovitch, E., and P. Verrell. 1993. Divergence of mate recognition systems among conspecific populations of the plethodontid salamander *Desmognathus santeetlah*. Biological Journal of the Linnean Society 49:19–29.

Markow, T. A. 1991. Sexual isolation among populations of *Drosophila mojavensis*. Evolution 45:1525–1529.

Markow, T. A., and E. C. Toolson. 1990. Temperature effects on epicuticular hydrocarbons and sexual isolation in *Drosophila mojavensis*. *In* J. S. F. Barker, W. T. Starmer, and R. J. MacIntyre, eds. Ecological and evolutionary genetics of *Drosophila*, pp. 315–331. Plenum Press, New York.

Martins, E. P. (ed). 1996. Phylogenies and the comparative method in animal behavior. Oxford University Press, New York.

Mayr, E. 1942. Systematics and the origin of species. Columbia University Press, New York.

Mayr, E. 1976. Evolution and the diversity of life. Harvard University Press, Cambridge, MA.

McLennan, D. A. 1996. Integrating phylogenetic and experimental analyses: The evolution of male and female nuptial coloration in the stickleback fishes (Gasterosteidae). Systematic Biology 45:261–277.

Meffert, L. M., and E. H. Bryant. 1991. Mating propensity and courtship behavior in serially bottlenecked lines of the housefly. Evolution 45:293–306.

Moller, A. P. 1993. Developmental stability, sexual selection and speciation. Journal of Evolutionary Biology 6:493–509.

Morton, E. S. 1975. Ecological sources of selection on avian sounds. American Naturalist 109:17–34.

Nei, M., T. Maruyama, and C.-I. Wu. 1983. Models of evolution of reproductive isolation. Genetics 103:557–579.

Nevo, E., and R. R. Capranica. 1985. Evolutionary origin of ethological reproductive isolation in cricket frogs, *Acris*. Evolutionary Biology 19:147–215.

Oden, N. L., and R. R. Sokal. 1992. An investigation of three-matrix permutation tests. Journal of Classification 9:275–290.

Otte, D. 1974. Effects and functions in the evolution of signaling systems. Annual Review of Ecology and Systematics 5:385–417.

Otte, D., and J. A. Endler (eds). 1989. Speciation and its consequences. Sinauer Associates, Sunderland, MA.

Partridge, L., and J. A. Endler. 1987. Life history constraints on sexual selection. *In* J. W. Bradbury and M. B. Andersson, eds. Sexual selection: Testing the alternatives, pp. 265–277. Wiley, New York.

Paterson, H. E. H. 1993. Evolution and the recognition concept of species. Johns Hopkins University Press, Baltimore, MD.

Pinto, J. D., R. Stouthamer, G. R. Platner, and E. R. Oatman. 1991. Variation in reproductive compatibility in *Trichogramma* and its taxonomic significance (Hymenoptera: Trichogrammatidae). Annals of the Entomological Society of America 84:37–46.

Powell, J. R. 1978. The founder-flush speciation theory: An experimental approach. Evolution 32:465–474.

Ritchie, M. G. 1991. Female preference for "song races" of *Ephippiger ephippiger* (Orthoptera: Tettigoniidae). Animal Behaviour 42:518–520.

Ryan, M. J., R. B. Cocroft, and W. Wilczynski. 1990. The role of environmental selection in intraspecific divergence of mate recognition signals in the cricket frog, *Acris crepitans*. Evolution 44:1869–1872.

Ryan, M. J., and A. Keddy-Hector. 1992. Directional patterns of female mate choice and the role of sensory biases. American Naturalist 139 (supplement):S4–S35.

Ryan, M. J., and A. S. Rand. 1993. Species recognition and sexual selection as a unitary problem in animal communication. Evolution 47:647–657.

Ryan, M. J., and W. Wilczynski. 1991. Evolution of intraspecific variation in the advertisement call of the cricket frog (*Acris crepitans*, Hylidae). Biological Journal of the Linnean Society 44:249–271.

Sakaluk, S. K. 1990. Sexual selection and predation: Balancing reproductive and survival needs. *In* D. L. Evans and J. O. Schmidt, eds. Insect defenses, pp. 63–90. State University of New York Press, Albany.

Sawada, S. 1963. Studies on the local races of the Japanese newt, *Triturus pyrrhogaster* Boie. II. Sexual isolation mechanisms. Journal of Science of Hiroshima University Series B 21:167–180.

Schluter, D., and T. Price. 1993. Honesty, perception and population divergence in sexually selected traits. Proceedings of the Royal Society of London B 253:117–122.

Shapiro, A. 1978. The assumption of adaptivity in genital morphology. Journal of Research on Lepidoptera 17:68–72.

Soans, A. B., D. Pimentel, and J. S. Soans. 1974. Evolution of reproductive isolation in allopatric and sympatric populations. American Naturalist 108:117–124.

Stratton, G. E., and G. W. Uetz. 1986. The inheritance of courtship behavior and its role as a reproductive isolating mechanism in two species of *Schizocosa* wolf spiders (Araneae: Lycosidae). Evolution 40:129–141.

Tilley, S. G., P. A. Verrell, and S. J. Arnold. 1990. Correspondence between sexual isolation and allozyme differentiation: A test in the salamander *Desmognathus ochrophaeus*. Proceedings of the National Academy of Science USA 87:2715–2719.

Tinbergen, N. 1953. Social behaviour in animals. Methuen, London, UK.

Toth, M., C. Lofstedt, B. W. Blair, T. Cabello, A. I. Farag, B. S. Hansson, B. G. Kovalev, S. Maini, E. A. Nesterov, I. Pajor, P. Sazanov, I. V. Shamshev, M. Subchev, and G. Szocs. 1992. Attraction of male turnip moths *Agrotis segetum* (Lepidoptera: Noctuidae) to sex pheromone components and mixtures at 11 sites in Europe, Asia and Africa. Journal of Chemical Ecology 18:1337–1347.

Uzendoski, K., and P. Verrell. 1993. Sexual incompatibility and mate-recognition systems: A study of two species of sympatric salamanders (Plethodontidae). Animal behaviour 46:267–278.

Verrell, P. A. 1989. An experimental study of the behavioral basis of sexual isolation between two sympatric plethodontid salamanders, *Desmognathus imitator* and *D. ochrophaeus*. Ethology 80:274–282.

Verrell, P. A. 1990a. Frequency of interspecific mating in salamanders of the plethodontid genus *Desmognathus*: Different experimental designs may yield different results. Journal of Zoology 221:441–451.

Verrell, P. A. 1990b. Sexual compatibility among plethodontid salamanders: Tests between *Desmognathus apalachicolae*, and *D. ochrophaeus* and *D. fuscus*. Herpetologica 46:415–422.

Verrell, P. A. 1990c. Tests for sexual isolation among sympatric salamanders of the genus *Desmognathus*. Amphibia-Reptilia 11:147–153.

Verrell, P. A. 1991. Illegitimate exploitation of sexual signalling systems and the origin of species. Ethology, Ecology and Evolution 3:272–283.

Verrell, P. A. 1992. Primate penile morphologies and social systems: Further evidence for an association. Folia Primatologica 59:114–120.

Verrell, P. A., and S. J. Arnold. 1989. Behavioral observations of sexual isolation among allopatric populations of the mountain dusky salamander, *Desmognathus ochrophaeus*. Evolution 43:745–755.

Verrell, P. A., and S. G. Tilley. 1992. Population differentiation in plethodontid salamanders: Divergence of allozymes and sexual compatibility among populations of *Desmognathus imitator* and *Desmognathus ochrophaeus* Caudata: Plethodontidae). Zoological Journal of the Linnean Society 104:67–80.

Vlijm, L. 1986. Ethospecies: behavioral patterns as an interspecific barrier. Actas X Congresso Internationale Aracnologia Jaca/Espana 2:41–45.

Weinberg, J. R., V. R. Starczak, and D. Jorg. 1992. Evidence for rapid speciation following a founder event in the laboratory. Evolution 46:1214–1220.

West-Eberhard, M. J. 1982. Sexual selection, social competition, and speciation. Quarterly Review of Biology 58:155–183.

Wu, C.-I. 1985. A stochastic simulation study on speciation by sexual selection. Evolution 39:66–82.

Zink, R. M., and J. V. Remsen, Jr. 1986. Evolutionary processes and patterns of geographic variation in birds. Current Ornithology 4:1–69.

Zouros, E., and C. J. D'Entremont. 1980. Sexual isolation among populations of *Drosophila mojavensis*: Response to pressure from a related species. Evolution 34:421–430.

13

Thoughts on Geographic Variation in Behavior

SUSAN A. FOSTER
JOHN A. ENDLER

In the past, behavior was assumed to be largely invariant within species, particularly those elements of behavior used as criteria of mate choice or in species recognition (see Magurran this volume, Verrell this volume). As is obvious from this volume, geographic variation could well be the common condition rather than the exception, and this applies to the full spectrum of behavioral phenotypes. Not only must students of behavior avoid typological thinking (Mayr 1963), but those wishing to infer similarity of behavior among populations must demonstrate the similarity just as surely as those interested in exploring population differentiation must demonstrate the differences.

Behavior is as much a phenotype as is morphology; it is the expression of the combined effects of genotype and environment. Like other traits, behavior varies geographically because it is subject to geographically varying conditions and, hence, to natural selection, gene flow, and genetic drift. The chapters in this book provide examples of this variation, of the underlying genetic bases for the differences, and in many cases, the causes of the geographic variation.

The study of geographic variation in behavior is in very early stages and lags well behind research on geographic variation in other kinds of traits (Endler 1977, 1986, 1995). Consequently, we cannot answer with assurance many of the questions we would like to be able to answer. However, we can take a first step using the insights offered by the research presented in this book. Before doing so, we briefly address some of the methodological issues that emerged over the course of the research because many are specific to the study of behavior or of geographic variation. We hope this will help others avoid problems encountered in these early studies.

Methodological Issues

Many of the methodological issues discussed in the chapters in this book are related to the difficulty of working with behavioral characteristics that are extremely labile and responsive to environmental conditions. The remainder are issues related to the interpretation of data collected to assess patterns and causes of geographic variation. We will examine them in turn.

The Measurement and Mismeasurement of Behavior

Evolutionists who have not tried to characterize behavior quantitatively often hold the view that the study of behavioral evolution is no different from research on the evolution of other aspects of phenotype. Those who have attempted to quantify behavior patterns often hold quite a different view. These views range from the idea that the intrinsic lability of behavioral characters simply can make them difficult to study in an evolutionary context (Barlow 1981) to the argument that there is little evidence that natural selection is a significant mechanism by which behavioral phenotypes are determined (Hailman 1982). The problem lies, to a large extent, in lability of behavioral expression, in the time frame over which it varies, and in the sensitivity of many aspects of behavior to many subtle environmental cues. Basically, behavioral characteristics of an individual can be difficult to characterize reliably without accounting for lability.

Sometimes differences among natural populations are large enough that they can be readily quantified and demonstrated. These are the differences that attract attention, and that beg for explanation. Even under such circumstances, difficulty often arises when an investigator wants to determine whether the differences have a genetic basis or how genotype and environment interact to produce the population differences (see Burghardt and Schwartz this volume for additional detail). Once environmental influences are removed, or altered, population differences may become far less pronounced and, consequently, more difficult to demonstrate and measure.

A major problem is that individuals can exhibit different responses to the same stimuli over short time scales. Therefore, assessment of the repeatability of a trait is essential. Low repeatability offers low confidence that the measurement accurately reflects an individual's phenotype. The low repeatability could be an outcome of unmeasured changes in the environment or of intervening experience. The problem is particularly acute when multiple traits are being measured and presentation order can affect behavioral expression. The importance of these issues in studies of behavioral genetics or of geographic variation in behavior cannot be stressed strongly enough (Boake 1994, Burghardt and Schwartz this volume).

Behavioral phenotypes also change during the life of an animal. The changes are caused by a range of factors, including maternal effects, changes in environment, and changes in the animal's physiology as it matures or seasons change. The simplest means of coping with these problems is to rear animals under uniform conditions and then to score behavioral phenotypes at a single appropriate developmental stage, unless the goal is specifically to understand differences in development and its causes across populations. If mate choice is to be scored, possible differences in outcome between sequential and simultaneous mate presentations need be considered (Verrell 1990 this volume).

Even with this relatively simple approach to the study of innate behavioral differences among populations, there are pitfalls. Because normal behavioral development often requires social interaction or experiences afforded by the animal's environment, behavioral abnormalities can arise from rearing in isolation or from rearing in depauperate laboratory environments (e.g., Barlow 1981, Boinski this volume, Burghardt and Schwartz this volume). As a consequence, assessment of population differences in behavior can prove frustrating, and the results can be unexpectedly misleading.

The extreme responsiveness of many behavioral phenotypes to environmental conditions means that, even under the best of conditions, sample sizes will have to be large if subtle population differences are to be detected. This is especially true if heritabilities or genetic covariances among traits are to be estimated. Often the requisite sample sizes will be prohibitive, especially with large or rare animals, or with those that are difficult or expensive to rear. Under these circumstances, relatively simple common garden experiments may provide adequate evidence of genetic differences among populations (see Carroll and Corneli this volume).

When, on the other hand, the question is one of differences in learning capabilities or of the pattern of phenotypically plastic responses to environmental differences, the appropriate paradigm is the behavioral norm of reaction (Carroll and Corneli this volume, Thompson this volume). When measuring norms of reaction, all of the concerns outlined above hold, but behavioral phenotypes are measured on the same individuals following different experiences or under differing (often reciprocal) sets of environmental conditions. From these, population mean norms of reaction are used to examine evolved differences in plastic responses to the environment. Norms of reaction have rarely been used in the study of animal behavior but hold great promise, particularly when population comparisons are used to evaluate evolutionary responses to different spatial and temporal patterns of environmental variation or when simple measures of learning capabilities are to be explored.

Interpreting Geographic Variation in an Evolutionary Context

As pointed out so well by Carroll and Corneli (this volume) and by Thompson (this volume), there can be three general causes of geographic variation in behavior. Behavioral differences across populations can reflect (1) differences in the way the same or similar mean genotypes are expressed in different environments, (2) genetic differences between the populations that are uninfluenced or little influenced by environmental variation, or (3) differences in the patterns of responses of the mean genotypes in different populations to environmental variation. In the first instance, the differences are due entirely to phenotypic plasticity, and arguments invoking Darwinian evolution are inappropriate. In the latter two cases, genetic differences exist between the populations, and efforts to understand the evolution of the differences are appropriate.

When genetic differences exist between the populations, they typically can be revealed by "common garden" rearing, in which individuals from both populations are reared under a single set of conditions. This will not reveal the extent to which genetic expression is dependent on environmental context, and critical insights into the potential adaptive nature of this interaction may be missed. Only when behavioral norms of reaction are examined can the evolution of strategic, or conditional, behavioral responses be explored. Especially when traits are responsive to environmental conditions, as is so often the case with behavioral traits, heritability or common garden studies performed under a single set of environmental conditions may miss the actual target of selection—the norm of reaction (e.g., Carroll and Corneli this volume, Thompson this volume).

Once a genetic basis for population differences in the behavior of interest has been established (an assumption we will make for the remainder of this discussion), other evolutionary questions can be addressed. The most common use of population comparison is to infer adaptation. As stressed in the introduction to this book, such comparisons are strongest when the character states are distributed in complex geographic patterns or when the populations are not connected by gene flow and have been derived independently from a single ancestral population (see also Endler 1977, 1983, 1986; Bell and Foster 1994). Although population comparisons can provide powerful insights into the causes of adaptive differentiation, direct methods must be used to determine whether the inferred selective regime does have the predicted outcome (Endler 1986; also see the Introduction to this volume).

Adaptive differentiation cannot be assumed when comparisons are made among populations. Gene flow can prevent populations from reaching adaptive optima, as can insufficient genetic variation in the founding population (see, e.g., Riechert this volume, Thompson this volume). Even if the requisite genetic variation exists within founding populations, insufficient time may have passed for the populations to have adapted fully to the local selective regime, or selection on correlated characters may have prevented or slowed adaptive differentiation (Thompson this volume). Finally, establishment of multiple local populations from different source populations can result in population differences in phenotype if the source populations comprise different arrays of genotypes. All of these possibilities must be addressed when apparently nonadaptive character states are observed and the research goal is to explain the pattern of geographic variation in behavior. On the other hand, if the correlation between environment and character state is strong, but not perfect, one may be able to proceed to direct tests of adaptive hypotheses on the strength of the correlation.

Knowledge of population histories, including past and present connections among them, can permit discrimination of these alternatives in some cases. For example, if populations with a nonadaptive phenotype are descended from a common ancestor that possessed that phenotype, a plausible explanation is phylogenetic constraint—particularly if there is no evidence of current or recent gene flow. Insights derived from knowledge of population phylogenies can also offer insight into the processes by which apparently adaptive differences arise. An excellent example is found in the differentiation of diversionary display behavior across populations of threespine stickleback (Foster 1988, 1994).

In lacustrine populations of sticklebacks in which groups cannibalize young at the nests of defending parental males, the approach of a group elicits a complex and conspicous display from the male, the function of which appears to be to divert the groups from approaching the nest. In some populations, the males simply swim out of their territories rapidly, rooting in the substratum vigorously as if feeding on an especially attractive resource. In other populations, the display is more complex, also incorporating motor patterns that appear to be derived from fright responses and courtship repertoires (Whoriskey and FitzGerald 1985, Foster 1988, 1994; Ridgway and McPhail 1988). In noncannibalistic populations, these displays are not elicited by groups. Instead, the males court females within them, attack them directly, or ignore them (Foster 1988, 1994).

The absence of the displays from these populations could be a consequence of loss of a complex ancestral repertoire or could itself be the ancestral state which has become elaborated in derived, cannibalistic populations. Because marine populations that are thought to be ancestral in the freshwater radiation of threespine stickleback are typically cannibalistic and possess the complex diversionary display repertoire, the former hypothesis is the best supported explanation (Foster 1988, 1994, 1995). Without knowledge of the phylogenetic relationships among populations in this radiation, these alternatives could not have been discriminated.

Knowledge of population phylogenies also has the potential to afford substantial insights into the speciation process. Perhaps the best example involves a comparative study of populations in a diploid–tetraploid cryptic species pair of gray tree frogs *Hyla chrysoscelis* and *H. versicolor* (Ptacek et al. 1994). Wasserman (1970) originally suggested that the tetraploid, *H. versicolor* ($2n = 48$) had arisen from *H. chrysoscelis* ($2n = 24$) by autopolyploidy. Both electorphoretic data (Ralin et al. 1983) and allele frequency data (Romano et al. 1987) have subsequently been interpreted as supporting a single origin for the tetrapolid species from the diploid.

This inference was also supported from patterns of advertisement call divergence. The tetraploid *H. versicolor* was first distinguished from its diploid progenitor on the basis of advertisement call pulse rate (Johnson 1966, Wasserman 1970). All tetraploid pulse rates are 50–60% lower than those of the diploids that occur in sympatry or parapatry, and the temperature-corrected pulse rates of tetraploid populations vary by no more than 10% (Gerhardt 1994, personal communication). In contrast to this relative uniformity within the tetraploid, call differences among diploid populations (fast pulse rate in the west and slow pulse rate along the east coast) provided the first evidence of at least two lineages of *H. chrysoscelis* (Gerhardt 1974), an inference confirmed on the basis of allozyme divergence and chromosome polymorphisms (Wiley 1983, Wiley et al. 1989).

A recent phylogenetic analysis of the group using the cytochrome *b* gene of mitochondrial DNA suggests instead that the tetraploid *H. versicolor* has arisen from the diploid *H. chrysoscelis* at least three, and possibly four times (Ptacek et al. 1994). From a behavioral perspective, this result is intriguing because it suggests remarkable similarity in the advertisement call of the independently derived tetraploid lineages and raises the question of how the call came to be so similar in those lineages. Because the diploids and tetraploids are syntopic through much of their ranges, several explanations for the patterns of call variation are possible. These include initial shifts to lower pulse rates (Ueda 1993) and shifts in female preference (e.g., Bogart and Wasserman 1972) associated with polyploidy and, possibly, character displacement in zones of sympatry (Gerhardt 1994). The advantages offered by the phylogenetic analysis to understanding the evolution of call variation within this group are that populations can be selected in such a way that these hypotheses can be discriminated without confounding effects due to historical differentiation (Gerhardt 1994). We also have good evidence that similar derived call characteristics have evolved repeatedly as a consequence of polyploidy. Without cladistic analyisis of molecular sequence data, the pattern of speciation in this group would remain obscure, and there would be little hope of resolving the causes of subtle geographic variation in the calls of these frogs.

The gray treefrog example is one of the few cases in which population phylogenies have been resolved using traditional cladistic methods. The problem is that populations often share common ancestry too recently to have permitted accumulation of many derived characters distinguishing them, and they often are, or recently have been, interconnected by gene flow. Both of these conditions limit the power of traditional cladistic methods. Recently, population geneticists have realized that the alleles at a polymorphic locus can also be related to one another in a branching relationship, commonly called a genealogy, which can be nested within an organismal phylogeny. Crandall and Templeton (1996) have developed methods based on the coalescent that permit population genetic data to be used in the inference of genealogies and potentially in the inference of population phylogenies. These methods are not well developed yet, and until we have a better understanding of their power, those interested in applying the phylogenetic comparative method to population level studies will need to rely on traditional biogeographic and cladistic inference where possible (see Foster and Cameron 1996 for a discussion).

Geographic Variation

As Riechert points out so aptly (1986, this volume), much of the geographic variation that captures the attention of biologists is ecotypic variation, or adaptive variation in phenotype that reflects genetic differentiation in response to local selective regimes (sensu Turesson 1922a,b). Given the pervasive influence of adaptation in studies of geographic variation in behavior, we begin here and then proceed to consider a series of issues that we believe either can be addressed to some extent by available studies of geographic variation in behavior or that should be addressed in future research.

Adaptive Geographic Variation

Nearly every chapter in this book provides evidence of ecotypic variation in behavior. In some instances the role of selection has been compellingly demonstrated. In other instances the evidence is primarily correlational, sometimes involving few populations. Overall, however, there seems to be a strong role of selection in the evolution of geographic variation in a wide array of behavior patterns across a diversity of taxa.

In many cases the relationship between environmental selection and behavior is straightforward. For example, differences in predation regime in different areas can favor the evolution of appropriate, divergent, behavioral defenses (reviewed by Magurran this volume). However, a striking feature of much of the adaptive differentiation described in this book, and in papers cited in the chapters, is that the divergence is not the product of a simple relationship between selective regime and behavior. Instead, a number of population differences in behavior appear to be the products of cascading effects of relationships between environmental variation and elements of phenotype that might not be associated intuitively with selection imposed by the observed differences in environmental conditions. The following are some examples of ways in which geographically varying selection can cause cascades of effects on behavior and other traits.

Boinski (this volume) showed how a difference in the distribution of food used by squirrel monkeys at two distant study sites has apparently produced differences in social structure, which have in turn affected stress physiology in a way that appears to feed back to maintain and induce further variation in social structure. Although Boinski's inferences are based on field studies at only two locations, and the stress reactivity studies involved animals from other sites, her chapter points out the value of examining divergence in the physiological systems that link genetic differentiation to behavior. There are few such studies (see also Bakker 1986, Garland and Adolph 1991). Following from Boinski's chapter, on a more general level, geographical differences in food availability can result in differences in social structure which lead to ancillary effects, including physiological effects, which can feedback to maintain and induce variation in social behavior. Geographical variation in patterns of food availability can also lead to variation in social structure. For example, more clumped food means that the food is more defendable, leading to more aggression/coalition, leading to more stress, more stress response, and consequent physiological and evolutionary effects.

Another example involving feeding differentiation is intriguing because it has been repeated so often, independently, across populations of threespine stickleback. In this instance, differences in the ecology of lakes have favored the evolution of populations of these small fish that are morphologically and behaviorally specialized for feeding on either plankton (deep, oligotrophic lakes) or benthic invertebrates (shallow, more eutrophic lakes) (see Schluter and McPhail 1993, Bell and Foster 1994, McPhail 1994 for reviews). Differences in foraging location and behavior across populations have resulted in loss of cannibalistic behavior in which large groups of foragers attack nests defended by males and consume the young within. The loss of this behavior appears to have resulted in turn in loss of an ancestral diversionary display and an enhancement of courtship conspicuousness relative to ancestral, marine populations (Foster 1994, 1995).

Geographical differences in predation regimes can also affect multiple traits. Antipredator behavior can take time from other essential tasks such as foraging or courtship or can modify the options available to individuals (Lima and Dill 1990, Magnhagen 1991, Godin 1995, Magurran this volume). Outcomes include population variation in female mating preferences (Endler and Houde 1996), male mating behavior (Magurran and Seghers 1994b), foraging patterns (Fraser and Gilliam 1987, Magurran and Seghers 1994a), levels of aggression at food patches (Magurran this volume), and functional relationships among traits (Ehlinger this volume).

Geographical differences in sex ratio can also result in behavioral variation because they cause differences in costs of mate searching, which result in differences in mate guarding, etc., changing many aspects of social behavior. Where sex ratio displays relatively great temporal variation, or where density is low, males may also exhibit tactical plasticity, consistent with more variable sex ratios. Natural selection then favors different norms of reaction, rather than affecting the trait itself, or it affects the trait less strongly than it does the norm of reaction (e.g., Carroll and Corneli this volume).

Different selective factors do not necessarily change independently with location, and even when they do, different combinations of selective regimes can again produce complex cascades of effects that may be difficult to disentangle (see Ehlinger this volume, Magurran this volume, Riechert this volume). For example, geographical variation in predation intensity and food availability may be strongly correlated. Regardless of the sign of the correlation, such correlations among selective factors will yield a cascade of effects on behavior (Endler 1995, Reichert this volume, Magurran this volume). An advantage of research involving geographic variation is that careful examination of a number of populations may permit discrimination of the relative effects of different selection pressures on phenotype evolution (Endler 1995).

Correlations among traits can also cause a cascade of geographical variation in a variety of traits. For example, there is geographical variation in sexual selection traits (Endler and Houde 1995, Verrell this volume, Ryan and Wilczynski this volume), and some of this variation may be favored by geographical variation in the signaling environment (Ryan et al. 1990, Ryan and Wilczynski 1991, Endler 1992, Endler and Houde 1995, Ryan and Wilczynsky this volume). Signals and responses both vary geographically, but not all components covary (Endler and Houde 1995, Verrell this volume). The concordance of some traits and non-concordance of others may generate additional geographical variation. Sexual incompatibility may increase with geographical distance, but not necessarily with genetic distance. Behavioral and morphological/genetic traits do not necessarily covary either (Endler 1995, Endler and Houde 1995, Verrell this volume).

Adding to this complexity, behavioral phenotypes can respond differently to the same selective regimes in different populations, and different combinations of traits may offer alternative evolutionary "solutions" to similar selective "problems" (Foster and Bell 1994, Magurran this volume). Equally, phenotypes may often reflect compromises between divergent selective regimes, a situation elegantly revealed by Ehlinger's demonstration of the differential effects of selection on the morphology and behavior of male and female bluegills (Ehlinger this volume). Thus, adaptive variation among populations may exist but may require comparisons among a large number of populations to be revealed.

Conditional and strategic behavior patterns can also evolve in an adaptive fashion, and may typically do so (Carroll and Corneli this volume, Thompson this volume). As pointed out by Carroll and Corneli, population differences in behavioral norms of reaction have typically been documented where appropriate methods of detection have been employed, and a failure to test for norms of reaction can lead to inappropriate interpretations of the nature of behavioral variation across populations. Because these studies demonstrate so clearly that conditional behavioral patterns do evolve across populations, comparisons of carefully selected populations should permit testing of models (e.g., evolutionarily stable strategy or optimality models) that predict the ways in which strategic conditionality should evolve in response to divergent selective regimes. Riechert's work (this volume) on the evolution of aggressive behavior in the spider, *Agelenopsis aperta*, provides an excellent example of the value of such an approach.

Norms of reaction offer unique tools for examining the relationships between within-population variation and that which evolves across populations. Although general similarities in the patterns of intra- and interpopulation variation in behavior can be taken to indicate that both levels of variation respond similarly to selection (e.g., Reynolds et al. 1993), we have few data with which to evaluate the generality of such assertions. This is an area in which research on the evolution of behavioral norms of reaction across populations could lead to substantial insights into the causes and patterns of behavioral evolution.

Plasticity itself may be geographically variable. For example, guppies and sticklebacks are more behaviorally plastic in high-predation populations (Huntingford et al. 1994, Endler 1995). Plasticity may be more common near boundaries between predation regimes, which are geologically unstable and hence subject to more frequent changes in predation, than sites well away from boundaries. Also, gene flow downstream might make high-predation populations more plastic. As Ehlinger says in chapter 6, "We cannot assume that the relationship between phenotype and fitness is the same among populations" (p. 133). Both the genetic and environmental components of phenotype can vary.

Given all the sources of complexity in nature, it is surprising that so much of the population variation in behavior can be interpreted in a functional or adaptive context when so few populations have been studied in each case. However, those phenotypes most often described in the literature tend to be those we think we best understand. Certainly, there are others like sneaking behavior and cannibalism that occur during courtship in stickleback for which there is no obvious explanation for the observed pattern of geographic variation (Foster 1995). In some cases, the difficulty may lie in a failure of populations to reach local adaptive optima because the populations do not harbor the requisite genetic variation, because sufficient time has not passed since a perturbation for local optima to be reached, or because gene flow has prevented local adaptation.

Nonadaptive Geographic Variation

Even when new populations are founded from a single ancestral population, the daughter populations are likely to possess different subsets of the genetic variation harbored in the ancestral population. The sampling error that produces these differences becomes increasingly pronounced as the size of the founding populations decreases. Thus, the daughter populations may, by chance alone, have different sets of alleles at their founding (founder effect). Chance loss of additional alleles or chance changes in gene frequencies that are particularly pronounced in small populations may further enhance nonadaptive differences among the daughter populations (genetic drift). Thus, even populations derived from a common genetic background may have different genetic compositions for selection to affect. Of course, initial differences among newly established populations are likely to be even greater when they are drawn from different populations.

If several populations are subject to similar selective regimes that are novel relative to those experienced by their ancestor population, they may change in parallel to reach similar adaptive "solutions" to selection, either because the same essential alleles were retained in each (Muller 1939), or because different alleles were favored in each to achieve similar adaptive phenotypes (Cohan 1984). In contrast, if some populations are missing the requisite quantitative genetic variation, they may be unable to reach the favored phenotype. Thus, differences among the populations exist, but some are due to chance or history.

The impact of genetic drift (including founder events) is difficult to determine, as there are no direct tests that can be applied except at founding or during subsequent periods when populations are small and the impact of drift can potentially be measured. There are some indirect methods such as the use of the coalescent (Nee et al. 1996), but they suffer from large estimation errors (Barton and Wilson 1996, Otto et al. 1996). Typically, an effect of chance is inferred when other explanations have failed. This kind of inference is problematic because of the possibility that unmeasured environmental variables (selective pressures) could explain the differences, but the possibility of nonadaptive population variation generated in this way must be kept in mind, especially in populations likely to have originated from few founders. One of the biggest problems is that, as the number of populations sampled increases, it is increasingly likely that they will be from places with different selective environments. But, if the direction of selection varies at random among these populations, the effect would resemble that of drift with random differentiation. Only long-term studies of effective population size and genetic demography (natural selection) and examination of mechanisms of selection can solve this problem.

Even when the founding populations possess sufficient heritability and size to permit population differentiation in response to novel, local selective regimes, they require time to respond to selection. If insufficient time has passed, populations may differ from one another, but not according to the predicted pattern. Thompson (this volume) suggests that a combination of genetic drift, weak selection, and insufficient response time since glaciation may provide the explanation for the weak match between climbing behavior and local environment across populations of the deer mouse, *Peromyscus maniculatus*. Similarly, these are among the explanations offered for unexpectedly low schooling tendencies of Trinidadian guppies, *Poecilia reticulata*, from the lower, high-risk reaches of the Quare and Oropuche Rivers (Magurran this volume). This possibility is supported by independent evidence of founder events in the differentiation of Trinidadian guppy populations, but the role of such events as constraints on the evolution of schooling behavior is circumstantial at best, as Magurran (this volume) notes.

Insufficient trait heritability is only one factor that can limit the potential of populations to respond to selection. Genetic correlations among traits may also constrain or direct evolution of a trait of interest. When selection pressures on correlated traits are in conflict, evolution is constrained. As long as genetic variances and covariances are maintained, they can be used to predict the direction of evolutionary change. The problem is that these values may themselves evolve, weakening their long-term predictive value (e.g., Turelli 1988, Shaw et al. 1995).

Some classes of correlations may be more resistant to evolutionary change than others. For example, Coss (this volume) suggests that phylogenetically ancient behavioral traits, like the recognition of a generalized predator sign stimulus, may be hard to lose because they are so embedded in neural organization that their loss would disrupt other essential behavioral functions. If this is true, behavioral traits appearing early in the evolutionary history of a group should be invariant across populations, whereas those that have appeared relatively recently should exhibit greater population differentiation. Behavioral traits under control of major physiological axes, like the gonadal–pituitary axis of vertebrates, might similarly prove resistant to evolutionarily change because a change in the expression of one behavioral character would be likely to disrupt a wide array of life-history, behavioral, and physiological characteristics of the animal (see Bakker 1994 for a discussion). Although we do not yet have answers to these questions, they can be addressed through population and higher order comparisons.

A final common reason that populations may often fail to reach local adaptive optima is gene flow. Both Thompson and Riechert (this volume) provide excellent examples of the effects of gene flow where barriers to gene flow do not exist. Both illustrate the principle that behavioral differentiation is most likely to be observed in population samples taken from large environmental patches and least likely to be observed in small environmental patches surrounded by expanses of an environment that favors a different behavioral phenotype. These results demonstrate clearly that genetically based behavioral variation is influenced by gene flow in the manner predicted by theory and demonstrated for other kinds of traits. Facile assumptions that all population differences in behavior are adaptive are clearly inappropriate.

Gene flow may produce mosaics of differentiated groups of populations when combined with natural selection (Endler 1977). The pattern becomes even more complex when several traits are involved, particularly in sexual selection systems.

For example, the Fisher process (Lande 1981, 1982) plus gene flow will not result in random patterns but in a patchwork of different preferences including many genetic neighborhoods. This may give rise to a hierarchy of resemblances among populations as it does in morphological traits (Endler 1977).

Although gene flow is most commonly viewed as a factor that retards adaptive evolution, it may sometimes generate variation that promotes evolutionary change. Because genetic drift and selection remove variation from populations, populations will tend to lose the potential to respond to environmental change, especially where population size is small or selection very strong. Where strong gene flow occurs, populations will fail to meet adaptive predictions but will tend to retain greater evolutionary potential, ultimately promoting the evolution of geographic variation in behavior. If gene flow is weak, it may not necessarily disrupt the general response to selection at a particular location, but it will contribute genetic variation, which may allow subsequent response to selection, especially if the environment changes.

Plasticity, Geographic Variation, or Geographic Variation in Plasticity?

When habitat patches are very large relative to the range of gene flow, populations may exhibit ecotypic differentiation. At the other extreme, if environmental patches are small relative to the range of gene flow, phenotypic plasticity may evolve (Bradshaw 1965), and there may be little evidence of ecotypic differentiation, even in behavioral norms of reaction. Finally, when environmental mosaics are interconnected by gene flow, behavioral norms of reaction may evolve that are different from those in other geographic regions where environmental mosaics differ in the combination of selective pressures imposed (Carroll and Corneli this volume, Riechert this volume, Thompson this volume). The importance of the relationship between patch size and the range of gene flow was clearly demonstrated by Sandoval (1994) and in the chapters by Riechert and Thompson. Although expected ecotypic differences were observed in populations of spiders and mice occupying large habitat patches, gene flow precluded attainment of the adaptive optimum in smaller patches bordered by large areas of different habitat. In contrast, as predicted, adaptive phenotypic plasticity was observed in the use of host plants by grasshoppers (Thompson this volume).

Adaptive phenotypic plasticity also can be favored by unpredictable temporal variation in environmental conditions within sites (Moran 1992). As expected, mate-guarding behavior was affected differently by sex ratio in two geographically disparate populations of soapberry bugs (Carroll and Corneli this volume). In the Oklahoma site, where natural variation in sex ratio was greatest, males altered guarding behavior in response to sex ratio, whereas those from the more stable Florida population did not. The Oklahoma population also possessed greater genetic variation, demonstrating that genetic variation and phenotypic plasticity are not mutually exclusive.

Rates of Behavioral Evolution

Carefully constructed research on geographic variation can provide unique insights into rates of evolutionary change. This is because in some species populations are of recent origin or are genetically isolated from all other populations. In some cases the exact age of a population is known because the date of introduction to a novel location is known (Berthold this volume; examples in Coss this volume, Magurran this volume). In other cases, maximum population age can be inferred from the age of the habitat patch in which the population resides or from the age of a physical barrier to dispersal (for examples see Coss this volume). Although such aging techniques only provide the outside age of the population when the dates are post-Pleistocene or more recent, they afford opportunities to evaluate rates of change that are more accurate than are most other methods. By examining a large number of appropriate populations in this way, we may be able to substantially improve our understanding of the potential speed of evolutionary change.

That population differences in behavior can evolve extremely rapidly in response to selection is apparent from Berthold's elegant demonstration of the rapid evolution of a novel migratory pattern in the blackcap warbler and from research on birds introduced to areas in which avian nest parasites are absent (Lotem and Rothstein 1995). In the latter cases, the introduced birds have rapidly lost the ability to recognize the eggs of the parasites with which they previously co-occurred. Similarly, an introduction of 200 Trinidadian guppies from a high-risk site to a guppy- and predator-free site 34 years ago has produced guppies that display a reduced tendency to school and greater willingness to approach predators than members of the ancestral population. Although descendants of a similar introduction made in 1976 by Endler display less pronounced antipredator behavior than the ancestral population in the field, but not when reared under uniform laboratory conditions, the overall evidence is that behavioral phenotypes can sometimes respond very rapidly, at least when selection favoring loss of ancestral behavioral phenotypes is strong (Magurran this volume).

Comparative studies of freshwater fishes in recently deglaciated areas also provide suggestions of rapid behavioral evolution, although the genetic bases of the behavioral differences have rarely been examined. In particular, rapid behavioral differentiation of lacustrine populations of stickleback suggests that most behavioral changes that appear in a time frame of 15,000 years or less are either quantitative changes in the expression of ancestral behavioral patterns or losses of these ancestral states (Foster et al. 1996). These changes have apparently contributed to speciation events which, in British Columbia, must have occurred in less than 11,000 years, and possibly in less than 2,000 years (McPhail 1994).

Speciation

Communication behavior elements are consistently involved in mate selection, a process critical to the evolution of species differences in many species. For this reason, research on geographic variation in behavior provides unique opportunities to examine the process of speciation in many animals as it occurs in the wild. Of course, differences in communication systems and patterns of mate choice can be examined across closely related, fully formed species, but such studies are limited in several respects. Perhaps the most serious is that the process itself is complete and cannot be reconstructed easily. Equally, genetic bases of divergence cannot be as readily examined as when partially formed species can be crossed, and the role of the environment in causing changes that result in speciation can no longer be examined directly. In essence, research on speciation that involves geographic variation allows investigators to examine the process in an ecological context, as it occurs under natural conditions. Although laboratory studies can offer insights into the means by which speciation can occur (Rice and Hostert 1993), they cannot tell us how the process actually occurs in nature. These are all points that have been made repeatedly (Mayr 1970, Lewontin 1974, Stratton and Uetz 1987, Endler 1989, McPhail 1994), yet there are remarkably few examples (Littlejohn this volume, Verrell this volume, Wilczynski and Ryan this volume).

Those research programs that have involved geographic comparisons have already provided substantial insights. We know, for example, that there is considerable geographic variation in both signal and receiver properties (although more is known about the former), and that in many cases, including the auditory communication system of the cricket frog and the visual communication system of Trinidadian guppies, receptor characteristics, environmental characteristics that influence signal transmission, and sexual selection all have contributed to the variation (Verrell this volume, Wilczynski and Ryan this volume). Equally, variation in illegitimate receivers across sites can cause geographic variation in communication systems (Endler 1980, 1988; Endler and Houde 1995). Often, but not always (e.g., Claridge and Morgan 1993, Endler and Houde 1995, Houde and Hankes 1997) the signal and the response covary. When they do not match, unique insights into mechanism and directionality of evolution leading to speciation may be gained (Kaneshiro 1976, Moodie 1982).

Both guppies (Endler 1980, Endler and Houde 1995) and cricket frogs (Ryan et al. 1990, Ryan and Wilczynski 1991, Wilczynski and Ryan this volume) show variation in both signals and receivers (mate choice) within and among populations which can be related to the physical environment. These results contrast strikingly with the view that stabilizing selection of mating systems prevents divergence (Paterson 1985, Templeton 1989). Stabilizing selection works within populations but does not prevent concomitant changes of both signals and receivers within populations. Consequently, populations can diverge even when under their own stabilizing selection. The only factor that would prevent populations from diverging is the presence of several populations in the same strong selective regime. But with different stabilizing selection optima, or no optima, populations can diverge in both mate preferences and male traits (Lande 1981, 1982; Pomiankowski 1988, Iwasa and Pomiankowski 1994). Populations can diverge concomitantly in mate preference and male traits for a variety of reasons (Houde 1993), with or without stabilizing selection, and this can easily lead to speciation (Lande 1981, 1982; Endler 1989).

Research on geographic variation in sexual incompatibility, particularly when combined with exploration of the causes of incompatibility, has great power for elucidating processes that can lead to speciation. Such studies are rare, however, because they are so labor intensive. The most complete study, conducted by Tilley, Verrell, and Arnold (reviewed with similar studies by Verrell this volume) offered the surprising insight that sexual incompatibility may increase with geographical distance but not necessarily with genetic distance, although this is not always the case (Coyne and Orr 1989, 1997). Examination of these attributes, in combination with those thought to be involved in mate choice in a diversity of taxa, ought to provide substantial insights into the causes and patterns of speciation.

A final class of studies designed explicitly to evaluate the dynamics of interactions following secondary contact are also geographical in nature; these are reviewed by Littlejohn (this volume). As he points out, after secondary contact, homogamy, if incomplete, can be eroded by gene flow. This is a process that can occur quickly when the divergent populations (incipient species) are sympatric throughout their range. However, when the two overlap in only a small part of their ranges, zones of hybridization can be maintained and reinforcement can occur if there is a cost to mismating. The nature of the transition from allopatry (and parapatry) to sympatry is poorly known, yet it is essential to understanding the dynamics of hybrid zones and the process of speciation.

Learning and Developmental Programs

Learning is a subset of the larger category of developmental programs in that animals are only capable of acquiring specific kinds of information or of responding to particular kinds of environmental stimuli once essential developmental stages are reached. In some cases learning must occur at a particular time in development and is strongly biased (imprinting); in other cases, learning increases progressively with age but may still be biased in the nature of the information acquired. These programs can evolve, just as can other less complex aspects of phenotype that are the products of interactions between genotype and environment.

The study of learning from the perspective of geographic variation is just beginning, and we have only enough examples to suggest the exciting potential of research in this field. Although the study of bird song has long fallen in the domain of geographic study, only recently has the possibility of genetically based differences in learning been considered as a contributor to song dialect production (Wilczynski and Ryan this volume). Nelson et al. (1996) have demonstrated that hand-reared males from a migratory population of the white-crowned sparrow, *Zonotrichia leucophrys oriantha*, learned far more tutorial material when a rich repertoire was offered than did males from a population of nonmigratory white-crowned sparrows, *Z. l. nuttalli*. They suggest that the reason for the difference could entail the need for song matching with neighbors. Because migratory birds will experience a wider range of song types among neighbors, they may require the broader song repertoire. Whatever the reason, this study suggests that differences in song learning abilities may be more common than previously suspected.

Research on antipredator behavior in minnows, guppies, and stickleback fish also offers evidence of geographic variation in ability to learn to avoid predators (Magurran this volume). The case of the stickleback is most surprising in that young in populations sympatric with predators learn to avoid predators as a consequence of efforts by their fathers to retrieve them if they stray from the nest. Young from populations devoid of predators simply do not learn avoidance skills to the same degree, although they interact similarly with the father (Huntingford and Wright 1993, Huntingford et al. 1994). This example illustrates the unexpected complexity that can be present in learning programs.

One of the most interesting developmental programs to display geographic variation is that by which male cowbirds (*Molothrus ater*) learn geographically appropriate song. King and West (1990) have shown that differences in the acoustic structure of males' songs in different regions is not a consequence of biased learning on the part of the males but is instead a reflection of vocal adjustments made in response to subtle visual cues offered by females in response to song. The genetic difference across populations lies in the females and in their responsiveness to songs of different types, not in genetically based differences among males. This remarkable system again offers insight into the diversity of developmental programs that may be detectable through study of geographic variation. This is an area of research likely to prove especially exciting in the future.

Conclusions

There can be little question that behavior evolves just as do other classes of traits. Consequently, it is not surprising to find widespread geographical variation in behavioral traits in a variety of taxa. Many of the behavioral differences observed among populations have proven to be heritable and subject to the same evolutionary forces as other aspects of the phenotype. This holds for a wide array of taxa, vertebrate and invertebrate. Certainly, behavioral phenotypes in many species are strongly influenced by environment and learning, and we expect that population differences in behavior will not always reflect genetic differentiation. However, the chapters in this book clearly demonstrate the fallacy of assuming that behavior is genetically uniform across all populations.

Because of their responsiveness to environmental conditions and their clear adaptive value, many kinds of behavior may serve to ameliorate the effects of environmental change on reproductive success. This very responsiveness may permit populations to survive in marginal environments. Indeed, more plastic organisms may respond to novel conditions with novel phenotypes as a consequence of genotype–environment interactions that will in turn increase the number of potential evolutionary trajectories available to the population (Morgan 1896, West-Eberhardt 1989, Wcislo 1989). Research on geographic variation may provide one of the most powerful means of testing this hypothesis. As with most evolutionary questions, the answers will not come quickly, but at least resolution is possible.

Although the existence of geographic variation in behavior might have been disturbing years ago when the focus of research was to predict exactly the behavioral outcomes of specific sensory stimuli, the existence of geographic variation now should excite the imaginations of those interested in understanding the evolution of behavior and of those interested in understanding the speciation process in animals that use behavioral criteria in mate choice. Even the ability to learn is now known to vary across populations, and the means by which behavioral variation develops has proven surprising and exciting in many instances. Clearly, research on geographic variation in behavior has enormous potential to resolve issues specific to the evolution of behavior, as well as those of broader theoretical interest.

References

Bakker, T. C. M. 1986. Aggressiveness in sticklebacks (*Gasterosteus aculeatus* L.): A behaviour-genetic study. Behaviour 98:1–144.

Bakker, T. C. M. 1994. Evolution of aggressive behaviour in the threespine stickleback. *In* M. A. Bell and S. A. Foster, eds. The evolutionary biology of the threespine stickleback, pp. 345–380. Oxford University Press, Oxford.

Barlow, G. W. 1981. Genetics and development of behavior, with special reference to patterned motor output. *In* K. Immelmann, G. W. Barlow, L. Petrinovich, and M. Main, eds. Behavioral development: The Bielefeld Interdisciplinary Project, pp. 191–251. Cambridge University Press, Cambridge.

Barton, N. H., and I. Wilson. 1996. Genealogies and geography. *In* P. H. Harvey, A. J. Leigh Brown, J. Maynard Smith, and S. Nee, eds. New uses for new phylogenies, pp. 23–56. Oxford University Press, Oxford.

Bell, M. A., and S. A. Foster. 1994. Introduction to the evolutionary biology of the threespine stickleback. *In* M. A. Bell and S. A. Foster, eds. The evolutionary biology of the threespine stickleback, pp. 1–27. Oxford University Press, Oxford.

Boake, C. R. B. 1994. Evaluation of applications of the theory and methods of quantitative genetics to behavioral evolution. *In* Quantitative genetic studies of behavioral evolution, pp. 305–326. Chicago University Press, Chicago.

Bogart, J. P., and A. O. Wasserman. 1972. Diploid-tetraploid species pairs: A possible clue to evolution by polyploidization in anuran amphibians. Cytogenetics 11:7–24.

Bradshaw, A. D. 1965. Evolutionary significance of phenotypic plasticity in plants. Advances in Genetics 13:115–155.

Claridge, M. F., and J. C. Morgan. 1993. Geographical variation in acoustic signals of the planthopper, *Nilaparvata bakeri* (Muir), in Asia: Species recognition and sexual selection. Biological Journal of the Linnean Society 48:267–281.

Cohan, F. M. 1984. Can uniform selection retard random genetic divergence between isolated conspecific populations? Evolution 39:495–504.

Coyne, J. A., and H. A. Orr. 1989. Patterns of speciation in *Drosophila*. Evolution 43: 362–381.

Coyne, J. A., and H. A. Orr. 1997. Patterns of speciation in *Drosophila* revisited. Evolution 51:295–303.

Crandall, K. A., and A. R. Templeton. 1996. Applications of intraspecific phylogenetics. *In* P. H. Harvey, A. J. Leigh Brown, J. Maynard Smith, and S. Nee, eds. New uses for new phylogenies, pp. 81–99. Oxford University Press, Oxford.

Endler, J. A. 1977. Geographic variation, speciation, and clines. Princeton University Press, Princeton, NJ.

Endler, J. A. 1980. Natural selection on color patterns in *Poecilia reticulata.* Evolution 34:76–91.

Endler, J. A. 1983. Testing causal hypotheses in the study of geographic variation. *In* J. Felsenstein, ed. Numerical taxonomy, pp. 424–443. Springer-Verlag, New York.

Endler, J. A. 1986. Natural selection in the wild. Princeton University Press, Princeton, NJ.

Endler, J. A. 1989. Conceptual and other problems in speciation. *In* D. Otte and J. A. Endler, eds. Speciation and its consequences, pp. 625–648. Sinauer Associates, Sunderland, MA.

Endler, J. A. 1988. Frequency-dependent predation, crypsis and aposematic colouration. Philosophical Transactions of the Royal Society of London B 319:505–523.

Endler, J. A. 1992. Signals, signal conditions, and the direction of evolution. American Naturalist 139:S125–153.

Endler, J. A. 1995. Multiple-trait coevolution and environmental gradients in guppies. Trends in Ecology and Evolution 10:22–29.

Endler, J. A., and A. E. Houde. 1995. Geographic variation in female preferences for male traits in *Poecilia reticulata*. Evolution 49:456–468.

Foster, S. A. 1988. Diversionary displays of paternal stickleback: Defenses against cannibalistic groups. Behavioural Ecology and Sociobiology 22:335–340.

Foster, S. A. 1994. Inference of evolutionary pattern: Diversionary displays of three-spined sticklebacks. Behavioral Ecology 5:114–121.

Foster, S. A. 1995. Understanding the evolution of behavior in threespine stickleback: The value of geographic variation. Behaviour 132:1107–1129.

Foster, S. A., and M. A. Bell. 1994. Evolutionary inference: The value of viewing evolution through stickleback-tinted glasses. *In* M. A. Bell and S. A. Foster, eds. The evolutionary biology of the threespine stickleback, pp. 472–486. Oxford University Press, Oxford.

Foster, S. A., and S. A. Cameron. 1996. Geographic variation in behavior: A phylogenetic framework for comparative studies. *In* E. Martins, ed. Phylogenies and the comparative method in animal behavior, pp. 138–165. Oxford University Press, Oxford.

Foster, S. A., W. A. Cresko, K. P. Johnson, M. U. Tlusty, and H. E. Willmott. 1996. Patterns of homoplasy in behavioral evolution. *In* M. J. Saunders and L. Hufford, eds. Homoplasy and the evolutionary process, pp. 245–269. Academic Press, New York.

Fraser, D. F., and J. F. Gilliam. 1987. Feeding under predation hazard: Response of the guppy and Hart's rivulus from sites with contrasting predation hazard. Behavioral Ecology and Sociobiology 21:203–209.

Garland, T., and S. C. Adolph. 1991. Physiological differentiation of vertebrate populations. Annual Review of Ecology and Systematics 22:193–228.

Gerhardt, H. C. 1974. Mating call differences between eastern and western populations of the gray tree frog *Hyla chrysoscelis.* Copeia 1974:534–536.

Gerhardt, H. C. 1994. Reproductive character displacement of female mate choice in the gray tree frog, *Hyla chrysoscelis.* Animal Behaviour 47:959–969.

Godin, J.-G. 1995. Predation risk and alternative mating tactics in male Trinidadian guppies (*Poecilia reticulata*). Oecologia 103:224–229.

Hailman, J. P. 1982. Evolution and behavior: An iconoclastic view. *In* H. C. Plotkin, ed. Learning, development, and culture, pp. 205–254. John Wiley and Sons, New York.

Houde, A. E. 1993. Evolution by sexual selection—what can population comparisons tell us? American Naturalist 141:796–803.

Houde, A. E., and M. A. Hankes. 1997. Evolutionary mismatch of mating preferences and male colour patterns in guppies. Animal Behaviour 53:343–351.

Huntingford, F. A., and P. J. Wright. 1993. The development of adaptive variation in predator avoidance in freshwater fishes. *In* F. A. Huntingford and P. Torricelli, eds. Behavioural ecology of fishes, pp. 45–61. Harwood, Chur, Switzerland.

Huntingford, F. A., P. J. Wright, and J. F. Tierney. 1994. Adaptive variation in antipredator behaviour in threespine stickleback. *In* M. A. Bell and S. A. Foster, eds. The evolutionary biology of the threespine stickleback, pp. 277–296. Oxford University Press, Oxford.

Iwasa, Y., and A. Pomiankowski. 1994. The evolution of sexual preferences for multiple handicaps. Evolution 48:853–867.

Johnson, C. F. 1996. Species recognition in the *Hyla versicolor* complex. Texas Journal of Science 18:361–364.

Kaneshiro, K. Y. 1976. Ethological isolation and phylogeny in the *Plantibia* subgroup of Hawaiian *Drosophila*. Evolution 30:740–745.

King, A. P., and M. J. West. 1990. Variation in species-typical behavior: A contemporary issue for comparative psychology. *In* D. A. Dewsbury, ed. Contemporary issues in comparative psychology, pp. 321–339. Sinauer Associates, Sunderland, MA.

Lande, R. 1981. Models of speciation by sexual selection on polygenic characters. Proceedings of the National Acadademy of Science USA 78:372–375.

Lande, R. 1982. Rapid origin of sexual isolation and character divergence in a cline. Evolution 36:213–223.

Lewontin, R. C. 1974. The genetic basis of evolutionary change. Columbia University Press, New York.

Lima, S. L., and L. M. Dill. 1990. Behavioural decisions made under the risk of predation: A review and a prospectus. Canadian Journal of Zoology 68:619–640.

Lotem, A., and S. I. Rothstein. 1995. Cuckoo-host coevolution: From snapshots of an arms race to the documentation of microevolution. Trends in Ecology and Evolution 10: 436–437.

Magnhagen, C. 1991. Predation risk as a cost of reproduction. Trends in Ecology and Evolution 6:183–186.

Magurran, A. E., and B. H. Seghers. 1994a. Predator inspection behaviour covaries with schooling tendency amongst wild guppy, *Poecilia reticulata*, populations in Trinidad. Behaviour 128:121–134.

Magurran, A. E., and B. H. Seghers. 1994b. Sexual conflict as a consequence of ecology: Evidence from guppy, *Poecilia reticulata*, populations in trinidad. Proceedings of the Royal Society of London B 255:31–36.

Mayr, E. 1963. Animal species and evolution. Belknap Press, Cambridge, MA.

Mayr, E. 1970. Populations, species, and evolution. Belknap Press, Cambridge, MA.

McPhail, J. D. 1994. Speciation and the evolution of reproductive isolation in the sticklebacks (*Gasterosteus*) of south-western British Columbia. *In* M. A. Bell and S. A. Foster eds. The evolutionary biology of the threespine stickleback, pp. 399–437. Oxford University Press, Oxford.

Moodie, G. E. E. 1982. Why asymmetric mating preferences may not show the direction of evolution. Evolution 36:1096–1097.

Moran, N. A. 1992. The evolutionary maintenance of alternative phenotypes. American Naturalist 139:971–989.

Morgan, C. L. 1896. On modification and variation. Science 4:733–740.

Muller, H. J. 1939. Reversibility in evolution considered fom the standpoint of genetics. Biological Reviews of the Cambridge Philosophical Society 14:261–279.

Nee, S., E. C. Holmes, A. Rambaut, and P. H. Harvey. 1996. Inferring population history form molecular phylogenies. *In* P. H. Harvey, A. J. Leigh Brown, J. Maynard Smith, and S. Nee, eds. New uses for new phylogenies, pp. 66–80. Oxford University Press, Oxford.

Nelson, D. A., C. Whaling, and P. Marler. 1996. The capacity for song memorization varies in populations of the same species. Animal Behaviour 52:379–387.

Otto, S. P., M. P. Cummings, and J. Wakeley. 1996. Inferring phylogenies from DNA sequence data: The effects of sampling. *In* P. H. Harvey, A. J. Leigh Brown, J. Maynard Smith, and S. Nee, eds. New uses for new phylogenies, pp. 103–115. Oxford University Press, Oxford.

Paterson, H. E. H. 1985. The recognition concept of species. *In* E. Vrba, ed. Species and speciation, pp. 21–29. Transvaal Museum Monograph 4, Pretoria, South Africa.

Pomiankowski, A. 1988. The evolution of female mate preferences for male genetic quality. Oxford Surveys in Evolutionary Biology 5:136–184.

Ptacek, M. B., H. C. Gerhardt, and R. D. Sage. 1994. Speciation by polyploidy in tree frogs: Multiple origins of the tetraploid, *Hyla versicolor*. Evolution 48:898–908.

Ralin, D. B., M. A. Romano, and C. W. Kilpatrick. 1983. The tetraploid tree frog *Hyla versicolor*: Evidence for a single origin from the diploid *H. chrysoscelis*. Herpetologica 39:212–225.

Reynolds, J. D., M. R. Gross, and M. J. Coombs. 1993. Environmental conditions and male morphology determine alternative mating behaviour in Trinidadian guppies. Animal Behaviour 45:145–152.

Rice, W. R., and E. E. Hostert. 1993. Laboratory experiments on speciation: What have we learned in 40 years? Evolution 47:1637–1653.

Ridgway, M. S., and J. D. McPhail. 1988. Raiding shoal size and a distraction display in male sticklebacks (*Gasterosteus*). Canadian Journal of Zoology 66:201–205.

Riechert, S. E. 1986. Spider fights as a test of evolutionary game theory. American Scientist 74:604–610.

Romano, M. A., D. B. Ralin, S. I. Guttman, and J. H. Skillings. 1987. Parallel electromorph variation in the diploid-tetraploid gray tree frog complex, *Hyla chrysoscelis* and *Hyla versicolor*. American Naturalist 130:864–878.

Ryan, M. J., R. B. Cocroft, and W. Wilczynski. 1990. The role of environmental selection in intraspecific divergence of mate recognition signals in the cricket frog, *Acris crepitans*. Evolution 44:1869–1872.

Ryan, M. J., and W. Wilczynski. 1991. Evolution of intraspecific variation in the advertisement call of a cricket frog (*Acris crepitans*, Hylidae). Biological Journal of the Linnean Society 44:249–271.

Sandoval, C. P. 1994. The effects of the relative geographic scales of gene flow and selection on morph frequencies in the walking-stick, *Timema cristinae*. Evolution 48: 1866–1879.

Schluter, D., and J. D. McPhail. 1993. Character displacement and replicate adaptive radiation. Trends in Ecology and Evolution 8:197–200.

Shaw, F. H., R. G. Shaw, G. S. Wilkinson, and M. Turelli. 1995. Changes in genetic variances and covariances: G whiz! Evolution 49:1260–1267.

Stratton, G. E., and G. W. Uetz. 1987. The inheritance of courtship behavior in *Schizocosa* wolf spiders (Araneae; Lycosidae). *In* M. D. Huettel, ed. Evolutionary genetics of invertebrate behavior, pp. 63–77. Plenum Press, New York.

Templeton, A. R. 1989. The meaning of species and speciation. *In* D. Otte and J. A. Endler, eds. Speciation and its consequences, pp. 3–27. Sinauer Associates, Sunderland, MA.

Turelli, M. 1988. Phenotypic evolution, constraint covariances, and the maintenance of additive variance. Evolution 42:1342–1348.

Turesson, G. 1922a. The genotypic response of the plant species to the habitat. Hereditas 3:211–350.

Turesson, G. 1922b. The species and the variety as ecological units. Hereditas 3:100–113.

Ueda, H. 1993. Mating calls of auto-triploid and auto-tetraploid males in *Hyla Japonica*. Scientific Reports of the Laboratory of Amphibian Biology, Hiroshima University 12:177–189.

Verrell, P. A. 1990. Frequency of interspecific mating in salamanders of the plethodontid genus Desmognathus: Different experimental designs may yield different results. Journal of Zoology 221:441–451.

Wasserman, A. O. 1970. Polyploidy in the common tree toad, *Hyla versicolor* Le Conte. Science 167:385–386.

Wcislo, W. T. 1989. Behavioral environments and evolutionary change. Annual Review of Ecology and Systematics 20:137–169.

West-Eberhart, M. J. 1989. Phenotypic plasticity and the origins of diversity. Annual Review of Ecology and Systematics 20:249–278.

Whoriskey, F. G., and G. J. FitzGerald. 1985. Sex, cannibalism and sticklebacks. Behavioral Ecology and Sociobiology 18:15–18.

Wiley, J. E. 1983. Chromosome polymorphism in *Hyla chrysoscelis.* Copeia 1983:273–275.

Wiley, J. E., M. A. Little, M. A. Romano, D. A. Blount, and G. R. Cline. 1989. Polymorphism in the location of the 18s and 28s rRNA genes on the chromosomes of the diploid tetraploid tree frogs *Hyla chrysoscelis* and *H. versicolor*. Chromosoma 97: 481–487.

Index